ENVIRONMENTAL SCIENCE, ENGINEERING AND TECHNOLOGY

GREENHOUSE GAS EMISSION SCENARIOS FOR USE IN CLIMATE BASED RESPONSE

ENVIRONMENTAL SCIENCE, ENGINEERING AND TECHNOLOGY

Additional books in this series can be found on Nova's website under the Series tab.

Additional E-books in this series can be found on Nova's website under the E-books tab.

CLIMATE CHANGE AND ITS CAUSES, EFFECTS AND PREDICTION

Additional books in this series can be found on Nova's website under the Series tab.

Additional E-books in this series can be found on Nova's website under the E-books tab.

ENVIRONMENTAL SCIENCE, ENGINEERING AND TECHNOLOGY

GREENHOUSE GAS EMISSION SCENARIOS FOR USE IN CLIMATE BASED RESPONSE

JACQUELINE L. AIGNER
EDITOR

Nova Science Publishers, Inc.
New York

For permission to use material from this book please contact us:
Telephone 631-231-7269; Fax 631-231-8175
Web Site: http://www.novapublishers.com

LIBRARY OF CONGRESS CATALOGING-IN-PUBLICATION DATA

Greenhouse gas emission scenarios for use in climate-based response /
editor, Jacqueline L. Aigner.
p. cm.
Includes bibliographical references and index.
ISBN 978-1-61209-610-0 (hardcover)
1. Greenhouse gas mitigation--United States. 2. Climatic
changes--Forecasting. I. Aigner, Jacqueline L.
TD885.5.G73G7325 2011
363.738'746--dc22
2011002547

Published by Nova Science Publishers, Inc. † New York

CONTENTS

PREFACE

This new book presents research scenarios which are designed to stabilize the influence of a suite of greenhouse gases. This research was conducted using computer-based research tools known as integrated assessment models. Three modeling groups each independently developed a reference scenario, in which all climate policies were assumed to expire in 2012, and then developed four stabilization scenarios as departures from their respective reference scenarios. The scenarios in this book are not predictions or best-judgment forecasts from the modeling groups; rather they constitute new research intended to advance understanding of the forces that lead to GHG emissions and ways in which to stabilize GHG concentrations and radiative forcing.

Chapter 1- This and a companion report constitute one of twenty-one Synthesis and Assessment Products called for in the *Strategic Plan for the U.S. Climate Change Science Program*. These studies are structured to provide high-level, integrated research results on important science issues with a particular focus on questions raised by decision-makers on dimensions of climate change directly relevant to the U.S. One element of the CCSP's strategic vision is to provide *decision support tools* for differentiating and evaluating response strategies. Scenario-based analysis is one such tool. The scenarios in this chapter explore the implications of alternative stabilization levels of anthropogenic greenhouse gases (GHGs) in the atmosphere, and they explicitly consider the economic and technological foundations of such response options. Such scenarios are a valuable complement to other scientific research contained in the twenty-one CCSP Synthesis and Assessment Products. The companion to the research reported here, *Global-Change Scenarios: Their Development and Use*, explores the broader strategic frame for developing and utilizing scenarios in support of climate decision making.

Chapter 2- A scenario is a description of potential future conditions produced to inform decision-making under uncertainty. Scenarios can help inform decisions that involve high stakes and poorly characterized uncertainty, which may thwart other, conventional forms of analysis or decision support. Originally developed to study military and security problems, scenarios are now widely used for strategic planning and assessment in businesses and other organizations, and increasingly to inform planning, analysis, and decision making for environmental issues, including climate change.

Scenarios can serve many purposes. They can help inform specific decisions, or can provide inputs to assessments, models, or other decision-support activities when these activities need specification of potential future conditions. They can also provide various forms of indirect decision support, such as clarifying an issue's importance, framing a decision agenda, shaking up habitual thinking, stimulating creativity, clarifying points of agreement and disagreement, identifying and engaging needed participants, or providing a structure for analysis of potential future decisions.

In: Greenhouse Gas Emission Scenarios For Use In Climate... ISBN: 978-1-61209-610-0
Editor: Jacqueline L. Aigner

Chapter 1

SCENARIOS OF GREENHOUSE GAS EMISSIONS AND ATMOSPHERIC CONCENTRATIONS*

United States Climate Change Science Program

ACKNOWLEDGMENT

This chapter has been peer reviewed in draft form by individuals chosen for their diverse perspectives and technical expertise. The expert review and selection of reviewers followed the OMB's Information Quality Bulletin for Peer Review.The purpose of this independent review is to provide candid and critical comments that will assist the Climate ChangeScience Program in making this published report as sound as possible and to ensure that the report meets institutionalstandards. The peer review comments, draft manuscript, and response to the peer review comments are publicly available at: www.climatescience.gov/ Library/sap/sap2-1/default.php.

We wish to thank the following individuals for their peer review of this chapter:
Joseph Aldy, Resources for the Future
Bill Chameides, James Wang, Environmental Defense
Russell Jones, American Petroleum Institute
David Rind, NASA Goddard Institute for Space Studies
Brent Sohngen, Ohio State University
Richard Tol, Hamburg, Vrije and Carnegie Mellon Universities
John Weyant, Stanford University

We would also like to thank the numerous individuals who provided their comments during the public comment pe-riod. The public review comments, draft manuscript, and response to the public comments are publicly available at:www.climatescience.gov/ Library/sap/sap2-1/default.php

* This is an edited, reformatted and augmented edition of a United States Climate Change Science Program publication, dated July 2007.

The authors also wish to acknowledge the contributions of Geoffrey Blanford, Josh Lurz, Sergey Paltsev, ThomasRutherford, and Marshall Wise in developing these scenarios.

EXECUTIVE SUMMARY

Introduction

This and a companion report constitute one of twenty-one Synthesis and Assessment Products called for in the *Strategic Plan for the U.S. Climate Change Science Program*. These studies are structured to provide high-level, integrated research results on important science issues with a particular focus on questions raised by decision-makers on dimensions of climate change directly relevant to the U.S. One element of the CCSP's strategic vision is to provide *decision support tools* for differentiating and evaluating response strategies. Scenario-based analysis is one such tool. The scenarios in this chapter explore the implications of alternative stabilization levels of anthropogenic greenhouse gases (GHGs) in the atmosphere, and they explicitly consider the economic and technological foundations of such response options. Such scenarios are a valuable complement to other scientific research contained in the twenty-one CCSP Synthesis and Assessment Products. The companion to the research reported here, *Global-Change Scenarios: Their Development and Use*, explores the broader strategic frame for developing and utilizing scenarios in support of climate decision making.

Study Design

The scenarios in this chapter were developed using integrated assessment models (IAMs). These analysis capabilities integrate computer models of socioeconomic and technological determinants of the emissions of GHGs with models of the natural science of Earth system response, including the atmosphere, oceans, and terrestrial biosphere. Three IAMs were applied in the scenario development:

- The Integrated Global Systems Model (IGSM) of the Massachusetts Institute of Technology's Joint Program on the Science and Policy of Global Change.
- The Model for Evaluating the Regional and Global Effects (MERGE) of GHG reduction policies developed jointly at Stanford University and the Electric Power Research Institute.
- The MiniCAM Model of the Joint Global Change Research Institute, a partnership between the Pacific Northwest National Laboratory and the University of Maryland.

Each modeling group first produced a reference scenario under the assumption that no climate policies are imposed beyond current commitments, namely the 2008-12 first period of the Kyoto Protocol and the U.S. goal of reducing reduce GHG emissions per unit of its gross domestic product by 18% by 2012. The resulting reference cases are not predictions or best-judgment forecasts but scenarios designed to provide clearly defined points of departure for studying the implications of alternative stabilization goals. As instructed in the Prospectus for the study, the modeling teams used model input assumptions they considered *meaningful* and *plausible*. The resulting reference scenarios provide insights into how the world might evolve *without* additional efforts to constrain GHG emissions, given various assumptions about principal drivers of these emissions such as population increase, economic growth, land and labor productivity growth, technological options, and resource endowments.

Each modeling group then produced four additional stabilization scenarios framed as departures from its reference scenario. The stabilization levels are common across the modeling groups and are defined in terms of the total long-term effect on the Earth's heat balance of the combined effect of the primary anthropogenic GHGs: carbon dioxide (CO_2), nitrous oxide (N_2O), methane (CH_4), hydrofluorocarbons (HFCs), perfluorocarbons (PFCs), and sulfur hexafluoride (SF_6). The potential for climate-related controls on other human emissions, such as aerosols and their precursors, was not incorporated into the stabilization constraints, although the participating models represent the emissions of many of these substances. With the exception of these stabilization levels, and a common hypothesis about the sharing among nations of the mitigation task, there was no direct coordination among the modeling groups either in the assumptions underlying the reference scenario or the precise paths to stabilization.

The results drawn from the simulations were selected to provide insight into questions such as the following:

- What emissions trajectories over time are consistent with meeting the four stabilization levels, and what are the key factors that shape them?
- What energy system characteristics are consistent with each of the four alternative stabilization levels, and how might these characteristics differ among stabilization levels?
- What are the possible economic consequences of meeting each of the four alternative stabilization levels?

With its focus on reducing emissions to meet various stabilization levels the study does not explore climate damages that might be avoided or ancillary benefits (such as lower air pollution) of emissions reduction. Thus, though the scenarios provide a useful input to climate-related decision making they address only one of several components of a cost-benefit analysis of climate policy. In addition , although these scenarios incorporate new thinking on GHG emissions and possible mitigation paths they were not designed to span the full range of possible futures or to provide an uncertainty analysis of key forces. They are intended, rather, to enhance understanding of the implications of different ways that the future might evolve without assigning likelihoods to outcomes.

Potential Applications

There are many potential applications of scenarios of this form, and to facilitate their use the numerical results are provided in a companion data set. Possible users include climate modelers and the science community; those involved in national public policy formulation; managers of Federal research programs; state and local government officials who face decisions that might be affected by climate change and mitigation measures; and individual firms, non-governmental organizations, and members of the public. Insights from the scenarios may be used directly as inputs to the decision-making processes, or the scenarios may serve as inputs to further analyses in support of climate decision making. A sample of possible further analyses would include the following:

- The scenarios can provide a basis for study of the climate implications of alternative stabilization levels, as an input to climate models, and then to follow-on studies of potential climate impacts.
- The scenarios can serve as a point of departure for exploring possible technology cost and performance goals, using information from the scenarios on energy prices and technology deployment levels.
- The scenarios can provide a foundation for analysis of the non-climate environmental implications of new energy sources at large scale.
- The scenarios could serve as an input to a more complete analysis of the economic effects of stablizing and the different radiative forcing levels, such as indicators of consumer impact in the U.S.
- The scenarios can be applied in comparative mode, extending the lessons to be learned from the three models in this research to those to be gained from scenarios developed using different approaches.

The varied clientele for these scenarios and the variety of questions they might inform implies a highly diverse set of possible needs, and no single scenario exercise can hope to fully satisfy all of them. Therefore these scenarios likely will stimulate further questions and the demand for more detailed analysis, some of which might be satisfied by further scenario development from models like those used here, but others demanding detail that can only be provided with alternative modeling and analysis techniques.

Several characteristics of these scenarios make them particularly valuable for these and other types of applications. One advantage is the update of economic and technology data and assumptions and the use of improved scenario development tools. It has been over a decade since the last emissions scenario development project of the Intergovernmental Panel on Climate Change (IPCC) – its *Special Report on Emissions Scenarios* (SRES) – and over five years since the subsequent CO2 stabilization scenarios in the IPCC's Third Assessment Report. Over this time, substantial advances have been made in both economic and natural science components of the IAMs used to simulate the various scenarios. A second advance of this research is its all-gas approach. Many other stabilization scenarios have focused on CO_2 with little attention to other human influences. The scenarios presented here consider stabilization in terms of the combined effect of all six categories of GHGs listed earlier so that the full range of policy options is considered simultaneously. Finally, there is great advantage in the simultaneous application to the task, and parallel presentation of results, by three independent modeling groups, applying IAMs each of which has its own special strengths. Comparison of scenarios across the models provides useful insights into the role of key assumptions, the realms of most fruitful technology development, and aspects of the natural science (particularly the carbon cycle) that have a substantial effect on the difficulty of the stabilization task.

Scenario Highlights

The report and supporting database provide many details of the implications for the U.S. and global economy, with particular focus on the energy sector, for the reference conditions and the four levels of possible atmospheric stabilization. Highlights of the picture that is found there include the following:

In the reference scenarios, economic and energy growth, combined with continued fossil fuel use, lead to changes in the Earth's radiation balance that are three to four times that already experienced since the beginning of the industrial age. By 2100, primary energy consumption increases from over 3 to nearly 4 times 2000 levels as economic growth outpaces improvements in the efficiency of energy use. Non-fossil energy use grows from over 4 to almost 9 times over the century, but this growth is insufficient to supplant fossil fuels as the major source of energy. As a result, global CO_2 emissions more than triple between 2000 and 2100, and emissions are rising at the end of the twenty-first century in all three reference scenarios. Combined with the effects of non-CO_2 GHGs, the increase in anthropogenic radiative forcing from preindustrial levels is substantial.

In the stabilization scenarios, CO_2 emissions peak and decline during the twenty-first century or soon thereafter. Emissions of non-CO_2 GHGs are also reduced.The timing of GHG emissions reductions varies substantially across the four radiative forcing stabilization levels. Under the most stringent stabilization levels, CO_2 emissions begin to decline immediately or within a matter of decades. Under the less stringent stabilization levels, CO_2 emissions do not peak until late in the century or beyond, and they are 1½ to over 2½ times today's levels in 2100.

In the stabilization scenarios, GHG emissions reductions require a transformation of the global energy system, including reductions in the demand for energy (relative to the reference scenarios) and changes in the mix of energy technologies and fuels. This transformation is more substantial and takes place more quickly at the more stringent stabilization levels. Fossil fuel use and energy consumption are reduced in all the stabilization scenarios due to increased consumer prices for fossil fuels. Use of shale oil, tar sands, and synthetic fuels from coal are greatly reduced or, under the most stringent stabilization levels, eliminated. Across the stabilization scenarios, CO_2 emissions from electric power generation are reduced at relatively lower prices than CO_2 emissions from other sectors, such as transport, industry, and buildings. Emissions are reduced from electric power by increased use of technologies such as CO_2 capture and storage, nuclear energy, and renewable energy. Other sectors respond to rising GHG prices by reducing demands for fossil fuels; substituting low- or non-emitting energy sources such as bioenergy and low-carbon electricity or hydrogen; and applying CO_2 capture and storage where possible.

Substantial differences in GHG emissions prices and associated economic costs arise among the modeling groups for each stabilization level. These differences are illustrative of some of the unavoidable uncertainties in long-term scenarios. Among the most important factors influencing the variation in economic costs are: (1) differences in assumptions – such as those regarding economic growth over the century, the behavior of the oceans and terrestrial biosphere in taking up CO_2, and opportunities for reduction in non-CO_2 GHG emissions – that determine the amount that CO_2 emissions that must be reduced to meet the radiative forcing stabilization levels; and (2) differences in assumptions about technologies, particularly in the second half of the century, to shift final demand to low-carbon sources such as biofuels and low- carbon electricity or hydrogen, in transportation, industrial, and buildings end uses. All other things being equal, scenarios with more low-cost technology options and lower required emissions reductions have lower economic costs.

Follow on Efforts

Generating scenarios is not a once-and-for-all activity. The scenarios in this chapter represent but one step in a long process of research and assessment, continuing an over 20-year tradition of research and analysis in the climate area. They will need to be updated as knowledge advances and conditions change. Indeed, the research presented here suggests several areas of potentially fruitful research:

- Analysis of the sensitivity of results to assumptions about the cost, performance and environmental issues surrounding key technologies such as nuclear power, carbon capture and storage, and biofuels.
- Consideration of scenarios based on different assumptions than used here about the way that the burdens of emissions mitigation may be shared among nations and over time.
- Expansion and improvement of analysis of human land use and the terrestrial carbon cycle.
- Inclusion of other anthropogenic emissions that affect the Earth's heat balance, such as the different types of aerosols, and the effect of the tropospheric ozone (another GHG) that results from urban air pollution.
- Addition of uncertainty analysis and consideration of decision-making under these conditions.

TECHNICAL SUMMARY

Highlights of the Report

Background

This chapter presents research from Synthesis and Assessment Product 2.1 a of the Climate Change Science Program (CCSP), *Scenarios of Greenhouse Gas Emissions and Atmospheric Concentrations*. The scenarios in this research product were designed to stabilize the influence of a suite of greenhouse gases (GHGs) – carbon dioxide (CO_2), methane (CH_4), nitrous oxide (N_2O), hydrofluorocarbons (HFCs), perfluorocarbons (PFCs) and sulfur hexafluoride (SF_6) – on the Earth's radiation balance, measured in terms of radiative forcing. Four radiative forcing stabilization levels are considered. The resulting atmospheric concentrations of the largest single contributor, CO_2, are roughly 450, 550, 650 and 750 parts per million by volume (ppmv). Responding to the Prospectus for this research product (CCSP 2005), this chapter focuses on (1) GHG emissions trajectories, (2) global and U.S. energy system implications, and (3) economic implications of stabilization.

This research was conducted using computer-based research tools known as integrated assessment models. Three modeling groups each independently developed a reference scenario, in which all climate policies were assumed to expire in 2012, and then developed four stabilization scenarios as departures from their respective reference scenarios. Idealized emissions- reduction measures – designed to achieve emissions reductions wherever, whenever, and using whichever GHG was most cost effective – were imposed to limit GHG emissions and meet the four radiative forcing stabilization levels. Evidence from previous literature suggests that if less idealized measures were employed to stabilize radiative forcing, the costs could be substantially higher. Further, this research considers only the costs of stabilization; it does not consider the benefits of potential climate change avoided or of possible ancillary benefits of emissions reduction, such as reduced air pollution.

The scenarios in this chapter are not predictions or best-judgment forecasts from the modeling groups. Rather, they constitute new research intended to advance understanding of the forces that lead to GHG emissions and that shape opportunities to stabilize GHG concentrations and radiative forcing. Although the future is uncertain and the scenarios are strongly dependent on many underlying assumptions, this research provides useful insights for those engaged in climate-related decision making.

Highlights of the Report

In the reference scenarios, economic and energy growth, combined with continued fossil fuel use, lead to changes in the Earth's radiation balance that are three to four times that already experienced since the beginning of the industrial age. By 2100, primary energy consumption increases from over three to nearly four times 2000 levels as economic growth outpaces improvements in the efficiency of energy use. Non-fossil energy use grows from over four to almost nine times over the century, but this growth is insufficient to supplant fossil fuels as the major source of energy. As a result, global CO_2 emissions more than triple between 2000 and 2100, and emissions are rising at the end of the twenty-first century in all three reference scenarios. Combined with the effects of non-CO_2 GHGs, the increase in anthropogenic radiative forcing from preindustrial levels is substantial.

In the stabilization scenarios, CO_2emissions peak and decline during the twenty-first century or soon thereafter. Emissions of non-CO_2 GHGs are also reduced. The timing of GHG emissions reductions varies substantially across the four radiative forcing stabilization levels. Under the most stringent stabilization levels, CO_2 emissions begin to decline immediately or within a matter of decades. Under the less stringent stabilization levels, CO_2 emissions do not peak until late in the century or beyond, and they are 1½ to over 2½ times today's levels in 2100.

In the stabilization scenarios, GHG emissions reductions require a transformation of the global energy system, including reductions in the demand for energy (relative to the reference scenarios) and changes in the mix of energy technologies and fuels. This transformation is more substantial and takes place more quickly at the more stringent stabilization levels. Fossil fuel use and energy consumption are reduced in all the stabilization scenarios due to increased consumer prices for fossil fuels. Use of shale oil, tar sands, and synthetic fuels from coal are greatly reduced or, under the most stringent stabilization levels, eliminated. Across the stabilization scenarios, CO_2 emissions from electric power generation are reduced at relatively lower prices than CO_2 emissions from other sectors, such as transport, industry, and buildings. Emissions are reduced from electric power by increased use of technologies such as CO_2 capture and storage (CCS), nuclear energy, and renewable energy. Other sectors respond to rising greenhouse gas prices by reducing demands for fossil fuels; substituting low- or non-emitting energy sources such as bioenergy and low-carbon electricity or hydrogen; and applying CCS where possible.

Substantial differences in GHG emissions prices and associated economic costs arise among the modeling groups for each stabilization level. These differences are illustrative of some of the unavoidable uncertainties in longterm scenarios. Among the most important factors influencing the variation in economic costs are: (1) differences in assumptions – such as those regarding economic growth over the century, the behavior of the oceans and terrestrial biosphere in taking up CO_2, and opportunities for reduction in non-CO_2 GHG emissions – that determine the amount that CO_2 emissions must be reduced to meet the radiative forcing stabilization levels; and (2) differences in assumptions about technologies, particularly in the second half of the century, to shift final demand to low-carbon sources such as biofuels and low- carbon electricity or hydrogen, in transportation, industrial, and buildings end uses. All other things being equal, scenarios with more low-cost technology options and lower required emissions reductions have lower economic costs.

Background

The Strategic Plan for the U.S. Climate Change Science Program (CCSP 2003) noted that "...sound, comprehensive emissions scenarios are essential for comparative analysis of how climate might change in the future, as well as for analyses of mitigation and adaptation options." The Plan includes Product 2.1, *Scenarios of Greenhouse Gas Emissions and Atmospheric Concentrations and Review of Integrated Scenario Development and Application,* which consists of two parts. This chapter presents the scenario development component (Product 2. 1A); the review of scenario methods (Product 2.1B) is the subject of a separate report (CCSP 2007).

Guidelines for producing these scenarios were set forth in a Prospectus, which specified that the new scenarios focus on alternative levels of atmospheric stabilization of the radiative forcing from the combined effects of a suite of the main anthropogenic GHGs. The Prospectus also set forth criteria for the facilities to be used in the analysis. Scenarios developed using three models that meet the Prospectus conditions are reported here.

The scenarios in this chapter are intended as one of many inputs to public and private discussions regarding climate change and what to do about it, and they may serve as a point of departure for further CCSP and other analyses that might inform these discussions in the future. The possible users of these scenarios are many and diverse. They include climate modelers and the science community; those involved in national public policy formulation; managers of Federal research programs; state and local government officials who face decisions that might be affected by climate change and mitigation measures; and individual firms, non-governmental organizations, and members of the public. Such a varied clientele implies an equally diverse set of possible needs, and no single scenario exercise can hope to fully satisfy all of these needs.

Each of the three modeling groups participating in this research first developed a no-climate policy scenario – referred to as a reference scenario – which serves as baseline for development of alternative scenarios with emissions control. Each modeling group then developed four control scenarios leading to stabilization of radiative forcing at four alternative levels. The resulting scenarios provide insight into questions such as the following:

- What emissions trajectories over time are consistent with meeting the four alternative stabilization levels, and what are the key factors that shape them?
- What energy system characteristics are consistent with each of the four alternative stabilization levels, and how might these characteristics differ among stabilization levels?
- What are the possible economic consequences of meeting each of the four alternative stabilization levels?

Although each of the models used to develop these scenarios represents the world as a set of interconnected nations and multi-nation regions, as specified in the Prospectus, this chapter focuses on the U.S. and world characteristics of the scenarios.

With the exception of the stabilization levels themselves and a common hypothesis about international burden sharing, there was no direct coordination among the modeling groups either in the assumptions underlying the reference scenario or the precise path to stabilization. Furthermore, the scenarios were not designed to span the full range of possible futures, and no explicit uncertainty analysis was called for. Although the future is uncertain and the scenarios depend on many underlying assumptions, this research illuminates a range of possible future developments and provides useful insights for those engaged in climate-related decision making.

> The scenarios in this chapter do not constitute a cost-benefit analysis of climate policy. They focus exclusively on the issues associated with reducing emissions to meet various stabilization levels; they do not consider the damages avoided through stabilization or ancillary benefits that could be realized by emissions reductions, such as reductions in local air pollution. Thus, although the scenarios should serve as a useful input to climate-related decision making, they address only one of several components of a benefit-cost analysis of climate policy.

Scenario research such as this continues a tradition of research and analysis that has gone on for over 20 years. This work will be continued and refined as the field advances, new information becomes available, and decision makers raise new questions and issues. Similar work is conducted by modeling groups in Europe and Asia. The scenarios developed here add to this larger body of scholarship and should be viewed as one additional piece of information in an ongoing and iterative process of scenario development.

Models Used to Develop the Scenarios

The Prospectus for this research set out the following criteria for participating models: they must (1) be global in scale, (2) be capable of producing global emissions totals for designated GHGs, (3) represent multiple regions, (4) be capable of simulating the radiative forcing from these GHGs and substances, (5) have technological resolution capable of distinguishing among major sources of primary energy (e.g., renewable energy, nuclear energy, biomass, oil, coal, and natural gas) as well as between fossil fuel technologies with and without carbon capture and storage systems, (6) be economics-based and capable of simulating macroeconomic cost implications of stabilization, and (7) look forward at least to the end of the twenty-first century. In addition, modeling groups were required to have a track record of publications in professional, refereed journals, specifically in the use of their models for the analysis of longterm GHG emission scenarios.

Application of these criteria led to the selection of three models:

- The Integrated Global Systems Model (IGSM) of the Massachusetts Institute of Technology's Joint Program on the Science and Policy of Global Change
- The Model for Evaluating the Regional and Global Effects (MERGE) of GHG reduction policies developed jointly at Stanford University and the Electric Power Research Institute.
- The MiniCAM Model of the Joint Global Change Research Institute, a partnership between the Pacific Northwest National Laboratory and the University of Maryland.

Each of these models has been used extensively for climate change analysis. The roots of each extend back more than a decade, during which time features and details have been refined, modified, and added. Research using each has appeared widely in peer-reviewed publications.

Approach

As directed by the Prospectus, each of the three modeling groups produced one reference scenario and four stabilization scenarios, for a total of 15 scenarios. First, the reference scenarios were developed under the assumption that no climate policy would be implemented beyond the set of policies currently in place (e.g., the Kyoto Protocol and the U.S. carbon intensity goal, each terminating in 2012 because goals beyond that date have not been identified). Each modeling group developed its own reference scenario. The Prospectus required only that each reference scenario be based on assumptions believed by the participating modeling groups to be *meaningful* and *plausible*. Each of the three reference scenarios is based on a different set of assumptions about how the future might unfold without additional climate policies. These assumptions are not intended as predictions or best-judgment forecasts of the future by the respective modeling groups. Rather, they represent possible paths that the future might follow to serve as a platform for examining how emissions might be reduced to achieve stabilization.

Each group then produced four stabilization scenarios by constraining the models to achieve four alternative radiative forcing levels. Stabilization was defined in terms of the total longterm radiative impact of a suite of GHGs including CO_2, N_2O, CH_4, HFCs, PFCs, and SF_6. These are the gases enumerated in the U.S. goal to reduce the intensity of GHG emissions relative to gross domestic product (GDP) as well as the Kyoto Protocol. Other substances with radiative impact, such as gases controlled under the Montreal Protocol, carbon monoxide (CO), ozone (O_3), and aerosols were not included in the radiative forcing levels.

Table TS.1. Greenhouse Gas Concentrations and Forcing

	Preindustrial Concentration (1750)	Current Concentration (1998)	Contribution to Radiative Forcing, (W/m^2, 1750 to 1998)
CO_2	278 ppmv	365 ppmv	1.46
CH_4	700 ppbv	1745 ppbv	0.48
N_2O	270 ppbv	314 ppbv	0.15
HFCs, PFCs, SF_6	0	various	≈ 0.02
Total	—	—	≈ 2.1

Source. IPCC 2001.

Concentrations of GHGs have increased since 1750(preindustrial), altering the radiative energy budget of the Earth's climate system.

Table TS.2. Radiative Forcing Stabilization Levels (W/m^2 from preindustrial) and Approximate Resulting CO_2 Concentrations (ppmv)

	Total Radiative Forcing from GHGs in this Research (W/m^2)	Approximate Contribution to Radiative Forcing from non-CO_2 GHGs (W/m^2)	Approximate Contribution to Radiative Forcing from CO_2(W/m^2)	Corresponding CO_2 Concentration (ppmv)
Level 1	3.4	0.8	2.6	450
Level 2	4.7	1.0	3.7	550
Level 3	5.8	1.3	4.5	650
Level 4	6.7	1.4	5.3	750
Year 1998	≈ 2.1	0.65	1.46	365
Preindustrial (1750)	—	—	—	278

The radiative forcing levels were constructed so that the CO_2 concentrations resulting from stabilization of total radiative forcing, after accounting for radiative forcing from the non-CO_2 GHGs, would be roughly 450 ppmv, 550 ppmv, 650 ppmv, and 750 ppmv.

The four radiative forcing stabilization scenarios were developed so that the combined radiative forcing from these GHGs since preindustrial times was constrained to no more than 3.4 W/m^2 for Level 1, 4.7 W/m^2 for Level 2, 5.8 W/m^2 for Level 3, and 6.7 W/m^2 for Level 4. Because radiative forcing was defined relative to preindustrial times, it includes the roughly 2.1 W/m^2 of radiative forcing from these substances that had already occurred from 1750 to 1998 (Table TS.1).

These radiative forcing stabilization levels were chosen so that the associated CO_2concentrations would be roughly 450 ppmv, 550 ppmv, 650 ppmv, and 750 ppmv after accounting for the contributions to radiative forcing from the non-CO_2 GHGs (Table TS.2). If these CO_2 concentrations were achieved exactly, the radiative forcing from CO_2 would be less than the radiative forcing stabilization levels because of the allowance for additional forcing from the nonCO_2 GHGs. Thus, the radiative forcing stabilization levels should not be interpreted as the "CO_2–equivalent" levels associated with the approximate CO_2 concentrations in Table TS.2. Because the stabilization exercises sought least- cost reductions among the gases, any correspondence between radiative forcing levels and CO_2 concentrations is necessarily approximate and differs among modeling groups because of differences in the treatment of the forces that influence emissions of GHGs, possibilities for emissions reductions, and tradeoffs between reductions among GHGs.

Overview of the Scenarios

This section provides an overview of the scenarios. The three reference scenarios are discussed in the next section, followed by a discussion of the twelve stabilization scenarios, four from each modeling group.

Reference Scenarios

The difficulty of achieving any specified level of atmospheric stabilization depends heavily on the emissions that would occur absent actions to address GHG emissions. In other words, the reference scenario strongly influences the stabilization scenarios. If the reference scenario has inexpensive fossil fuels and high-economic growth, then larger changes to the energy sector and other parts of the economy may be required to stabilize radiative forcing. On the other hand, if the reference scenario shows lower economic growth and emissions, and perhaps increased exploitation of non-fossil sources even in the absence of climate policy, then the effort required to stabilize radiative forcing will not be as great.

Energy production, transformation, and consumption are central features in all of these scenarios, although non-CO_2 gases and changes in land use also make a significant contribution to aggregate GHG emissions. Demand for energy over the coming century will be driven by economic growth and will also be strongly influenced by the way that energy systems respond to depletion of resources, changes in prices, and improvements in technology. Demand for energy in developed countries remains strong in all the scenarios and is even stronger in developing countries, where millions of people seek greater access to commercial energy. These developments strongly influence the emissions of GHGs, their disposition, and the resulting change in radiative forcing in the reference scenarios.

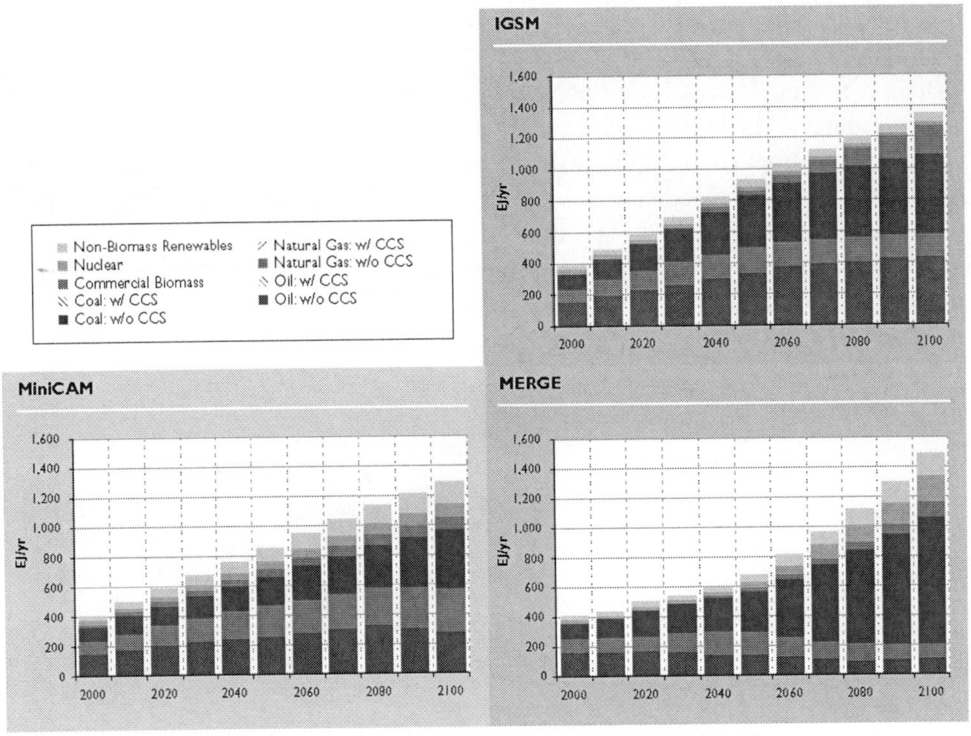

Global primary energy consumption rises in all three reference scenarios, from about 400 EJ/yr in 2000 to between roughly 1275 EJ/yr and 1500 EJ/yr in 2100.Dependence on conventional oil resources gradually decreases. However, a range of alternative fossil-based resources, such as syntheticfuels from coal and unconventional oil resources (e.g., tar sands and oil shales) are available and become economically viable. Fossil fuels provided almost 90% of global primary energy consumption in the year 2000, and they remain the dominant energy source in the three reference scenarios throughout the twenty-first century, supplying 70% to 80% of primary energy in 2100. Non-fossil fuel energy use grows over the century in all three reference scenarios. The range of contributions in 2100 is from 250 EJ/yr to 450 EJ/yr – an amount equaling roughly one-half to a little over global primary energy consumption today.

[Notes. i. Oil consumption includes that derived from tar sands and oil shales, and coal consumption includes that used to produce synthetic liquid and gaseous fuels. ii. Primary energy consumption from nuclear power and non-biomass renewable electricity are accounted for at the average efficiency of fossil-fired electric facilities, which vary over time and across scenarios. This long-standing convention means that, all other things being equal, increasing efficiency of fossil-electric energy lowers the contribution to primary energy from these sources.]

Figure TS.1. Global Primary Energy Consumption Across Reference Scenarios (EJ/yr)

The three reference scenarios show the implications of this increasing demand and the improved access to energy, with the ranges reflecting the variation among the scenarios from the three modeling groups. Global primary energy consumption rises substantially in all three reference scenarios, from about 400 EJ/yr in 2000 to between roughly 1275 EJ/yr and 1500 EJ/yr in 2100 (Figure TS.1). U.S. primary energy consumption also grows substantially, about 1¼ to 2½ times present levels by 2100 (Figure TS.2). Primary energy growth occurs despite continued improvements in the efficiency of energy use and energy production technologies. For example, the U.S. energy intensity – the ratio of primary energy consumption to economic output – declines 60% to 75% between 2000 and 2100 across the three reference scenarios.

All three reference scenarios include an eventual reduction in the consumption of conventional oil resources. However, in all three, a range of alternative fossil-based resources, such as synthetic fuels from coal and unconventional oil resources (e.g., tar sands and oil shales), are available and become economically viable. Fossil fuels provided almost 90% of global primary energy in the year 2000, and they remain the dominant energy source in the three reference scenarios throughout the twenty-first century, supplying 70% to 80% of total primary energy in 2100.

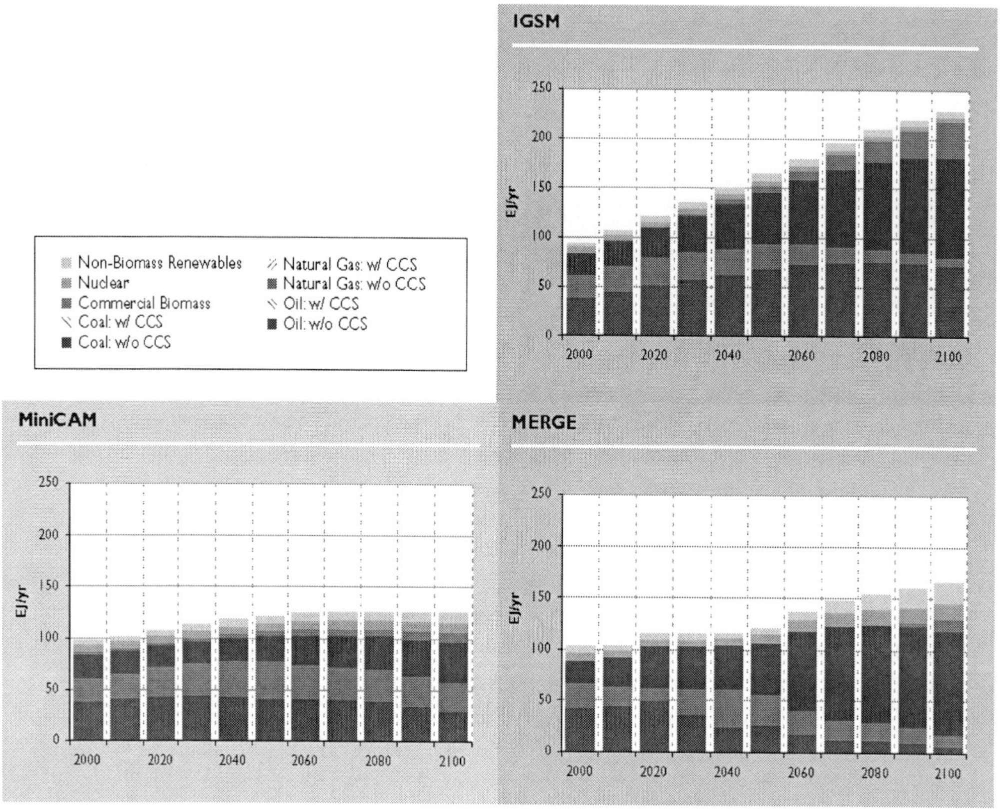

U.S. primary energy consumption rises in all three reference scenarios, to roughly 1¼ to 2½ times present levels by 2100. This growth occurs despite continued improvements in the efficiency of energy use and production. U.S. energy intensity declines 60% to 75% between 2000 and2100 in the reference scenarios.

[Notes. i. Oil consumption includes that derived from tar sands and oil shales, and coal consumption includes that used to produce synthetic liquid and gaseous fuels. ii. Primary energy consumption from nuclear power and non-biomass renewable electricity are accounted for at the average efficiency of fossil-fired electric facilities, which vary over time and across scenarios. This long-standing convention means that, all other things being equal, increasing efficiency of fossil-electric energy lowers the contribution to primary energy from these sources.]

Figure TS.2. U.S. Primary Energy Consumption Across Reference Scenarios (EJ/yr)

However, non-fossil fuel energy use also grows over the century in all three reference scenarios. Contributions to primary energy consumption in 2100 range from 250 EJ to 450 EJ – an amount equaling roughly ½ times to a little over global primary energy consumption today. Despite this growth, these sources never supplant fossil fuels, although they provide an increasing share of the total, particularly in the second half of the century.

Consistent with the characteristics of primary energy consumption, global and U.S. electricity production continues to rely on coal, although the contribution of coal varies among the reference scenarios (Figure TS.3 and Figure TS.4). The contribution of renewable and nuclear energy varies considerably in the different reference scenarios, depending on resource availability, technology, and non-climate policy considerations. For example, global nuclear generation in the reference scenarios ranges from about 1½ times current levels (if non-climate concerns such as safety, waste, and proliferation constrain its growth as is the case in one reference scenario), to an expansion of almost an order of magnitude assuming relative economics as the only constraint.

In the reference scenarios, oil and natural gas prices rise through the century relative to year 2000 levels, whereas coal and electricity prices remain relatively stable. It should be emphasized, however, that the models used in this research were not designed to simulate short- term, fuel-price spikes, such as those that occurred in the 1970s, early 1980s, and more recently in 2005. Thus, price trends in the scenarios should be interpreted as multi-year averages.

As a combined result of all these influences, CO_2 emissions from fossil fuel combustion and industrial processes in the reference scenarios increase from approximately 7 gigatonne carbon per year (GtC/yr) in 2000 to between 22.5 GtC/yr and 24.0 GtC/yr in 2100; that is, roughly 3 to 3½ times current levels (Figure TS.5). (Note that one tonne C is equivalent to 3.67 tonnes CO_2. See Box 3.2 for more on converting between units of carbon and units of CO_2.)

It is instructive to see how emissions are divided between industrialized countries (Annex 1) and developing countries (Non-Annex 1). Developing country emissions overtake those of developed countries in the 2020 to 2030 timeframe in the reference scenarios (Figure TS.6). This suggests the difficulty of stabilizing radiative forcing without developing-country participation. Indeed, even if developed countries were to reduce their emissions to zero, global involvement would still be necessary for stabilization.

The capacity of the ocean to absorb CO_2 differs among the three models. The ocean is a major sink for CO_2, and the rate at which the oceans take up CO_2 generally increases in the reference scenarios as concentrations rise early in the century. However, processes in the ocean can slow this rate of increase at high concentrations late in the century. Ocean uptake in the three reference scenarios is roughly 2 GtC/yr in 2000, rising to about 5 GtC/yr to 11 GtC/yr by 2100. The three ocean models behave more similarly in the stabilization scenarios; for example, the difference in ocean uptake between models, is less than 1 GtC/yr in 2100 under the most stringent stabilization level.

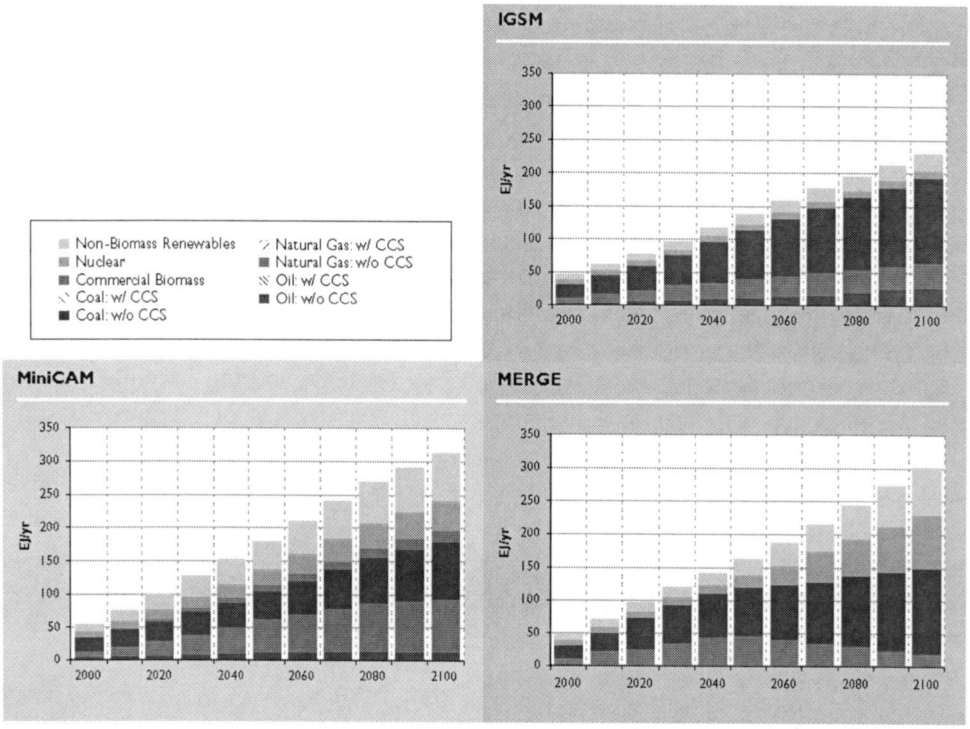

Global electricity production grows to over four times production levels in 2000 in all the reference scenarios. Global electricity production shows continued reliance on coal, although this contribution varies among the reference scenarios. The contribution of renewable energy and nuclear power varies considerably among the reference scenarios, depending on assumptions about resource availability, technology, and non-climate policy considerations. For example, global production of electricity from nuclear power in the reference scenarios ranges from about 1½ times current levels (if non-climate concerns such as safety, waste, and proliferation constrain its growth as is the case in one reference scenario), to an expansion of almost an order of magnitude assuming relative economics as the only constraint.

Figure TS.3. Global Electricity Production Across Reference Scenarios (EJ/yr)

Two of the three participating models include sub-models of the exchange of CO_2 with the terrestrial biosphere, including the net uptake by plants and soils and the emissions from deforestation. In the reference scenarios from these modeling groups, the terrestrial biosphere acts as a small annual net sink (less than 1 GtC/yr) in 2000, increasing to an annual net sink of roughly 2 GtC/yr to 3 GtC/yr by the end of the century. The third modeling group assumed a zero net exchange. Changes in emissions from terrestrial systems over time in the reference scenarios reflect assumptions about human activity (including a decline in deforestation) as well as increased CO_2 uptake by vegetation as a result of the positive effect of CO_2 on plant growth. There remains substantial uncertainty about this carbon fertilization effect and its evolution under a changing climate.

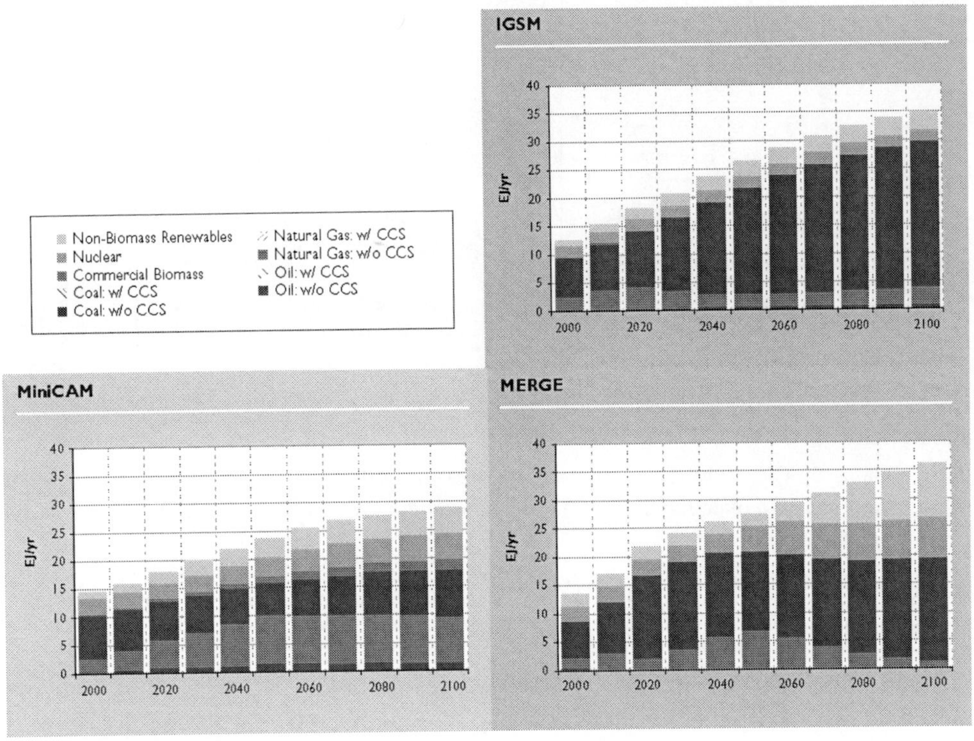

Continued dependence on coal for electricity generation is a feature of all three reference scenarios, with the degree of dependence varying among scenarios. Differences in the use of nuclear power reflect differing assumptions about the degree to which issues of safety, waste, and proliferation constrain its growth.

Figure TS.4. U.S. Electricity Production AcrossReference Scenarios (EJ/yr).

Although this Technical Summary focuses on the most important anthropogenic GHG, CO_2, the scenarios considered a number of other GHGs (CH_4 , N_2O SF_6, PFCs, and HFCs), which are emitted from various sources, including agriculture, waste management, biomass burning, fossil fuel production and consumption, and a number of industrial activities. Future global anthropogenic emissions of CH_4 and N_2O vary widely among the reference scenarios, ranging from flat or declining emissions to increases of 2 to 2½ times present levels. These differences reflect differing assumptions about technological opportunities and about whether current emissions rates will be reduced significantly for non-climate reasons, such as air pollution control and/or higher natural gas prices that would further stimulate the capture of CH_4 emissions for its fuel value.

Increases in emissions from the global energy system and other human activities lead to higher atmospheric GHG concentrations and radiative forcing. These increases are moderated by natural biogeochemical removal processes. As a result, GHG concentrations rise substantially over the century in the reference scenarios. By 2100, CO_2 concentrations range from about 700 ppmv to 900 ppmv, up from 365 ppmv in 1998. CH_4 concentrations in 2100 range from 2000 ppbv to 4000 ppbv, up from 1745 ppbv in 1998, and N_2O concentrations in 2100 range from about 375 ppbv to 500 ppbv, up from 314 ppbv in 1998.

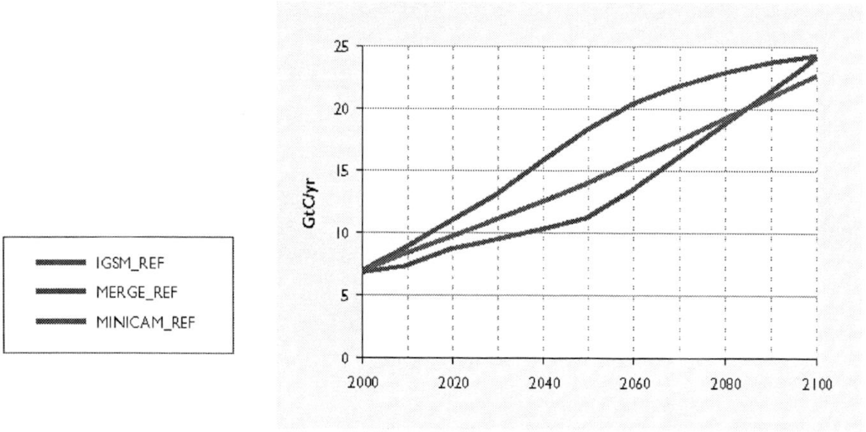

Global emissions of CO_2 from fossil fuel combustion and other industrial sources, mainly cement production, increase over the century in all three reference scenarios. By 2100 emissions reach 22.5 GtC/yr to 24.0 GtC/yr.

Figure TS.5. Global Emissions of CO_2 from Fossil Fuels and Industrial Sources [CO_2 from land-use change excluded] Across Reference Scenarios (GtC/yr).

As a result, radiative forcing in 2100 ranges from 6.4 W/m^2 to 8.6 W/m^2 from preindustrial, up from a little over 2 W/m^2 today. The non-CO_2 GHGs account for about 20% to 25% of radiative forcing by the end of the century (Figure TS.7).

Stabilization Scenarios

Important assumptions underlying the stabilization scenarios include the flexibility that exists in a policy design, as represented by the modeling groups, to seek out least cost options for emissions control regardless of where they occur, what substances are controlled, or when they occur. This set of conditions is referred to as *where*, *what*, and *when* flexibility. Equal marginal costs of abatement among regions, across time (taking into account discount rates and the lifetimes of substances), and among substances (taking into account their relative warming potential and different lifetimes) will, under specified conditions, lead to least-cost abatement. Each modeling group applied an economic instrument that priced GHGs in a manner consistent with the group's interpretation of *where*, *what* and *when* flexibility. The economic characteristics of the scenarios therefore assume a policy designed with the intent of achieving the required reductions in GHG emissions in a least-cost way. Key implications of these assumptions are that: (1) all nations proceed together in restricting GHG emissions from 2012 and continue together throughout the century, and that the same marginal cost is applied across sectors (*where* flexibility); (2) the marginal cost of abatement rises over time based on each modeling group's interpretation of *when* flexibility, with the effect of linking emissions mitigation efforts over the time horizon of the scenarios; and (3) stabilization of radiative forcing is achieved by combining control of all GHGs, with differences in how modeling groups compared them (*what* flexibility).

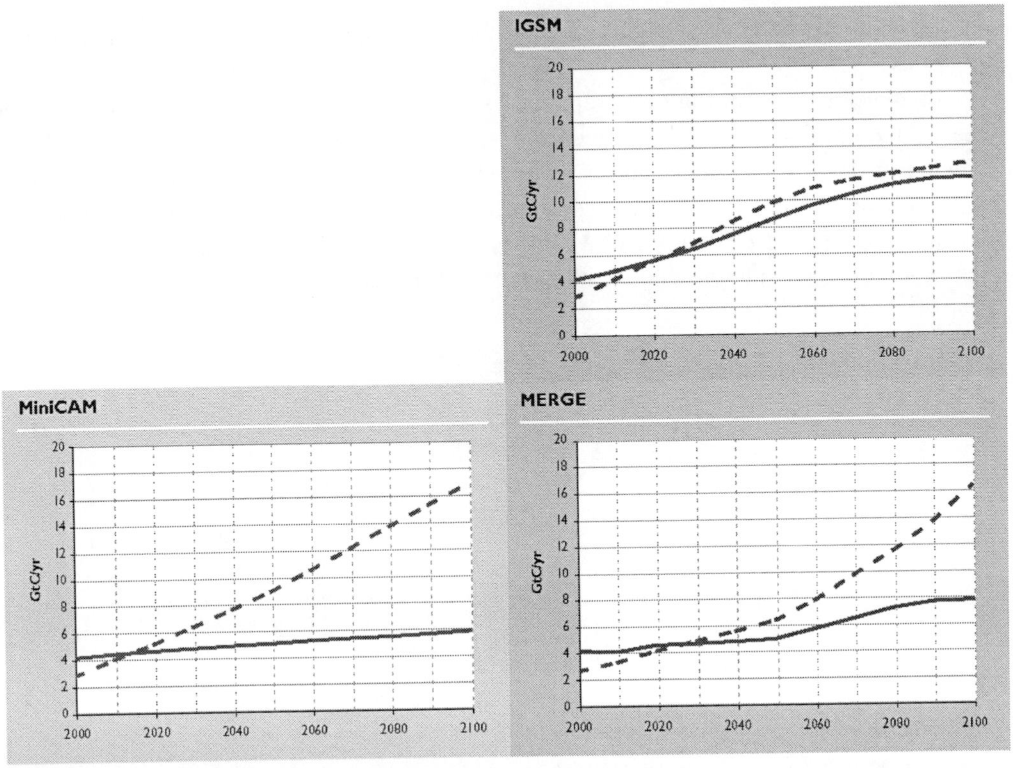

Emissions of fossil fuel and industrial $CO_2$from the Non-Annex I countries exceed those from the Annex I countries in all three reference scenarios by2030 or earlier. Non-Annex 1emissions continue to grow rapidly in two of the reference scenarios, such that their emissions are on the order of twice the level of Annex I by 2100.Emissions do not continue to diverge in the third reference scenario, due in part to relatively slower economic growth in Non-Annex I regions, faster growth in Annex I, and increased emissions in Annex I as they become producers and exporters of shale oil, tar sands, and synthetic fuels from coal.

Figure TS.6. Global Emissions of Fossil Fuel and Industrial $CO_2$by Annex I and Non-Annex I Countries AcrossReference Scenarios (GtC/yr).

Although these assumptions are convenient for analytical purposes, to gain an impression of the implications of stabilization, they are idealized versions of possible outcomes. For the abatement costs in these scenarios to be representative of actual abatement costs would require, among other things, that a negotiated international agreement include these flexibility mechanisms. Failure in that regard could have a substantial effect on the difficulty of achieving any of the stabilization levels considered in this research. For example, a delay in the participation of some large countries would require greater effort by the others, and policies that impose differential burdens on different sectors without mechanisms to allow for equalizing marginal costs across sectors can result in a many-fold increase in the cost of any environmental gain. Therefore, *it is important to view these result as scenarios under specified conditions, not as predictions or best-judgment forecasts of the most likely outcome within the national and international political system.* Further, none of the scenarios considered the extent to which variation from these least-cost rules might be improved upon given interactions with existing taxes, technology spillovers, or other non-market externalities.

If the developments in the three reference scenarios were to occur, concerted efforts to reduce GHG emissions would be required to stabilize radiative forcing at the levels considered in this research. Such limits would shape technology deployment throughout the century and have important economic consequences. The stabilization scenarios demonstrate that there is no single technology pathway consistent with a given level of radiative forcing. Furthermore, there are other possible pathways than those considered in this research.

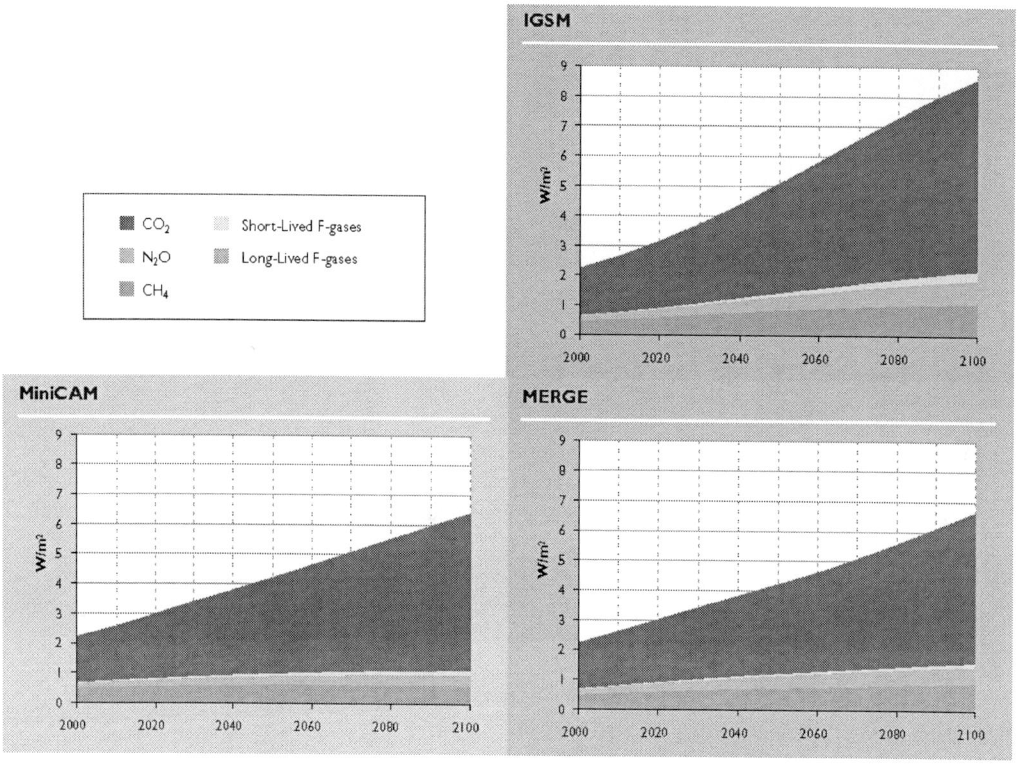

CO_2 accounts for 75% to 80% of the radiative forcing in 2100 in the three reference scenarios. Total radiative forcing in 2100 from all the GHGs considered in this research ranges from about 6.4W/m^2 to 8.6 W/m .

Figure TS.7. Radiative Forcing by Gas Across Reference Scenarios (W/m^2 from preindustrial)

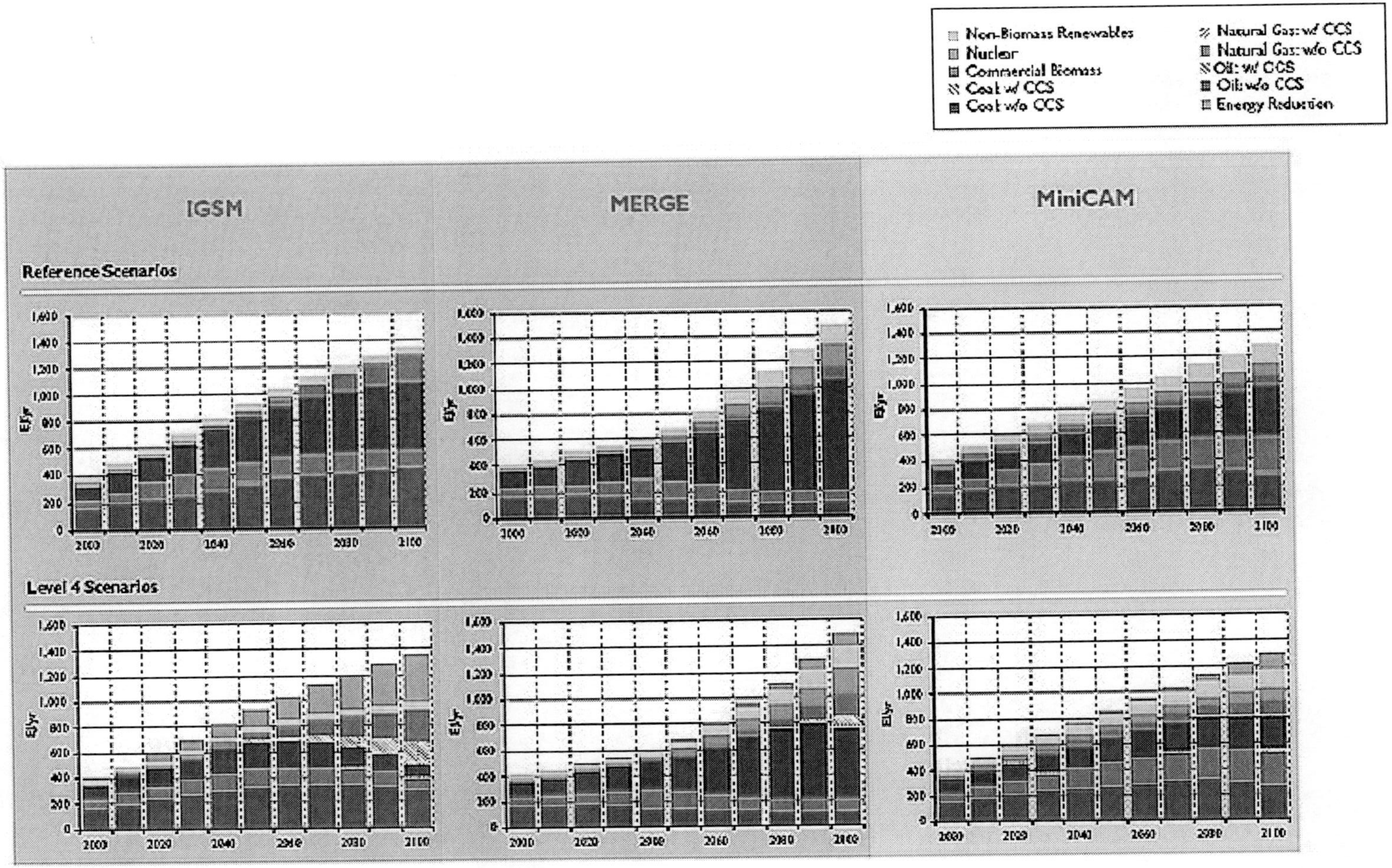

Figure TS. 8 continued

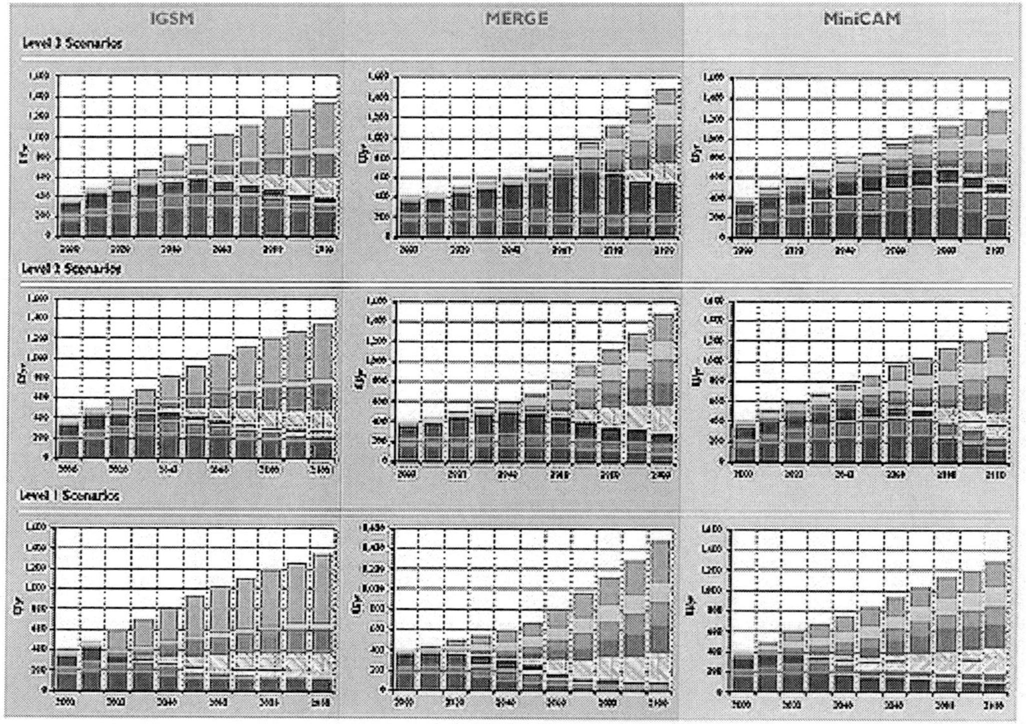

The global energy system undergoes a significant transformation in the stabilization scenarios from all three modeling groups. This transformation begins earlier the more stringent the radiative forcing stabilization level, and would continue into the next century for all stabilization levels. The transformation includes: reduction in energy use, increased use of carbon-free sources of energy such as biomass, other renewables, and nuclear power; and the addition of CCS. The contribution of each of these varies among the models reflecting different assumptions about cost and performance, policy, and resource limits.

[Notes. i. Oil consumption includes that derived from tar sands and oil shales, and coal consumption includes that used to produce synthetic liquid and gaseous fuels. ii. Primary energy consumption from nuclear power and non-biomass renewable electricity are accounted for at the average efficiency of fossil-fired electric facilities, which vary over time and across scenarios. This long-standing convention means that, all other things being equal, increasing efficiency of fossil-electric energy lowers the contribution to primary energy from these sources.]

Figure TS.8. Global Primary Energy Consumption by Fuel Across Scenarios (EJ/yr).

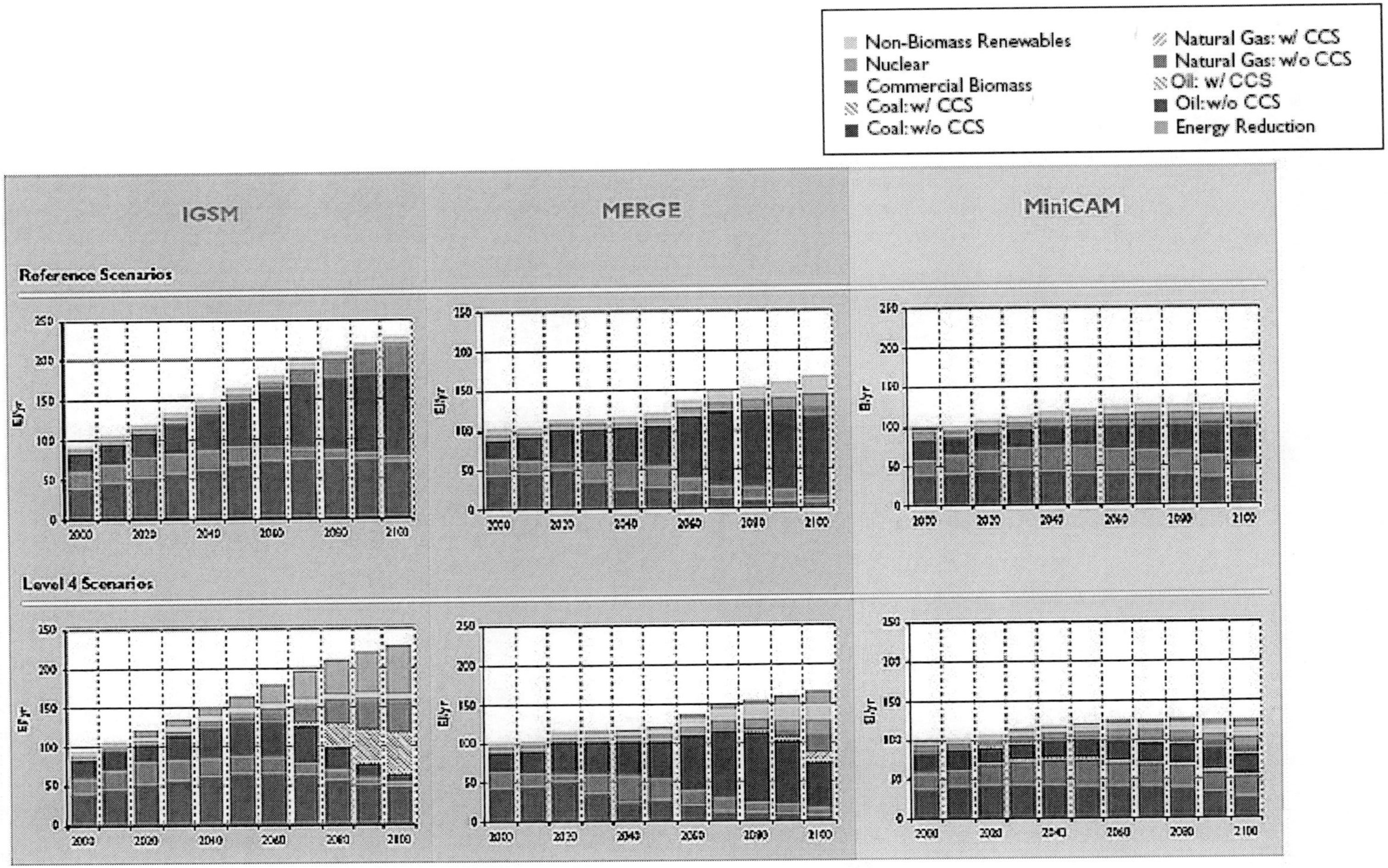

Figure TS. 9. continued

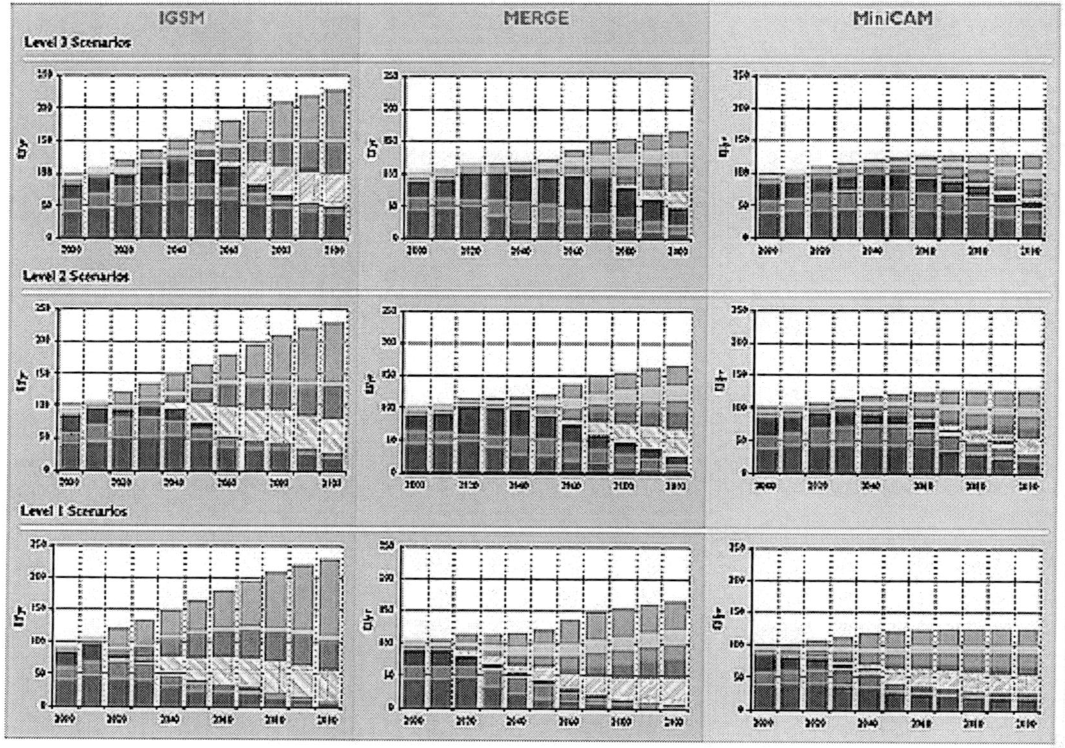

The U.S. energy system undergoes a significant transformation in the stabilization scenarios similar to the transformation in the global energy system. One difference, not obvious in this figure, is the transformation from conventional oil and gas to synthetic fuel production derived from shale oil or coal. One model (IGSM) includes heavy use of shale oil in the reference with some coal gasification, whereas another (MERGE) includes primarily synthetic liquid and gaseous fuels derived from coal. The third (MiniCAM) includes an intermediate mix of both. [Notes. i. Oil consumption includes that derived from tar sands and oil shales, and coal consumption includes that used to produce synthetic liquid and gaseous fuels. ii. Primary energy consumption from nuclear power and nonbiomass renewable electricity are accounted for at the average efficiency of fossil-fired electric facilities, which vary over time and across scenarios. This long-standing convention means that, all other things being equal, increasing efficiency of fossil-electric energy lowers the contribution to primary energy from these sources.]

Figure TS.9. U.S. Primary Energy Consumption by Fuel Across Scenarios (EJ/yr).

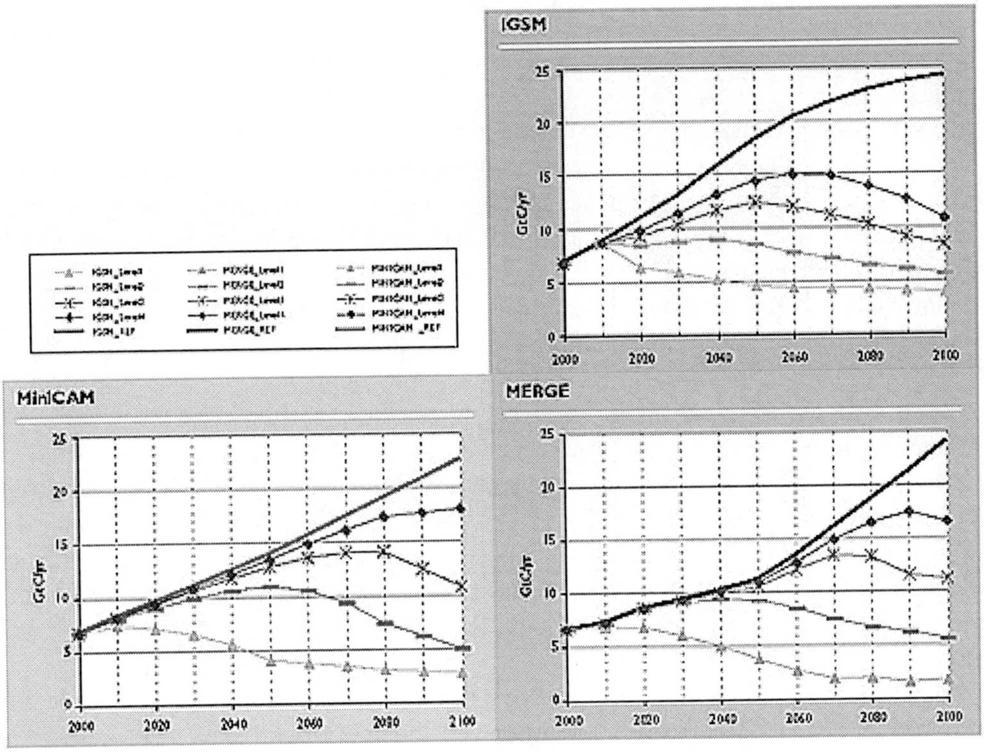

The tighter the constraint on radiative forcing, the faster carbon emissions must decline from those in the reference scenarios. This is because the stabilization level defines a long-term carbon budget;t hat is, the remaining amount of carbon that can be emitted in the future. The gradual deflection of the emissions from the reference reflects the assumption of when flexibility, with carbon prices rising gradually. Under the most stringent radiative forcing stabilization levels, CO_2 emissions begin to decline immediately or within a matter of decades. Under less stringent radiative forcing stabilization levels, CO_2emissions do not peak until late in the century or beyond, and they are 1½ to over 2½ times today's levels in 2100.

Figure TS.10. Global Emissions of $CO_2$from Fossil and other Industrial Sources Across Scenarios (GtC/yr).

Stabilization of radiative forcing at the levels examined in this research would require a substantially different energy system globally, and in the U.S., than what emerges in the reference scenarios. The degree and timing of change in the global energy system depends on the level at which radiative forcing is stabilized (Figure TS.8 and Figure TS.9). The lower the radiative forcing stabilization level, the larger the scale of change in the global energy system relative to the reference scenario required over the coming century and the sooner those changes would need to occur.

Across the stabilization scenarios, the energy system relies more heavily on non-fossil energy sources, such as nuclear, solar, wind, biomass, and other renewable energy forms, than in the associated reference scenarios. The stabilization scenarios differ in the degree to which these technologies are deployed, depending on assumptions about: technological improvements; the ability to overcome obstacles, such as intermittency in the case of solar and wind power, or safety, waste, and proliferation issues in the case of nuclear power; and the policy environment surrounding these technologies. Energy consumption, while still higher than today's levels, is lower in the stabilization scenarios than in the reference scenarios.

CCS is widely deployed in the stabilization scenarios because each modeling group assumed that the technology can be successfully developed and that concerns about storing large amounts of carbon do not impede its expansion. Removal of this assumption would make the stabilization levels more difficult to achieve and would lead to greater demand for low-carbon sources such as renewable energy and nuclear power, to the extent that growth of these other sources is not otherwise constrained.

Significant fossil fuel use continues across the stabilization scenarios, because stabilization allows for some level of carbon emissions through 2100, and because of the presence of CCS technology in all the stabilization scenarios.

Increased use is made of biomass energy crops in all the stabilization scenarios, the contribution of which is ultimately limited by competition with agriculture and forestry. One modeling group examined the importance of valuing terrestrial carbon similarly to the way fossil fuel carbon is valued in stabilization scenarios. It was found that important interactions between large-scale deployment of commercial bioenergy crops and land use occurred to the detriment of unmanaged ecosystems when no economic value was placed on carbon in terrestrial systems.

Across the stabilization scenarios, the scale of the emissions reductions required relative to the reference scenario increases over time, with the bulk of emissions reductions taking place in the second half of the century. But emissions reductions occur in the first half of the century in every stabilization scenario (Figure TS. 10).

The 2100 time horizon of this research limited examination of the ultimate stabilization requirements. Further reductions in CO_2 emissions after 2100 would be required in all of the stabilization scenarios, because stabilization of radiative forcing at any of the levels considered in this research requires human emissions of CO_2 in the long term to be essentially halted. Despite the fact that much of the carbon emissions will eventually make its way into oceans and terrestrial sinks, some will remain in the atmosphere for thousands of years. Only CCS can allow continued burning of fossil fuels. Higher radiative forcing limits can delay the point in time at which emissions must be reduced toward zero, but this requirement must ultimately be met.

Fuel sources and electricity generation technologies change substantially, both globally and in the U.S., in the stabilization scenarios compared to the reference scenarios. There are a variety of technological options in the electricity sector that reduce carbon emissions in these scenarios (Figure TS.1 1 and Figure TS.12).

By the end of the century, electricity produced by conventional fossil technology that freely emits CO_2 is reduced in the stabilization scenarios relative to reference scenarios. Electricity production from technologies that emit CO_2 varies substantially with the stabilization level; in the most stringent stabilization scenarios, electricity production from these technologies is reduced toward zero.

The economic effects of stabilization are substantial in many of the stabilization scenarios, although much of this cost is borne later in the century. As noted earlier, each of the modeling groups assumed that a global policy was implemented after 2012, with universal participation by the world's nations, and that the time path of reductions approximated a least-cost solution.

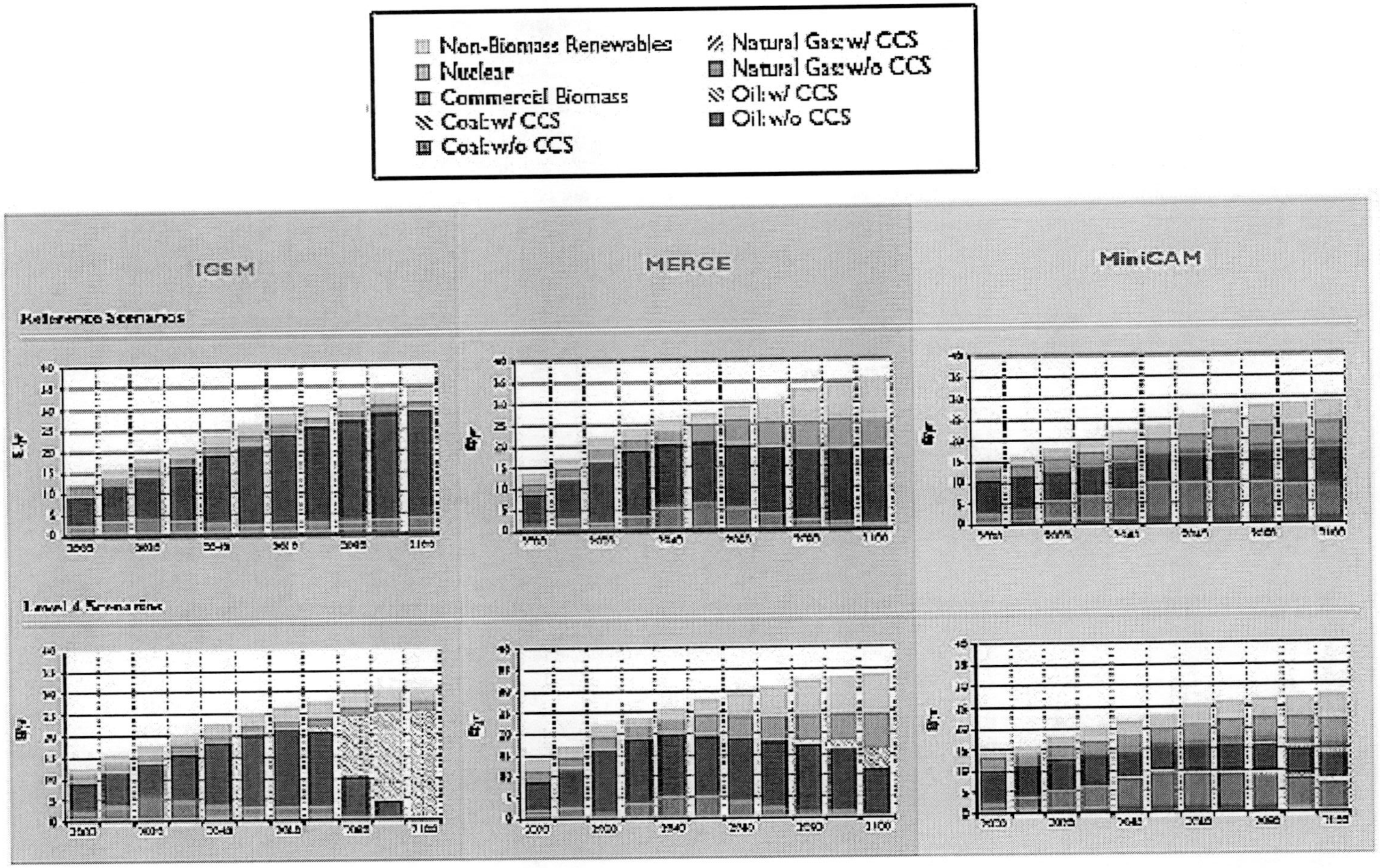

Figure TS. 11 continued

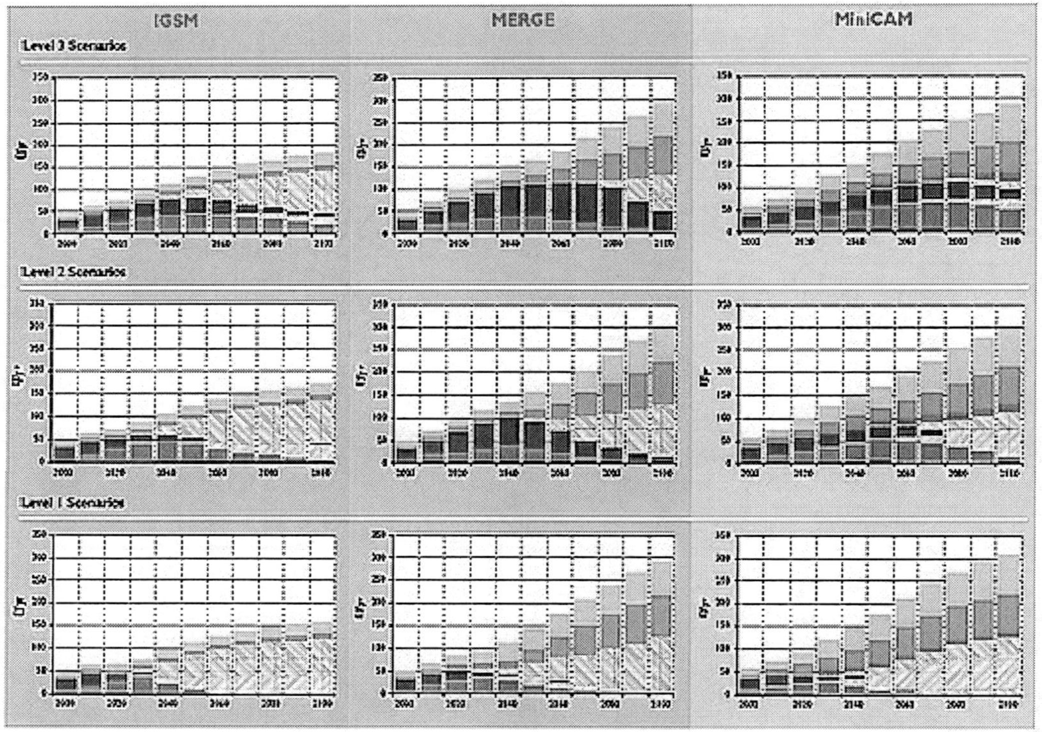

Various electricity technology options could be competitive in the future, and different assumptions regarding their relative economic viability, reliability, and resource availability lead to considerably different scenarios of the global electricity sector in reference and stabilization scenarios across modeling groups. One reference scenario includes relatively little change in the electricity sector mix in the reference scenario. The other two reference scenarios include more substantial transformations from the present. In all scenarios, large changes from reference are required to stabilize radiative forcing at the levels considered in this research. In most cases, the relative proportion of electricity in energy consumption increases in the stabilization scenarios, so the relative reductions in electricity production are generally smaller than for primary energy.

Figure TS.11. Global Electricity Production by Fuel Across Scenarios (EJ/yr)

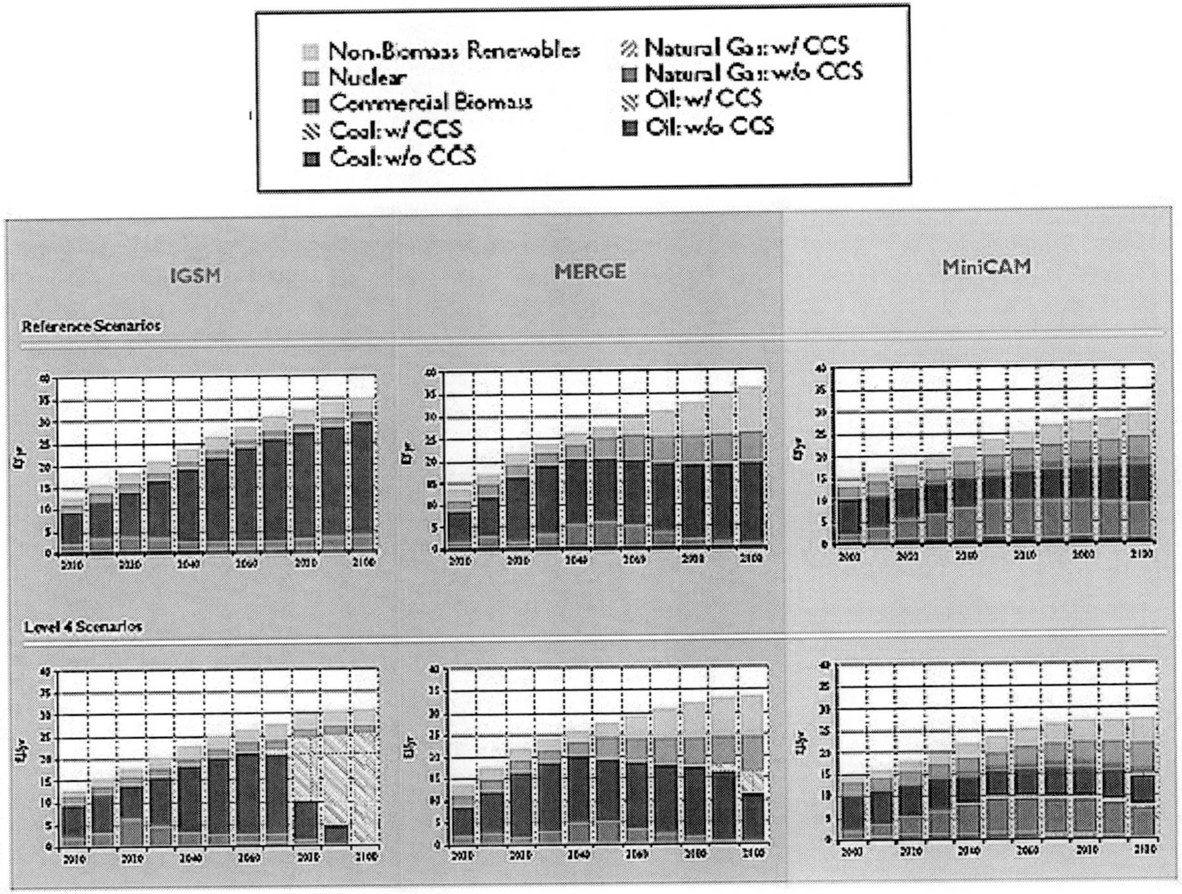

Figure TS. 12 continued

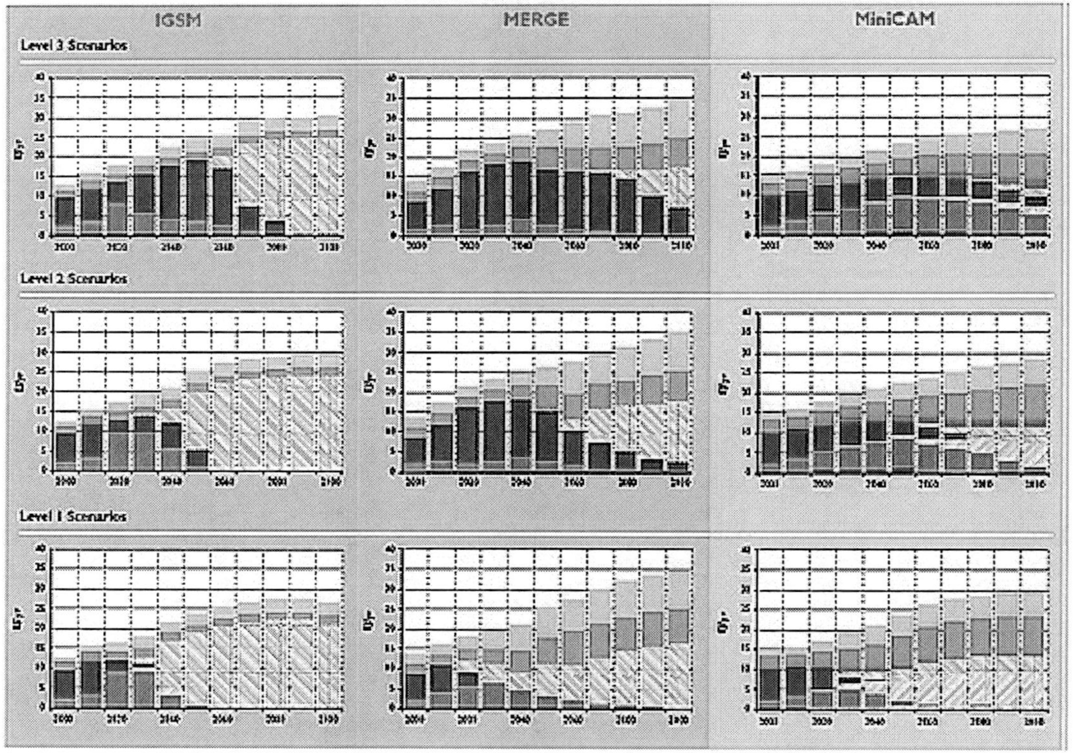

U.S. electricity generation sources and technologies are substantially transformed to meet the four radiative forcing stabilization levels. CCS figures in the stabilization scenarios from all three modeling groups, but the contribution of other sources and technologies and the total amount of electricity used differ substantially. In most cases, the relative proportion of electricity in energy consumption increases in the stabilization scenarios, so the relative reductions in electricity production are generally smaller than for primary energy. In one scenario (MiniCAM Level 1), electricity production in the U.S. increases under stabilization in the second half of the century.

Figure TS.12. U.S. Electricity Production by Fuel Across Scenarios (EJ/yr).

Table TS.3. Carbon Prices at Various Points in Time for the Stabilization Scenarios

	2020 ($/tonne C)			2030 ($/tonne C)		
Stabilization Level	**IGSM**	**MERGE**	**MiniCAM**	**IGSM**	**MERGE**	**MiniCAM**
Level 4	$18	$1	$1	$26	$2	$2
Level 3	$30	$2	$4	$44	$4	$7
Level 2	$75	$8	$15	$112	$13	$26
Level 1	$259	$110	$93	$384	$191	$170
	2050 ($/tonne C)			**2050 ($/tonne C)**		
Stabilization Level	**IGSM**	**MERGE**	**MiniCAM**	**IGSM**	**MERGE**	**MiniCAM**
Level 4	$58	$6	$5	$415	$67	$54
Level 3	$97	$11	$19	$686	$127	$221
Level 2	$245	$36	$69	$1,743	$466	$420
Level 1	$842	$574	$466	$6,053	$609	$635

These assumptions of where, when, and *what* flexibility lower the economic consequences of stabilization relative to what they might be with other implementation approaches.

The stabilization scenarios follow a pattern where, in most scenarios, the carbon price rises steadily over time (Table TS.3), providing an opportunity for the energy system to adjust gradually. Although the general shape of the carbon price trajectory over time is similar across the models, the carbon prices vary substantially across the models. For example, for the less stringent stabilization levels two of the modeling groups produced scenarios with carbon prices of $10 or below per tonne of carbon in 2020, with carbon prices rising to roughly $100 per tonne in 2020 at the most stringent stabilization level. The scenarios from the third modeling group show higher initial carbon prices in 2020, ranging from around $20 for the least stringent stabilization level to over $250 for the most stringent stabilization level. (Note that $100/tonne C is equivalent to $27/tonne CO_2.

See Box 3.2 for more on converting between units of carbon and units of CO_2.)

These differences in carbon prices, along with other model features, lead to similar variation in the costs of stabilization. Under the most stringent radiative forcing stabilization level, for example, gross world product (aggregating country figures using market exchange rates) is reduced in 2050 by around 1% in the scenarios from two of the modeling groups and approximately 5% in the scenario from the third. In 2100 it is reduced by less than 2% in two of the scenarios and over 16% in the third.

Table TS.4. Cumulative EmissionsReductions from the Reference Scenarios Across Models in the Stabilization Scenarios (GtC through 2100)

	IGSM	MERGE	MiniCAM
Level 4	472	112	97
Level 3	674	258	267
Level 2	932	520	541
Level 1	1172	899	934

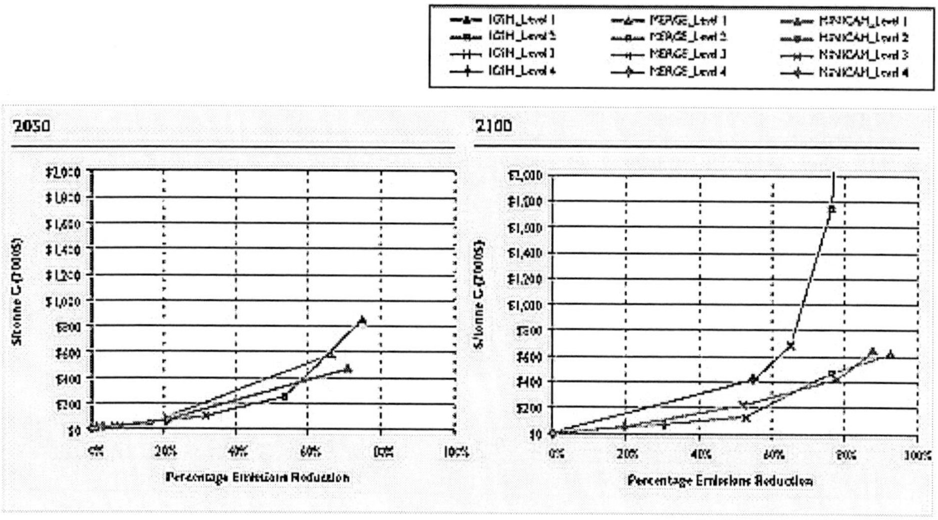

The relationship between carbon price and percentage reduction in emissions is similar among the models in 2050. In 2100, the relationship between carbon price and percentage reduction in emissions diverges across the models, due in large part to different assumptions regarding the technologies available to facilitate emissions reductions late in the century.

[Note. CO_2 emissions vary across the reference scenarios from the three modeling groups, so that similar percentage reductions, as shown in this figure, imply differing levels of total emissions reduction.]

Figure TS. 13. Relationship Between Carbon Price and Percentage Emissions Reductions in 2050 and 2100

The variation in carbon prices and reductions in gross world product is attributable to many factors, but two are most prominent. First, the amount that CO_2 emissions must be reduced to achieve stabilization differs between the scenarios from the different modeling groups (Table TS.4), because of differing assumptions regarding economic growth and other factors that determine emissions in the reference scenarios; levels of CO_2 uptake by the oceans and terrestrial biosphere; and availability of control for non-CO_2 GHGs.

Second, the modeling groups chose different assumptions regarding the technologies available for emissions reductions, particularly in the second half of the century. Most prominent are differences in assumptions about technologies to shift final energy demand to low-carbon sources such as biofuels and low-carbon electricity or hydrogen, in transportation, industrial and buildings end uses. The differences in technological assumptions among the modeling groups is reflected in the relationship between carbon prices and percentage abatement (Figure TS.13), a form of marginal abatement cost curve, for the three models in 2050 and 2100. The scenarios from the three modeling groups exhibit very similar behavior through 2050, but different assumptions about technological options lead to a divergence among the models by 2100.

In all of the scenarios, emissions reductions in electric power sector come at relatively lower prices than in other sectors (e.g. buildings, industry, and transport) so that the electricity sector is essentially decarbonized in the most stringent scenarios from all three modeling groups (Figure TS.14). At somewhat higher cost, other sectors can respond to rising carbon prices by reducing demands for fossil fuels, applying CCS technologies where possible, and substituting low-carbon energy sources such as bioenergy and low-carbon electricity or hydrogen. The amount of electricity used per unit of total primary energy increases in all of the stabilization scenarios (Figure TS. 15), but those scenarios with the highest relative use of electricity tend to exhibit lower stabilization costs in part because of the larger role of decarbonized power generation. Assumptions regarding costs and performance of technologies to facilitate these adjustments, particularly in the post-2050 period, play an important role in determining stabilization costs.

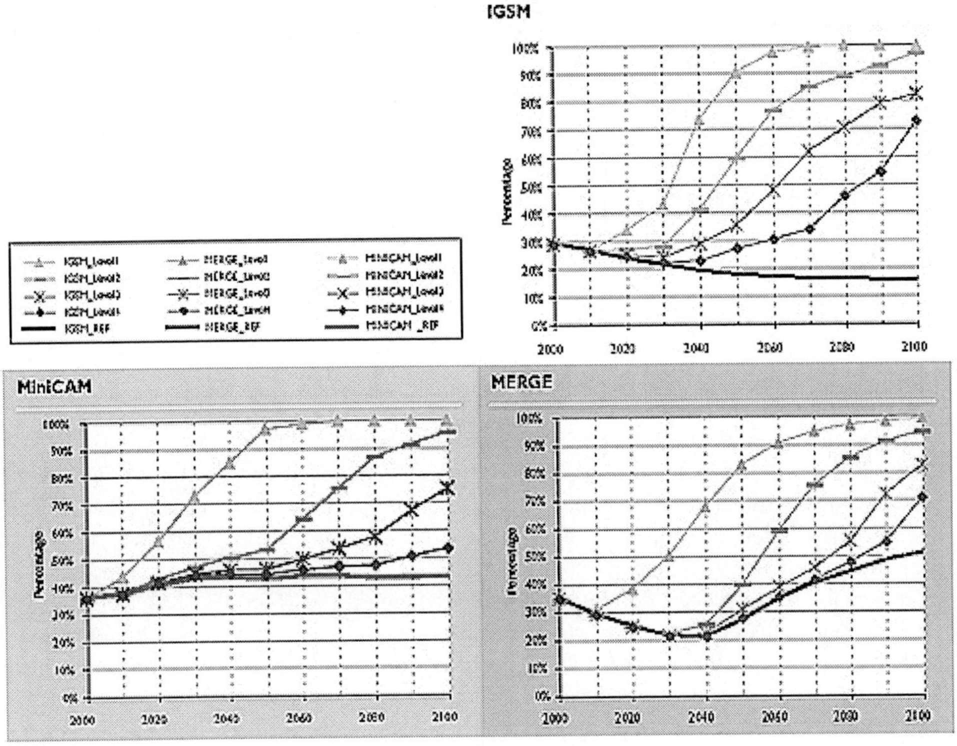

All three modeling groups assumed sufficient technological options to allow for substantial reductions in carbon emissions from electricity production. Options include fossil power plants with CCS, nuclear power, and renewable energy such as hydro electric power, wind power, and solar power. In all of the Level 1scenarios, the electricity sector is almost fully de carbonized by the end of the century.

Figure TS.14. Percentage of World Electricity Production from Low- or Zero-Emissions Technologies Across Scenarios (percentage).

The assumption of *when* flexibility links elements of each stabilization scenario through time. This in turn means that in addition to near- term technology availability, differences in assumptions about technology in the post-2050 period are reflected in near-term emissions re-ductions and GHG prices.

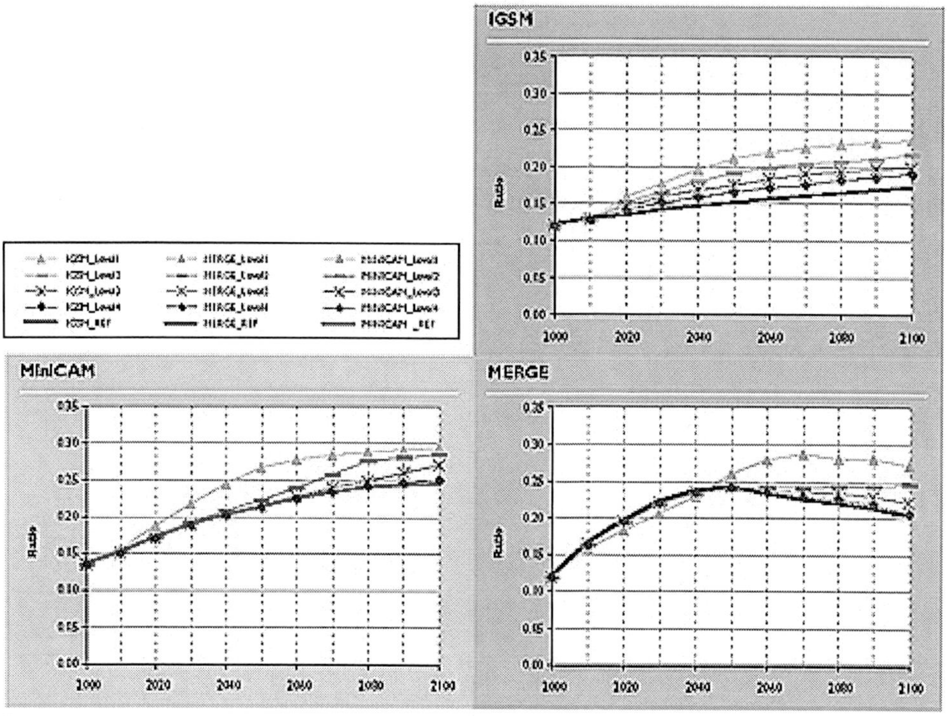

Efforts to constrain CO_2 emissions result in increased use of electricity as a fraction of total primary energy in the scenarios from all three modeling groups. This is because all three modeling groups assumed lower-cost technology options for reductions in emissions from electricity production than for substitution away from fossil fuels in direct uses such as transportation. The scenarios from two of the modeling groups (MERGE and MiniCAM) generally include greater electrification than the scenarios from the third modeling group (IGSM). Greater opportunities to electrify reduce the economic impacts of stabilization. [Note. Primary energy consumption from nuclear powerand non-biomass renewable electricity are accounted for at the average efficiency of fossil-fired electric facilities, which vary over timeand across scenarios.]

Figure TS.15. Ratio of Global Electricity Production to Primary Energy Consumption Across Scenarios.

> As noted earlier, the overall cost levels are strongly influenced by the idealized policy scenario that has all countries participating from the start, the assumption of *where* flexibility, an efficient pattern of emissions reductions over time, and integrated reductions in emissions of the different GHGs. Assumptions in which policies are implemented in a less efficient manner would lead to higher costs. Thus, these scenarios should not be interpreted as applying beyond the particular conditions assumed.

Constraints on GHG emissions also affect fuel prices. Generally, producer prices for fossil fuels fall as demand for them is depressed by the stabilization measures. Consumers of fossil fuels, on the other hand, pay for fuel plus a carbon price if the CO_2 emissions are freely released to the atmosphere (Table TS.5). There fore, consumer costs of energy rise with more stringent stabilization levels in these scenarios.

Table TS.5. Relationship Between a$100/tonne Carbon Price and Fuel Prices

Fuel	Base Cost ($2005)	Added Cost ($)	Added Cost (%)
Crude Oil ($/bbl)	$60.0	$12.2	20%
Regular Gasoline ($/gal)	$2.39	$0.26	11%
Heating Oil ($/gal)	$2.34	$0.29	12%
Wellhead Natural Gas ($/tcf)	$10.17	$1.49	15%
Residential Natural Gas ($/tcf)	$15.30	$1.50	10%
Utility Coal ($/short ton)	$32.6	$55.3	170%
Electricity (c/kWh)	9.6¢	1 .76¢	1 8%

Source. Bradley et al. 1991, updated with U.S. average prices for the 4th quarter of 2005 as reported in DOE 2006.

(In most cases, stabilization depresses producer prices and so the percentage rise in the fuel cost seen by consumers would be less than indicated here. The change in producer price is highly scenario- and model-dependent.)

Non-CO_2 gases play an important role in shaping the degree of change in the energy system. Scenarios that assume relatively better performance of technologies for reducing non-CO_2 emissions allow a given radiative forcing stabilization level to be met with greater radiative forcing from CO_2 and, all other things being equal, less extensive changes to the energy system. Differences in GHG concentrations among the three models reflect differences in assumed mitigation opportunities for non-CO_2 GHGs relative to CO_2. For example, lower CH_4 and N_2O emissions in the scenarios from one of the modeling groups reflects a greater market penetration of technologies that reduce CH_4 and N_2O emissions with positive profits even in the reference scenario, and significant abatement in the stabilization scenarios. With lower levels of CH_4 and N_2O than is the case in the scenarios from the other two modeling groups, higher levels of CO_2 are still consistent with the overall radiative forcing levels (Figure TS. 16).

Achieving stabilization of atmospheric GHGs poses a substantial technological and policy challenge. It would require important transformations of the global energy system. The cost and feasibility of such a goal depends on the evolution of technology and its ability to overcome existing limits and barriers to adoption, and it depends on the efficiency and effectiveness of the policy instruments employed to achieve stabilization. These scenarios provide a means to gain insight into the challenge of stabilization and the implications of technology.

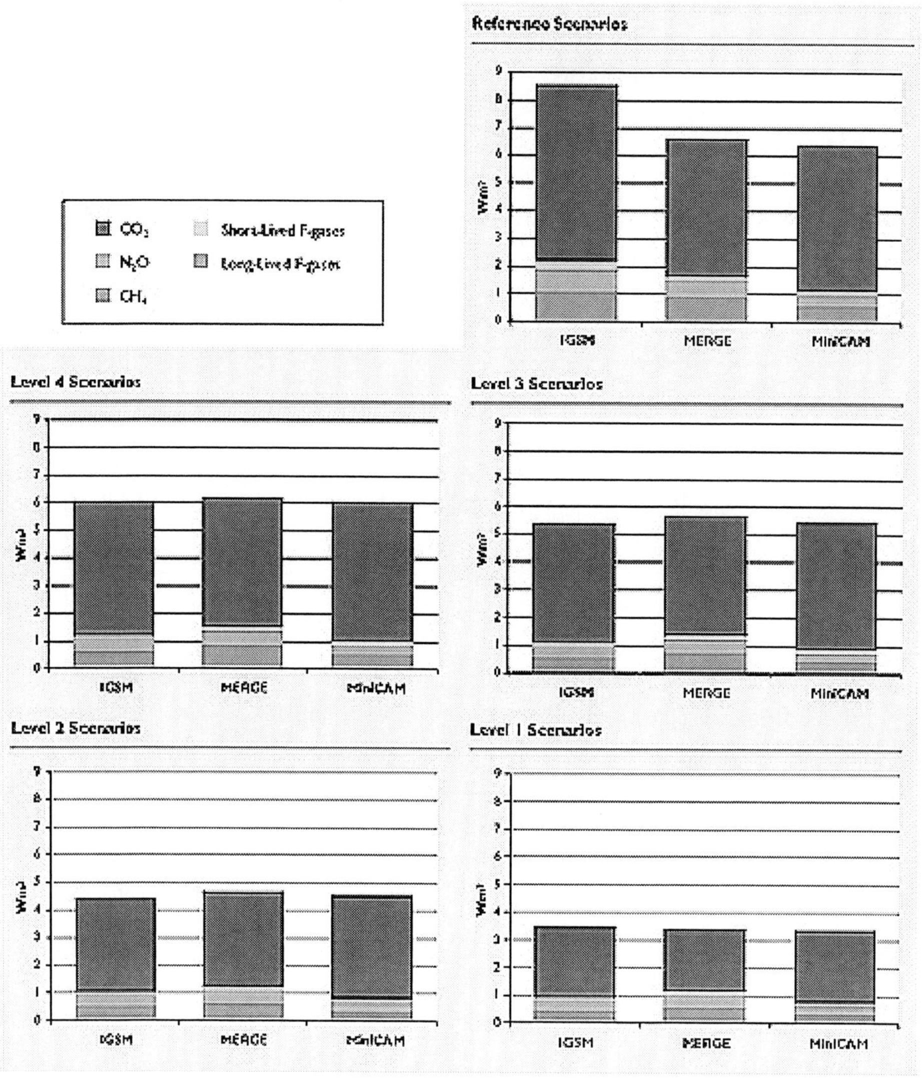

CO_2 is the main contributor to radiative forcing in the year 2100 in all of the scenarios. The opportunities to reduce control emissions from non-CO_2 GHGs influence the CO_2emissions reductions required to meet a given radiative forcing stabilization level. At any stabilization level, scenarios with lower contributions to radiative forcing from non-CO_2GHGs allow for greater radiative forcing from CO_2

Figure TS.16. Total Radiative Forcing in 2100 Across Scenarios (W/m2from preindustrial)

Using the Scenarios and Future Work

The scenarios in this chapter are intended as one of many inputs to public and private discussions regarding the threat of climate change, and they are also intended to serve as a point of departure for further CCSP and other analyses. A range of such analyses are possible. For example, the scenarios could be applied as the basis for assessing the climate implications of alternative stabilization levels, and then follow-on studies of potential climate impacts. They might also be used in studies exploring possible technology cost and performance goals, using information from the scenarios on energy prices and technology deployment levels. Similarly, the scenarios might inform analyses of the non-climate environmental implications of implementing potential new energy sources at a large scale. Another possibility is that the scenarios could serve as an input to a more complete analysis of the economic effects of stabilizing at the different radiative forcing levels, such as indicators of consumer impact in the U.S. (The reader is reminded, however, that these effects do not include the benefits that alternative stabilization levels might yield in reduced climate change risk or ancillary effects, such as effects on air pollution). The scenarios could also be compared against past and future scenarios analyses.

The scenarios in this chapter represent but one step in a long process of research and assessment, and the scenarios and their underlying models will benefit from further work. The review process has identified at least five different areas that hold the promise of potentially fruitful research: (1) technology sensitivity analysis, (2) consideration of non-idealized policy architectures, (3) expansion and improvement of the land use and terrestrial carbon cycle linkages to the energy and economic model components, (4) inclusion of other radiatively-important substances such as emissions affecting tropospheric ozone and aerosols, and (5) decision-making under uncertainty. These needs for additional research and analysis are elaborated in Chapter 5.

1. Introduction and Overview

Introduction

The Strategic Plan for the U.S. Climate Change Science Program (CCSP 2003) calls for the preparation of 21 synthesis and assessment products. Noting that "sound, comprehensive emissions scenarios are essential for comparative analysis of how climate might change in the future, as well as for analyses of mitigation and adaptation options," the Plan includes Product 2.1, *Scenarios of Greenhouse Gas Emissions and Atmospheric Concentrations and Review of Integrated Scenario Development and Application*. This chapter presents the scenarios created in the scenario-development component of Product 2.1; the review of scenario methods is the subject of a separate report (CCSP 2007). The guidelines for the development of these scenarios are set forth in the *Final Prospectus for Synthesis and Assessment Product 2. 1* (CCSP 2005). Consistent with the Prospectus and the nature of the climate change issue, these scenarios were developed using long-term models of global energy-agriculture-land-use-economy systems coupled to models of global atmospheric composition and radiation.

This chapter discusses the overall design of scenarios (Chapter 1); describes the key features of the participating models (Chapter 2); presents and compares the newly prepared scenarios (Chapters 3 and 4); and discusses emerging insights from these new scenarios, the uses and limitations of the scenarios, and avenues for further research (Chapter 5). Scenario details are available in a separate data archive.

The scenarios in this chapter are intended as one of many inputs to public and private discussions regarding climate change and what to do about it, and they may also serve as a point of departure for further Climate Change Science Program (CCSP) and other analyses that might inform these discussions in the future. The possible users of these scenarios are many and diverse. They include climate modelers and the science community; those involved in national public policy formulation; managers of Federal research programs; state and local government officials who face decisions that might be affected by climate change and mitigation measures; and individual firms, non-governmental organizations, and members of the public. Such a varied clientele implies an equally diverse set of possible needs, and no single scenario research product can hope to fully satisfy all of these needs. The Prospectus for this research highlighted three particular areas in which the scenarios might provide valuable insights:

- ***Emissions Trajectories.*** What emissions trajectories over time are consistent with meeting the four stabilization levels, and what are the key factors that shape them?
- ***Energy Systems.*** What energy system characteristics are consistent with each of the four alternative stabilization levels, and how might these characteristics differ among stabilization levels?
- ***Economic Implications.*** What are the possible economic consequences of meeting each of the four alternative stabilization levels?

It should be emphasized that there are issues of climate change decision making that these scenarios do not address. For example, they were not designed for use in exploring the role of aerosols in climate change. Also, they lack the regional detail that may be desired for many aspects of local or regional decision-making.

In addition, the scenarios in this chapter do not constitute a cost-benefit analysis of climate policy. They focus exclusively on the issues associated with reducing emissions to meet various stabilization levels; they do not consider the damages avoided through stabilization or ancillary benefits that could be realized by emissions reductions, such as reductions in local air pollution. Thus, although the scenarios should serve as a useful input to climate- related decision making, they address only one of several components of a benefit- cost analysis of climate policy.

Three analytical models, all meeting the criteria set forth in the Prospectus, were used in preparing the new scenarios. As also directed in the Prospectus, fifteen scenarios are presented in this document, five from each of the three modeling groups. First, each group produced a unique reference scenario based on the assumption that no climate policy would be implemented either nationally or globally beyond the current set of policies in place (e.g., the Kyoto Protocol and the President's greenhouse gas emissions intensity target for the U.S.). These reference scenarios were developed independently by the modeling groups, so they provide three separate visions of how the future might unfold across the globe over the 21st century without additional climate policies.[1]

Each group then produced four additional stabilization scenarios, which are departures from each group's reference scenario. The Prospectus specified that stabilization levels, common across the groups, be defined in terms of the total long-term radiative impact of the suite of greenhouse gases (GHGs) that includes carbon dioxide (CO_2), nitrous oxide (N_2O), methane (CH_4), hydrofluorocarbons (HFCs), perfluorocarbons (PFCs), and sulfur hexafluoride (SF_6). This radiative impact is expressed in terms of radiative forcing associated with increases from preindustrial concentrations of this suite of GHGs (Box 1.1).

Although stabilization is defined in terms of radiative forcing, the stabilization levels were constructed so that the resulting CO_2 concentrations, after accounting for radiative forcing from the non-CO_2 GHGs, would be roughly 450 parts per million by volume (ppmv), 550 ppmv, 650 ppmv, and 750 ppmv. The radiative forcing limits therefore are higher than the forcing from CO_2 alone at these concentrations. Based on this requirement, the four stabilization levels were chosen as 3.4 watts per meter squared (W/m^2) (Level 1), 4.7 W/m^2 (Level 2), 5.8 W/m^2 (Level 3), and 6.7 W/m^2 (Level 4). In comparison, radiative forcing relative to preindustrial levels for this suite of gases stood at roughly 2.1 W/m^2 in 1998. Details of these stabilization assumptions are elaborated in Section 1.3 and Chapter 4.

The production of emissions scenarios consistent with these stabilization goals required analysis beyond the study of the emissions themselves because of physical, chemical, and biological feedbacks within the Earth system. Scenarios focused only on emissions of GHGs and other substances generated by human activity (anthropogenic sources) can rely exclusively on energy-agriculture-economic models that represent human activity and the emissions that result. However, relating emissions paths to concentrations of GHGs in the atmosphere requires models that account for both anthropogenic and natural sources as well as the sinks for these substances.

Models that attempt to capture these complex interactions and feedbacks must, because of computational limits, use simplified representations of individual components of the Earth system. These simplified representations are typically designed to mimic the behavior of more complex models but cannot represent all of the elements of these systems. Thus, while the scenario research undertaken here uses models that represent both the anthropogenic sources (the global energy-industrial-agricultural economy) and the Earth system processes (ocean, atmosphere, and terrestrial systems), it is not intended to supplant detailed analysis of these systems using full scale, state-of-the-art models and analytic techniques. Rather, these scenarios provide a common point of departure for more complex analyses of individual components of the Earth system as it is affected by human activity. These might include detailed studies of sub-components of the energy sector, regional scenarios of climate change using three-dimensional general circulation models (GCMs) and further downscaling techniques, and assessment of the implications of climate change under various stabilization goals for economic activity and natural ecosystems.

The remainder of this chapter is organized into four sections. Section 1.2 provides an overview of scientific aspects of the climate issue as background for interpretation of these scenarios. Section 1.3 then presents the research design with a focus on the characteristics of the stabilization scenarios to be investigated in Chapter 4. Section 1.4 briefly discusses how scenarios of this type have been used to examine the climate change issue and the intended uses and limits of the new scenarios, focusing on interpretation of these scenarios under conditions of uncertainty. Section 1.5 provides a guide to the structure of the remaining chapters.

Background: Human Activities, Emissions, Concentrations, and Climate Change

Materials that influence the Earth's radiation balance come in various forms, and most have natural as well as anthropogenic sources. Some are gases which remain in the atmosphere from days to millennia, trapping heat. They are known as GHGs because, while transparent to incoming short-wave radiation (the visible spectrum that people commonly perceive as light), they capture and reflect back to Earth long-wave radiation, thus increasing the temperature of the lower atmosphere. These naturally occurring GHGs, plus clouds and water vapor (the most important GHG of all), are responsible for creating a habitable climate on Earth. Without them, the average temperature at the Earth's surface would be colder than it is today by roughly 55°F (~30°C).

GHGs are not the only influences on the Earth's radiative balance. Other gases such as oxides of nitrogen (NO_x) have no direct greenhouse effect, but they are components of the atmospheric chemistry that determine the lifetime of some of the heat-trapping GHGs and are involved in the reactions that produce tropospheric ozone, another GHG. Aerosols (non-aqueous particles suspended in air) may have positive or negative effects, depending on their relative brightness. Some present a white surface and reflect the sun's energy back to space; others are black and absorb solar energy, adding to the solar warming of the atmosphere. Aerosols also have an indirect effect on climate in that they influence the character and lifetime of clouds, which have a strong influence on the radiation balance and on precipitation. Humans also alter the land surface, changing its reflective properties, and these changes can have climate consequences with effects most pronounced at a local scale (e.g., urban heat islands) and regional levels (e.g., large-scale changes in forest cover). In addition, the climate itself has positive and negative feedbacks, such as the decrease in global albedo that would result from melting land and sea ice or the potential release of GHGs, such as CH_4 from wetlands.

Climate policy concerns are driven by the fact that emissions from human activities (mainly combustion of fuels and biomass, industrial activities, and agriculture) are increasing the atmospheric concentrations of these substances. Climate policy discussions have focused heavily on CO_2, CH_4, N_2O, and a set of fluorine-containing industrial chemicals – SF_6 and two families of substances that do not exist naturally, hydrogenated halocarbons (including hydrochlorofluorocarbons [HCFCs] and HFCs)[2] and PFCs. Some of these substances remain in the atmosphere for decades (CH4 and most HFCs), others for about 100 years (CO_2 and N_2O), and some for thousands of years (PFCs and SF_6).

Other naturally occurring substances whose levels have also been greatly enhanced by human activities remain in the atmosphere for days to months. With such short lifetimes, they are not well mixed in the atmosphere, so their effects have a regional pattern as well as global consequences. These substances include aerosols such as black carbon and other particulate matter; sulfur dioxide, which is the main precursor of the reflecting aerosols; and other gases such as volatile organic compounds, nitrogen dioxide, other oxides of nitrogen, and carbon monoxide. All are important components of atmospheric chemistry.

This suite of substances with different radiative potency and different lifetimes in the atmosphere presents a challenge in defining what is meant by atmospheric stabilization. Specification in terms of quantities of the substances themselves is problematic because there is no simple way to add them together in their natural units, such as tonnes or ppmv. Thus, a meaningful metric is needed to combine the effects of different GHGs.

One approach is to define stabilization in terms of some ultimate climate measure, such as the change in the global average temperature. One drawback of such measures is that they interject large uncertainties into the consideration of stabilization because the ultimate climate system response to added GHGs is uncertain. Climate models involve complex and uncertain interactions and feedbacks, such as increasing levels of water vapor, changes in reflective polar ice, cloud effects of aerosols, and changes in ocean circulation that determine the ocean's uptake of CO_2 and heat.

For the design of these scenarios, the Prospectus called for an intermediate, less uncertain measure of climate effect. The Prospectus directed that stabilization "be defined in terms of the radiative forcing resulting from the longterm combined effects of carbon dioxide (CO_2), nitrous oxide (N_2O), methane (CH_4), hydrofluorocarbons (HFCs), perfluorocarbons (PFCs), and sulfur hexafluoride (SF_6)." Radiative forcing (Box 1.1) is a measure of the instantaneous imbalance in the radiative energy budget of the Earth's climate system (energy in versus energy out) resulting from an externally imposed perturbation such as increasing GHG concentrations. It is measured in terms of W/m^2 at the Earth's shell and a positive value means a warming influence. For these scenarios, radiative forcing is measured against the concentrations of the GHGs considered in this research in preindustrial times, taken to be 1750.

Box 1.1. Radiative Forcing

Most of the Sun's energy that reaches the Earth is absorbed by the oceans and land masses and radiated back into the atmosphere in the form of heat or infrared radiation. Some of this infrared energy is absorbed and re-radiated back to the Earth by atmospheric gases, including water vapor, CO_2, and other substances. As concentrations of GHGs increase, there are direct and indirect effects onthe Earth's energy balance. The direct effect is often referred to as a radiative forcing, a subset of amore general set of phenomena referred to as climate forcings. The National Research Council(NRC 2005) offers the following set of definitions:

Factors that affect climate change are usefully separated into forcings and feedbacks. ... A climate forcing is an energy imbalance imposed on the climate system either externally or by human activities. Examples include changes in solar energy output, volcanic emissions, deliberate land modification, oranthropogenic emissions of greenhouse gases, aerosols, and their precursors. A climate feedback is aninternal climate process that amplifies or dampens the climate response to an initial forcing. An example is the increase in atmospheric water vapor that is triggered by an initial warming due to rising carbon dioxide (CO_2) concentrations, which then acts to amplify the warming through the greenhouseproperties of water vapor.

Climate forcing: An energy imbalance imposed on the climate system either externally or by human activities.

- **Direct radiative forcing:** A climate forcing that directly affects the radiative budget of the Earth'sclimate system; for example, added carbon dioxide (CO_2) absorbs and emits infrared radiation. Direct radiative forcing may be due to a change in concentration of radiatively active gases, a changein solar radiation reaching the Earth, or changes in surface albedo. Radiative forcing is reported inthe climate change scientific literature as a change in energy flux at the tropopause, calculated in unitsof watts per square meter (W/m2); model calculations typically report values in which the stratospherewas allowed to adjust thermally to the forcing under an assumption of fixed stratospheric dynamics.

- **Indirect radiative forcing:** A climate forcing that creates a radiative imbalance by first alteringclimate system components (e.g., precipitation efficiency of clouds), which then almost immediatelylead to changes in radiative fluxes. Examples include the effect of solar variability on stratosphericozone and the modification of cloud properties by aerosols.
- **Nonradiative forcing:** A climate forcing that creates an energy imbalance that does not immediately involve radiation. An example is the increasing evapotranspiration flux resulting from agricultural irrigation.

For purposes of this chapter, the radiative forcing stabilization levels are defined in terms of the direct radiative forcing caused by increases from preindustrial concentrations of CO_2, CH_4, N_2O,PFCs, HFCs, and SF_6. The indirect radiative effects are not included in calculating whether the radiative forcing stabilization level levels are met, nor are the direct radiative effects (positive or negative) of other substances such as ozone, CFCs, or aerosols, although emissions of these substancesand their radiative and climatic effects are part of these integrated system models.

Figure 1.1 shows estimates of how increases in GHGs, aerosols, and other changes have influenced radiative forcing since 1850. The GHGs considered in these scenarios are collected in the left-most bar and together they have had the biggest effect, with CO_2 being the largest of this group. Increased tropospheric ozone has also had a substantial warming effect. The reduction in stratospheric ozone has had a slight cooling effect. Changes in aerosols have had both warming and cooling effects. Aerosol effects are highly uncertain because they depend on the nature of the particles; how the particles are distributed in the atmosphere; and the concentrations of the particles, which are not as well understood as the GHGs. Land-use change and its effect on the reflectivity of the Earth's surface, jet contrails and changes in high-level (cirrus) clouds, and the natural change in intensity of the sun have also had effects.

Another important aspect of the climate effects of these substances, not captured in the W/m^2 measure, is the persistence of their influence on the radiative balance – a characteristic discussed in Box 1.2. The W/m^2 measure of radiative forcing accounts for only the effect of a concentration in the atmosphere at a particular instant. The GHGs considered here have influences that may last from a decade or two (e.g., the influence of CH_4) to millennia, as noted earlier.

An important difference between GHGs and most of the other substances in Figure 1.1 is their long lifetimes. In contrast to GHGs, aerosols remain in the atmosphere only for a few days to a couple of weeks. Once an aerosol emission source is eliminated, its effect on radiative forcing disappears very quickly. Tropospheric ozone lasts for a few months. Moreover, relatively short-lived substances are not well mixed in the atmosphere. Levels are very high near emissions sources and much lower in other parts of the world, so their climate effect has a different spatial pattern than that of long-lived substances. The regional differences and much shorter lifetimes of non-GHG substances make comparisons among them more difficult than among GHGs. The radiative effects of these substances also subject to more uncertainty, as shown in Figure 1.1.

Research Design

The broad elements of the research design for these scenarios are set forth in the Prospectus, including (1) selection of models, (2) guidance to the modeling groups for development of a reference scenario, and (3) guidance for the development of stabilization scenarios.

Box 1.2. Atmospheric Lifetimes of Greenhouse Gases

The atmospheric lifetime concept is more appropriate for CH_4, N_2O, HFCs, PFCs, and SF6 than it is for CO_2. These non-CO_2 gases are destroyed via chemical processes after some time in the atmosphere. In contrast, CO_2 is constantly cycled between pools in the atmosphere, the surface layer of the ocean, and vegetation, so it is (for the most part) not destroyed. Very slow processes lead to some removal of carbon from oceans, vegetation, and the atmosphere as calcium carbonate. Also, over long geological periods, carbon from vegetation is stored as fossil fuels, which is a permanent removal process as long as the fossil fuels are not burned to produce energy.

Although the lifetime concept is not strictly appropriate for CO_2 (see Box 2.2 in Chapter 2), the molecules in a kilogram of emissions can be thought of as residing in the atmosphere, exercising their radiative effect, for around 100 years. This approximation allows a rough comparison with the other gases: CH4 at 12 years, N_2O at 114 years, and SF6 at 3200 years. HFCs are a family of gases with varying lifetimes from less than a year to over 200 years; those predominantly in use now have lifetimes mostly in the range of 10 to 50 years. Similarly, the PFCs have various lifetimes, ranging from 2,600 to 50,000 years.

The lifetimes are not constant, as they depend to some degree on other Earth system processes. The lifetime of CH_4 is the most affected by the levels of other pollutants in the atmosphere.

Model Selection

The Prospectus set forth the model capabilities required to develop the desired stabilization scenarios. As stated in the Prospectus, participating models must:

1. Be global in scale
2. Be capable of producing global emissions totals for, at a minimum, CO_2, N_2O, CH_4, HFCs, PFCs, and SF_6 that may serve as inputs to global GCMs, such as the National Center for Atmospheric Research (NCAR) Community Climate System Model and the Geophysical Fluid Dynamics Laboratory climate model

3. Be capable of simulating the radiative forcing from CO_2, N_2O, CH_4, HFCs, PFCs, and SF_6
4. Represent multiple regions
5. Have technological resolution capable of distinguishing among major sources of primary energy (e.g., renewable energy, nuclear energy, biomass, oil, coal, and natural gas) as well as between fossil fuel technologies with and without CO_2 capture and storage (CCS) systems
6. Be economics based and capable of simulating macroeconomic cost implications of stabilization
7. Look forward to the end of the century or beyond.

In addition, the Prospectus required that the modeling groups have a track record of publications in professional, refereed journals, specifically in the use of their models for the analysis of long-term GHG emission scenarios.

Selection by these criteria led to the three models used in this research: (1) The Integrated Global Systems Model (IGSM) of the Massachusetts Institute of Technology's Joint Program on the Science and Policy of Global Change; (2) the Model for Evaluating the Regional and Global Effects of GHG reduction policies (MERGE), developed jointly at Stanford University and the Electric Power Research Institute; and (3) the MiniCAM Model of the Joint Global Change Research Institute, which is a partnership between the Pacific Northwest National Laboratory and the University of Maryland.

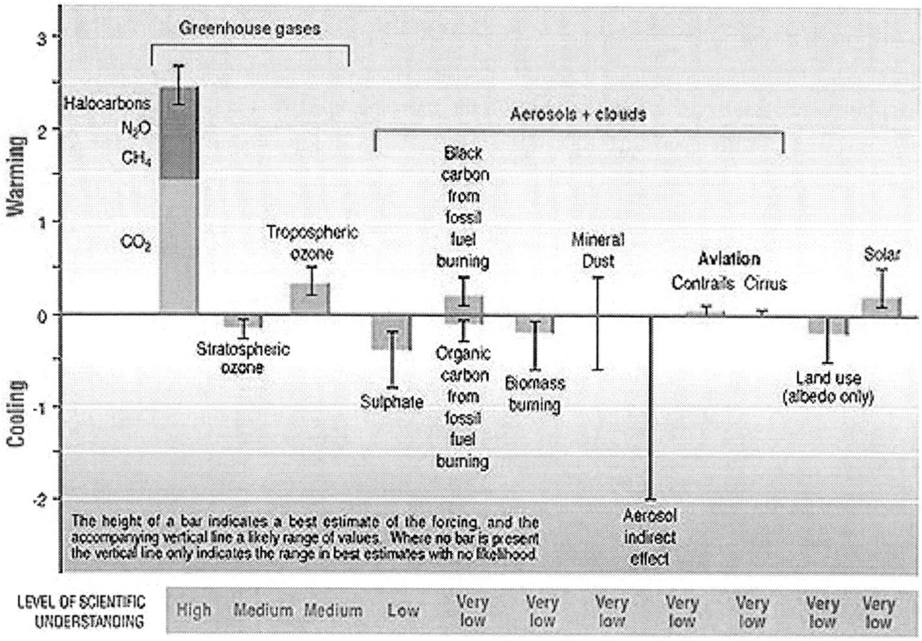

Source: IPCC 2001.

Figure 1.1. Estimated Influences of Atmospheric Gases on Radiative Forcing, 1750-Present

Each of these models has been used extensively for climate change analysis. The roots of each extend back more than a decade, during which time features and details have been added. Analyses using each have appeared widely in peer-reviewed publications. The features of the models are described in Chapter 2 with references to publications and reports that provide complete documentation.

These models fall into a class that has come to be known as Integrated Assessment Models (IAMs). There are many ways to define IAMs and to characterize the motivations for developing them (IPCC 1996). A particularly appropriate definition of their primary purposes, provided by Parson and Fisher-Vanden (1997), is "evaluating potential responses to climate change, structuring knowledge and characterizing uncertainty, contributing to broad comparative risk assessments, and contributing to scientific research."

Development of Reference Scenarios

As required by the Prospectus, each participating modeling group first produced a reference scenario that assumes no policies specifically intended to address climate change beyond implementation of any existing policies to the end of their commitment periods, including the Kyoto Protocol and the policy of the U.S. to reduce greenhouse gas emissions intensity by 18% by 2012. For purposes of the reference scenario (and for each of the stabilization scenarios), it was assumed that these policies are successfully implemented through 2012 and their goals are achieved. (This assumption could only be approximated within the models because their time steps did not coincide exactly with the period from 2002 to 2012. However, such approximation is a minor consideration as slight differences in emissions for a few years have little impact on long term concentrations.) As directed by the Prospectus, after 2012 these existing climate policies expire and are not renewed or replaced. This is not a prediction or a best-judgment forecast, but a scenario designed to provide a clearly defined point of departure for illuminating the implications of alternative stabilization goals. The paths toward stabilization are implemented to start after 2012 as discussed further in the following section. The reference scenarios and assumptions underlying them are detailed in Chapter 3.

The reference scenarios serve two main purposes. First, they provide insight into how the world might evolve without additional efforts to constrain GHG emissions, given various assumptions about principal drivers of the economy, energy use, and emissions. These assumptions include those concerning population increase, land and labor productivity growth, technological options, and resource endowments. These forces govern the supply and demand for energy, industrial goods, and agricultural products – the production and consumption activities that lead to GHG emissions. The reference scenarios are a thought experiment in that they assume that even as emissions increase and climate changes nothing is done to reduce emissions. The specific levels of GHG emissions and concentrations are not predetermined but result from the combination of assumptions made.

Second, the reference scenarios serve as points of departure for analysis of the changes brought about by stabilization of radiative forcing, and the underlying assumptions have a large bearing on the characteristics of the stabilization scenarios. For example, all other things being equal, the lower the economic growth and the higher the availability and competitiveness of low-carbon energy technologies in the reference scenario, the lower will be the GHG emissions and the easier it will be to reach stabilization. On the other hand, if a reference scenario assumes that fossil fuels are abundant, and fossil fuel technologies will become cheaper over time while low- or zero-carbon alternatives remain expensive, the scenario will show consumers having little reason to conserve, adopt more efficient energy equipment, or switch to non-fossil sources. Under such a reference scenario, emissions will grow rapidly, and stronger economic incentives will be required to achieve stabilization.

Finally, the Prospectus specified that the modeling groups develop their reference scenarios independently[3], applying *meaningful* and *plausible* assumptions for key drivers. Similarities and differences among the reference scenarios are useful in illustrating the uncertainty inherent in long-run treatment of the climate challenge. At the same time, with only three participating models, the range of scenario assumptions produced does not span the full range of possibilities.

Development of the Stabilization Scenarios

Although the model groups were required to independently develop their modeling assumptions, the Prospectus specified that a common set of four stabilization targets be used across the participating models. Also, whereas much of the literature on atmospheric stabilization focuses on concentrations of CO_2 only, an important objective of this research was to expand the range of coverage to include other GHGs. Thus, the Prospectus required that the stabilization levels be defined in terms of the combined effects of CO_2, N_2O, CH_4, HFCs, PFCs, and SF_6. This suite of GHGs forms the basis for the U.S. GHG-intensity-reduction policy, announced by the President on February 14, 2002; it is the same set subject to control under the Kyoto Protocol. These gases are included in the left-most bar of Figure 1.1. The stabilization targets specified in the Prospectus explicitly omit the aerosol, ozone, land surface, and other effects shown in Figure 1.1, which may be influenced by measures taken to achieve the stabilization goal. Table 1.1 shows the change in concentration levels for these gases from 1750 to 2000. The left-most bar in Figure 1.1 shows radiative forcing of roughly 2.4 W/m^2 compared with a sum of 2.1 W/m^2 in Table 1.1. The difference exists because Figure 1.1 includes roughly 0.3 W/m^2 of forcing from chlorofluorocarbons (CFCs) not in Table 1.1. CFCs, important in the historical data, are already being phased out under the Montreal Protocol because of their stratospheric ozone-depleting properties, so they are not expected to be a significant source of additional increased forcing in the future. The HFCs, which do not contribute to stratospheric ozone depletion, were developed as substitutes for the CFCs, but are of concern because of their radiative properties. Table 1.2 shows the specific radiative forcing targets chosen.

As noted earlier, the Prospectus instructed that the stabilization levels be constructed so that the CO_2 concentrations resulting from stabilization of total radiative forcing, after accounting for radiative forcing from the non-CO_2 GHGs, would be roughly 450 ppmv, 550 ppmv, 650 ppmv, and 750 ppmv. This correspondence was achieved by (1) calculating the increased radiative forcing from CO_2 at each of these concentrations, (2) adding to that amount the radiative forcing from the non-CO_2 gases from 1750 to present, and (3) adding an estimate of the change in radiative forcing from the non-CO_2 GHGs under each of the stabilization levels. Each of the models represents the emissions and abatement opportunities of the non-CO_2 gases somewhat differently and takes a different approach to representation of the tradeoffs among them, so an exact correspondence between overall radiative forcing and CO_2 levels that would fit all three models was not possible.

The Prospectus also specified that, beyond the implementation of any existing policies, the stabilization scenarios should be based on universal participation by the world's nations. This guidance was implemented by assuming a climate regime with simultaneous global participation in emissions mitigation and in which the marginal costs of emission controls are equalized across countries and regions. Under this assumption, known as *where* flexibility, emissions will be reduced where it is cheapest to do so regardless of their geographical location. One important implication of this assumption is that the stabilization scenarios produce estimates of stabilization costs that are systematically lower than what might be expected in a world in which some major countries remain out of an emissions mitigation regime for an extended period of time, some economies use more costly regulatory mechanisms, or emissions mitigation regimes within nations are incomplete either in terms of GHG or sectoral coverage. On the other hand, possible ancillary benefits, tax interaction effects, or effects of carbon policies on technical change were not considered, which in some cases can lower costs. These issues are discussed in more detail in Chapter 4.

Table 1.1. GreenhouseGas Concentrationsand Forcing

	Preindustrial Concentration (1750)	Current Concentration (1998)	Contribution to Radiative Forcing, (W/m^2, 1750 to 1998)
CO_2	278 ppmv	365 ppmv	1.46
CH_4	700 ppbv	1745 ppbv	0.48
N_2O	270 ppbv	314 ppbv	0.15
HFCs, PFCs, SF_6	0	various	≈ 0.02
Total	—	—	≈ 2.1

Source. IPCC 2001.

Concentrations of GHGshave increased since 1750(preindustrial), altering the radiative energy budget of the Earth's climate system.

Table 1.2. Radiative Forcing Stabilization Levels (W/m^2) and Approximate CO_2 Concentrations (ppmv)

	Total Radiative Forcing from GHGs (W/m^2)	Approximate Contribution to Radiative Forcing from non-CO_2 GHGs (W/m^2)	Approximate Contribution to Radiative Forcing from CO_2 (W/m^2)	Corresponding CO_2 Concentration (ppmv)
Level 1	3.4	0.8	2.6	450
Level 2	4.7	1.0	3.7	550
Level 3	5.8	1.3	4.5	650
Level 4	6.7	1.4	5.3	750
Year 1998	≈ 2.1	0.65	1.46	365
Preindustrial (1750)	—	—	—	278

The radiative forcing levels were constructed so that the CO2 concentrations resulting from stabilization of total radiative forcing, after accounting for radiative forcing from the non-CO_2 GHGs, would be roughly 450 ppmv, 550 ppmv, 650 ppmv, and 750 ppmv.

In addition, the Prospectus required that stabilization be defined as long term. Because of the inertia in the Earth system, largely attributable to the ocean, perturbations to the climate and atmosphere have effects for thousands of years. Economic models have little credibility over such timeframes. The Prospectus, therefore, instructed that the participating modeling groups report scenario information only up through 2100. Each group then had to address how to relate the level in 2100 to the long-term goal. The chosen approaches were generally similar, but with some differences in implementation. This and other details of the stabilization scenario design are addressed more completely in Chapter 4.

Interpreting Scenarios: Uses, Limits, and Uncertainty

Emissions scenarios have proven to be useful aids to understanding climate change, and there is a long history of their use (see Box 1.3). Scenarios are descriptions of future conditions, often constructed by asking *what if* questions, such as what if events were to unfold in a particular way? Informal scenario analysis is part of almost all decision making. For example, families making decisions about big purchases, such as a car or a house, might plausibly construct a scenario in which changes in employment forces them to move. Scenarios addressing major public-policy questions perform the same purpose, helping decision makers and the public to understand the consequences of actions today in the light of plausible future developments.

Models assist in creating scenarios by showing how assumptions about key drivers, such as economic and population growth or policy options, lead to particular levels of GHG emissions. Model-based scenario analysis is designed to provide quantitative estimates of multiple outcomes and to assure consistency among them that is difficult to achieve without a formal structure. Thus, a main benefit of such model simulation of scenarios is that they ensure basic accounting identities: the quantity demanded of fuel is equal to the quantity supplied, imports in one region are balanced by exports from other regions, cumulative fuel used does not exceed estimates of the available resources, and expenditures for goods and services do not exceed income. The approach complements other ways of thinking about the future, ranging from formal uncertainty analysis to narratives. Also, such model analyses offer a set of macro- scenarios that users can build on, adding more detailed assumptions about variables and decisions of interest to them.

The possible users of these scenarios are many and diverse, and a single scenario research product cannot hope to provide the details needed by all potential users or to address their specific questions. Thus, these scenarios are an initial set offered to potential user communities. If successful, they will generate further questions and the demand for more detailed analysis, some of which might be satisfied by further scenario development from models like those used here, but more often demanding detail that can only be provided with other modeling and analysis techniques. As such, this effort is one step in an ongoing and iterative process of producing and refining climate-related scenarios and scenario tools.

Box 1.3 Emissions Scenarios and Climate Change

Emissions scenarios that describe future economic growth and energy use have been importanttools for understanding the long-term consequences of climate change. They were used in assessments by the U.S. National Academy of Sciences in 1983 and by the Department of Energy in 1985(NAS 1983, US DOE 1985). Previous emissions scenarios have evolved from simple projectionsthat extrapolated a 1% per year increase in CO_2emissions to scenarios that incorporate assumptions about population, economic growth, energy supply, and controls on GHG emissions and CFCs(Leggett et al. 1992, Pepper et al. 1992). They played an important role in the reports of the Inter-governmental Panel on Climate Change (IPCC 1991, IPCC 1992, IPCC 1996). The IPCC Special Report on Emissions Scenarios (SRES) (Nakicenovic et al. 2000) was the most recent major effortundertaken by the IPCC to expand and update earlier scenarios. This set of scenarios was basedon storylines of alternative futures, updated with regard to the variables used in previous scenarios and with additional detail on technological change and land use.

The Energy Modeling Forum (EMF) has been an important venue for intercomparison of emissionsscenarios and IAMs. The EMF, managed at Stanford University, includes participants from academic,government, and other modeling groups from around the world. It has served this role for the energy-modeling community since the 1970s. Individual EMF studies run over a course of about twoyears, with scenarios designed by the participants to provide insight into the behavior of the participating models. Scenarios are often published in the peer-reviewed literature. A recent study, EMF21, focused on multi-gas stabilization scenarios (de la Chesnaye and Weyant 2006)

Although the required long-term perspective demands scenarios that stretch into the distant future, any such scenarios carry with them considerable uncertainty. Inevitably, the future will hold surprises. Scientific advances will be made, new technologies will be developed, and the direction of the economy will change, making it necessary to reassess the issues examined here. The Prospectus called for development of a limited number of scenarios, without a formal treatment of likelihood or uncertainty, requiring as noted earlier, only that the modeling groups use assumptions that they believe to be *meaningful* and *plausible*. Formal uncertainty analysis has much to offer and could be a useful additional follow-on or complementary research task. Here, however, the range of outcomes from the different modeling groups help to illustrate, if incompletely, the range of possibilities.

The scenarios developed here take the best information available now and assess what it may mean for the future. Any such research, however, will necessarily be incomplete and will not foresee all possible future developments. The best planning must prepare for changes in course later as new information becomes available.

Report Outline

Chapter 2 of this chapter provides an overview of the three models used in development of the scenarios. Chapter 3 describes the assumptions about key drivers in each of the models and reports the reference scenarios. Chapter 4 provides greater detail on the design of the stabilization scenarios and then presents these scenarios. Chapter 5 provides concluding observations, including possible avenues for additional research.

The chapters seek to show how the models and the assumptions used by the modeling groups to develop the scenarios differ and, to the degree possible, to relate where these differences matter and how they shape the scenarios. The models have their own respective areas of focus, and each offers its own reasonable representation of the world. The authors have distilled general conclusions common to the scenarios generated by the three modeling groups, while recognizing that other plausible representations could well lead to quite different scenarios. The scenarios are presented primarily in the figures. Associated with the report is a database with quantitative information available for those who wish to further analyze and use these scenarios. A description of the database, directions for use, and its location can be found in the appendix.

2. Models Used in This Research

Overview of the Models

The computer models used in this research are referred to as integrated assessment models because they combine, in an integrated framework, the socioeconomic and physical processes and systems that define the human influence on, and interactions with, the global climate. They integrate computer models of socioeconomic and technological determinants of the emissions of GHGs and other substances influencing the Earth's radiation balance with models of the natural science of Earth system response, including those of the atmosphere, oceans, and terrestrial biosphere. Although they differ in their specific design objectives and details of their mathematical structures, each of these IAMs was developed for the purpose of gaining insight into economic and policy issues associated with global climate change.

To create scenarios of sufficient depth, scope, and detail, a number of model characteristics were deemed critical for development of these scenarios. The criteria set forth in the Prospectus for this research led to the selection of three IAMs: IGSM, MERGE, and MiniCAM. These three are among the most detailed models of this type of IAM, and each a has long history of development and application.

- **IGSM** of the Massachusetts Institute of Technology's Joint Program on the Science and Policy of Global Change is an Earth system model that comprises a multi-sector, multi-region economic component and a science component, including a two-dimensional atmosphere, a three- dimensional ocean, and a detailed biogeochemical model of the terrestrial biosphere (Sokolov et al. 2005). Because this research focuses on new emissions scenarios, elements of the scenarios emerging from the economic model component of IGSM, the Emissions Prediction and Policy Analysis (EPPA) model (Paltsev et al. 2005), are featured in the discussion below. EPPA is a recursive-dynamic computable general equilibrium (CGE) model of the world economy and greenhouse-relevant emissions, solved on a five-year time step.

Previous applications of IGSM and its EPPA component system can be found at http://web

- **MERGE** was developed jointly at Stanford University and the Electric Power Research Institute (Manne and Richels 2005). It is an inter-temporal general equilibrium model of the global economy in which the world is divided into nine geopolitical regions. It is solved on a ten-year time step. MERGE is a hybrid model, combining a bottom-up representation of the energy supply sector with a top-down perspective on the remainder of the economy.[4] Savings and investment decisions are modeled as if each region maximizes the discounted utility of its consumption, subject to an inter-temporal wealth constraint. Embedded within this structure is a reduced-form representation of the physical Earth system. MERGE has been used to explore a range of climate-related issues, including multi-gas strategies, the value of low-carbon-emitting energy technologies, the choice of near-term hedging strategies under uncertainty, the impacts of learning-by-doing, and the potential importance of *when* and *where* flexibility. To support this scenario research, the multi-gas version has been revised by adjustments in technology and other assumptions. The MERGE code and publications describing its structure and applications can be found at http://www.stanford.edu/group/MERGE/.
- **MiniCAM** is an integrated assessment model (Brenkert et al. 2003) that combines a technologically detailed global energyeconomy-agricultural-land-use model with a suite of coupled gas-cycle, climate, and ice-melt models, integrated in the Model for the Assessment of Greenhouse-Gas Induced Climate Change (MAGICC). MiniCAM was developed and is maintained at the Joint Global Change Research Institute, a partnership between the Pacific Northwest National Laboratory and the University of Maryland, while MAGICC was developed and is maintained at NCAR. MiniCAM is solved on a 15-year time step. MiniCAM has been used extensively for energy, climate, and other environmental analyses conducted for organizations that include the U.S. Department of Energy (DOE), the U.S. Environmental Protection Agency, the IPCC, and several major private sector energy companies. Its energy sector is based on a model developed by Edmonds and Reilly (1985). The model is designed to examine long-term, large-scale changes in global and regional energy systems, focusing on the impact of energy technologies. Documentation for MiniCAM can be found in Brenkert et al. (2003).

Because these models were designed to address an overlapping set of climate change issues, they are similar in many respects. All three have social science-based components that capture the socioeconomic and technology interactions underlying the emissions of GHGs, and each incorporates models of physical cycles for GHGs and other radiatively important substances and other aspects of the natural science of global climate. The differences among them lie in the detail and construction of these components and in the ways they are modeled to interact. Each was designed with somewhat different aspects of the climate issue as a main focus. IGSM includes the most detailed representation of the chemistry, physics, and biology of the atmosphere, oceans, and terrestrial biosphere; thus, its EPPA component is designed to provide the emissions detail that these natural science components require. MERGE has its origins in an energy-sector model that was initially designed for energy technology assessment. It was subsequently modified to explore the influence of expectations (and uncertainty regarding expectations) about future climate policy on the economics of current investment and the cost-minimizing allocation of emissions mitigation over time. Its focus requires a forward- looking structure, which in turn employs simplified non-energy components of the economy. MiniCAM is a technology-rich IAM. It features detailed representations of energy technologies, energy systems, and energy markets and their interactions with demographics, the economy, agricultural technologies, markets, land use, and the terrestrial carbon cycle.

Each of these IAMs has unique strengths and areas of special insight. In this research, the simultaneous application of different model structures is useful in revealing different aspects of the task of stabilizing radiative forcing. The differences among the scenarios prepared by the three modeling groups, presented in Chapters 3 and 4, are an indication of the limits of the knowledge about future GHG emissions and the challenges in stabilizing atmospheric conditions. Indeed, differences among the emissions characteristics of the reference scenarios and in the implications of various stabilization targets are likely within the range that would be realized from an uncertainty analysis applied to any one of the three, as indicated by the analysis of the IGSM model by Webster et al. (2003).

Table 2.1 provides a cross-model overview of some of the key characteristics to be compared in the following sections of this chapter. Section 2.2 focuses on social science components, describing similarities and differences and highlighting the assumptions that have the greatest influences on the scenarios. Section 2.3 does the same for the natural science sub-models of each IAM, which in this research make the connection between the emissions of GHGs and the resulting atmospheric conditions.

Socioeconomic and Technology Components

Equilibrium, Expectations, and Trade

As can be seen in Table 2.1, the three participating models represent economic activity and associated emissions in a similar way; each divides the world economy into several regions, and further divides each region into economic sectors. In all three, the greatest degree of disaggregation is applied to the various components of energy supply and demand.

The models differ, however, in their representations of the equilibrium structure, the role of future expectations, and in the goods and services traded. MERGE and the EPPA component of IGSM are CGE models, which solve for a consistent set of supply-demand and price equilibria for each good and factor of production that is distinguished in the analysis. In the process, CGE models ensure a balance in each period of income and expenditure and of savings and investment for the economy, and they maintain a balance in international trade in goods and emissions permits. MiniCAM is a partial-equilibrium model, solving for supply-demand and price equilibria within linked energy and agricultural markets. Other economic sectors that influence the demand for energy and agricultural products and the costs of factors of production in these sectors are represented through exogenous assumptions.

Table 2.1. Characteristics of the Models

Feature	IGSM (with EPPA Economics Component)	MERGE	MiniCAM
Regions	16	9	14
Time Horizon, Time Steps	2100, 5-year steps	2200, 10-year steps	2095, 15-year steps
Model Structure	General equilibrium	General equilibrium	Partial equilibrium
Solution	Recursive dynamic	Inter-temporal optimization	Recursive dynamic
Final Energy Demand Sectors in Each Region	Households, private transportation, commercial transportation, service sector, agriculture, energy intensive industries, and other industry	A single, non-energy production sector	Buildings, transportation, and industry (including agriculture)
Capital Turnover	Five vintages of capital with a depreciation rate	A putty clay approach wherein the input-output coefficients for each cohort are optimally adjusted to the future trajectory of prices at the time of investment	Vintages with constant depreciation rate for all electricity-sector capital; capital structure not explicitly modeled in other sectors
Goods in International Trade	All energy and non-energy goods as well as emissions permits	Energy, energy intensive industry goods, emissions permits, and representative tradable goods	Oil, coal, natural gas, biomass, agricultural goods, and emissions permits
Emissions	CO_2, CH_4, N2O, HFCs, PFCs, SF_6, CO, NOX, SOX, NMVOCs, BC, OC, NH_3	CO_2, CH_4, N_2O, long-lived F-gases, short-lived F-gases, and SOX	CO_2, CH4, N_2O, CO, NOX, SO2, NMVOCs, BC, OC, HFC245fa, HFC 1 34a, HFC 125, HFC 143a, SF_6, C2F_6, and CF_4
Land Use	Agriculture (crops, livestock, and forests), biomass land use, and land use for wind and/or solar energy	Reduced-form emissions from land-use; no explicit land use sector; assume no net terrestrial emissions of CO_2	Agriculture (crops, pasture, and forests) as well as biomass land use and unmanaged land; the agriculture-land-use module directly determines land-use change emissions and terrestrial carbon stocks.
Population	Exogenous	Exogenous	Exogenous
GDP Growth	Exogenous productivity growth assumptions for labor, energy, and land; exogenous labor force growth determined from	Exogenous productivity growthassumptions for labor andenergy; exogenous labor	Exogenous productivitygrowth assumptions for labor; exogenous labor force growthbased

Table 2.1. (Continued)

Feature	IGSM (with EPPA Economics Component)	MERGE	MiniCAM
	population growth;endogenous capital growththrough savings and investment	forcegrowth determined frompopulation growth; endogenouscapital growth through savingsand investment	on populationdemographics
Energy Efficiency Change	Exogenous	Proportional to the rate ofGDP growth in each region	Exogenous
Energy Resources	Oil (including tar sands), shaleoil, gas, coal, wind and/or solar,land (biomass), hydro, andnuclear fuel	Conventional oil,unconventional oil (coal-basedsynthetics, tar sands, and shaleoil), gas, coal, wind, solar,biomass, hydro, and nuclear fuel	Conventional oil,unconventional oil (includingtar sands and shale oil), gas, coal, wind, solar, biomass(waste and/or residues and crops), hydro, and nuclear fuel(uranium and thorium);includes a full representation of the nuclear fuel cycle
Electricity Technologies	Conventional fossil (coal, gas,and oil), nuclear, hydro, naturalgas combined cycle (NGCC)with and without capture,integrated coal gasificationwith capture, and wind and/or solar, biomass	Conventional fossil (coal, gas, and oil), nuclear, hydro, new coal and gas with and without CCS, other renewables.	Conventional fossil (coal, gas,and oil) with and without capture; integrated gasification combined cycles (IGCCs) with and without capture; NGCC with and without capture; GenII, III, and IV reactors and associated fuel cycles; hydro, wind, solar, and biomass(traditional and modern commercial)
Conversion Technologies	Oil refining, coal gasification,and bio-liquids	Oil refining, coal gasification and liquefaction, bio-liquids, and electrolysis	Oil refining, natural gas processing, natural gas to liquids conversion, coal, and biomass conversion to synthetic liquids and gases; hydrogen production using liquids, natural gas, coal, biomass; and electrolysis, including direct production from wind and solar, and nuclear thermal conversion
Atmosphere-Ocean	2-dimensional atmosphere with a 3-dimensional ocean general circulation model,resolved at 20	Parameterized ocean thermal lag	Global multi-box energy balance model with upwelling-diffusion ocean heat transport

Table 2.1. (Continued)

Feature	IGSM (with EPPA Economics Component)	MERGE	MiniCAM
	minute time steps, 4° latitude, 4 surface types, and 12 vertical layers in the atmosphere		
Carbon Cycle	Biogeochemical models of terrestrial and ocean processes; depends on climate and/or atmospheric conditions with 35 terrestrial ecosystem types	Convolution ocean carbon cycle model assuming a neutral biosphere	Globally balanced carbon-cycle with separate ocean and terrestrial components, with terrestrial response to land-use changes
Natural Emissions	CH_4, N_2O, and weather and/or climate dependent as part of biogeochemical processmodels	Fixed natural emissions over time	Fixed natural emissions overtime
Atmospheric fate of GHGs,pollutants	Process models of atmospheric chemistry resolved for urban and background conditions	Single box models with fixeddecay rates. No considerationof reactive gases	Reduced form models forreactive gases and theirinteractions
Radiation Code	Radiation code accounting for all significant GHGs and aerosols	Reduced form, top-of-the-atmosphere forcing	Reduced form and top-of-the-atmosphere forcing; including indirect forcing effects

The models also differ in how expectations about the future affect current decisions. The EPPA component of IGSM and MiniCAM are recursive-dynamic, meaning they are solved one period at a time with economic agents modeled as responding to conditions in that period. This behavior is also referred to as myopic because these agents do not consider expected future market conditions in their decisions. The underlying behavioral assumption is that consumers and producers maximize their individual utilities or profits. In MiniCAM, this process is captured through the use of demand and supply functions that evolve over time as a function of evolving economic activity and regional economic development. In IGSM, explicit representative-agent utility and sector production functions ensure that consumer and producer decisions are consistent with welfare and profit maximization. In both of these models, the patterns of emissions mitigation over time in the scenarios that stabilize radiative forcing are imposed through assumptions intended to capture the features of a strategy that, as explained in Section 2.4, would be cost efficient. MERGE, on the other hand, is an inter-temporal optimization model, meaning that all periods are solved simultaneously such that resources and mitigation effort are allocated optimally over time as well as among sectors. Inter-temporal models of this type are often referred to as forward-looking or perfect foresight models because actors in the economy base current decisions not only on current conditions but on future ones, which are assumed to be known with certainty. Simultaneous solution of all periods ensures that agents' expectations about the future are realized in the model solution. MERGE's forward-looking structure allows it to explicitly solve for cost-minimizing emissions pathways, in contrast to MiniCAM and IGSM, which exogenously prescribe emissions mitigation policies over time.

Although all three models also represent international trade in goods and services and include exchange in emissions permits, they differ in the combinations of goods and services traded. In IGSM, all goods and services represented in the model are traded, with electricity trade limited to geographically contiguous regions to the extent that it occurs in the base data. MiniCAM models international trade in oil, coal, natural gas, agricultural goods, and emission permits. MERGE models trade in oil and natural gas, emissions permits, energy-intensive industrial goods, and a single non-energy good representing all other tradable goods and services.

Population and Economic Growth

An increase in the overall scale of economic activity is among the most important drivers of GHG emissions. However, economic growth depends, in part, on growth in population, which in all three models is an exogenously determined input. Although economic activity is an output of the models, its level is largely determined by assumptions about labor productivity and labor force growth, which are also model inputs. Policies to reduce emissions below those in the reference scenarios also affect economic activity, which may be measured as changes in gross domestic product (GDP) or in national consumption. (See Chapter 4, which provides a discussion of the interpretation and limitations of GDP and other welfare measures.)

In MiniCAM, labor productivity and growth in the labor force are the main drivers of GDP growth. GDP is calculated as the product of labor force and average labor productivity modified by an energy-service cost feedback elasticity. The labor force and labor productivity are both exogenous inputs to MiniCAM, but were developed for these scenarios from detailed demographic analysis. Starting with the underlying population scenario, the labor force was estimated from age- and gender-specific labor force participation rates applied to the relevant cohorts, then summed and adjusted by a fixed unemployment rate. Trends were explicitly considered, such as the increasing rate of labor force participation by females in the U.S. economy, the aging of the baby boomers, and evolving labor participation rates in older cohorts, reflecting the consequences of changing health and survival rates. Labor force productivity growth rates vary over time and across region to represent these evolving demographics.

In MERGE and the EPPA component of IGSM, the labor force and its productivity, while extremely important, are not the only factors determining GDP. Savings and investment and productivity growth in other factors (e.g., materials, land, labor, and energy) variously contribute as well. IGSM and MERGE use population directly as a measure of the labor force and apply assumptions about labor productivity change that are appropriate for that definition.

Energy Demand

In all three models, energy demands are represented regionally and driven by regional economic activity. As a region's economic activity increases, its corresponding demand for energy services rises. Energy demand is also affected by assumptions about changes in technology, in the structure of the economy, and in other economic conditions (see Section 2.2.5). Similarly, all the models represent the way demand will respond to changes in price.

The formulation of price response is particularly important in the construction of stabilization scenarios because the imposition of a constraint on carbon emissions will require the use of more expensive energy sources with lower emissions and will, therefore, raise the consumer price of all forms of energy.

The demand for energy is derived from demands for other goods and services in all three IAMs. However, the models differ in the way they derive their energy demands. In IGSM each good- or service-producing sector demands energy. The production sector is an input-output structure in which every industry (including the energy sector) supplies its outputs as inputs to intermediate production in other industries and for final consumption. Households have separate demands for automobile fuel and for all other energy services. Each final demand sector can use electricity, liquid fuels (petroleum products or biomass liquids), gas, and coal; fuel for automobiles is limited to liquids. MiniCAM is similar in that each MiniCAM sector demands energy. Energy is demanded by both final consumers and transforming sectors. In MiniCAM, there are three final energy consumption sectors – buildings, industry, and transport – which consume electricity and energy products such as coal, biomass, refined liquid fuels, methane, and hydrogen. In addition, energy is demanded by energy-producing and refining sectors, power generators, and hydrogen producers, whose demands in turn are derived from the demands arising in the final energy consumption sectors. MERGE is similar to IGSM except that its interindustry transactions are aggregated into a single, non-energy-production sector for each region from which demands for fuels (oil, gas, coal, and bioenergy) and electricity are derived. The power generation sector's demands for energy are derived from the economy's demand for electricity.

Energy Resources

The future availability of energy resources, particularly of exhaustible fossil fuels, is an important determinant of energy use and emissions, so all three of the participating models provide explicit treatments of the underlying resource base. All three include empirically based estimates of in-ground resources of oil, coal, and natural gas that might ultimately be available, along with a model of the costs of extraction. The levels of detail in the different models are shown in Table 2.1. Each of the models includes both conventional and unconventional sources in its resource base and represents the process of exhaustion of resources by an increasing cost of exploitation. That is, lower-cost resources are utilized first so that the costs of extraction rise as the resources are depleted. The models differ, however, in the way they represent the increasing costs of extraction. MiniCAM divides the resource base for each fossil fuel into discrete grades with increasing costs of extraction, along with an exogenous technological change parameter that lowers extraction costs over time. MERGE has similar differential grades for oil and gas, but assumes that the coal base is more than sufficient to meet potential demand and that exogenous technological improvements in extraction will be minimal. For these reasons, MERGE represents coal as having a constant cost over time irrespective of utilization. IGSM models resource grades with a continuous function, separately identifying conventional oil, shale oil, natural gas, and coal. Fuel-producing sectors are subject to economy-wide technical progress (e.g., increased labor productivity growth), which partly offsets the rise in extraction costs. The models all incorporate tar sands and unconventional gas (e.g., tight gas and coal-seam gas) in the grade structure for oil and natural gas, and each also includes the potential development of shale oil.

The models seek to represent all resources that could be available as technology and economic conditions vary over time and across simulations. Thus, they represent conditions under which currently unused resources could be economically exploited due to advances in technology or higher prices driven by increasing demands. Generally, then, the modeling groups define a resource base that is more expansive than, for example, that of the U.S. Geological Survey, which estimates technological and economic feasibility only at current technology and prices. However, differences exist in the treatments of potentially available resources. Mini- CAM includes a detailed representation of the nuclear power sector, including uranium and thorium resources; nuclear fuel fabrication; reactor technology options; and associated fuel- cycle cycles, including waste, storage, and fuel reprocessing. IGSM and MERGE assume that the uranium resources used for nuclear power generation are sufficient to meet likely use and, therefore, do not explicitly model their depletion.

The treatment of wind and solar resources also differs among the models. IGSM represents the penalty for intermittent supply by modeling wind and solar as imperfect substitutes for central station generation, where the elasticity of substitution implies a rising cost as these resources supply a larger share of electricity supply. Land is also an input, and the regional cost of wind and solar energy is based on estimates of regional resource availability and quality. MERGE represents these resources as having a fixed cost, but it applies upper limits on the proportion of these resources in the electricity system, representing limits on the integration of these resources into the grid. MiniCAM represents wind and solar technologies as extracting power from graded, regional, renewable resource bases. Variation in resource availability across diurnal and annual cycles affects market penetration of these technologies. As wind and solar technologies achieve larger fractions of the total power generation system, storage and ancillary power production capacity are required, which in turn affects the cost of power generation and technology choice.

IGSM and MiniCAM model biomass production as competing for agricultural land. Increasing production leads to increasing land rent, representing the scarcity of agricultural land, and thus, to increasing cost of biomass as production expands. MiniCAM also has a separate set of regional supply functions for biomass supplied from waste and residue sources. In these scenarios, MERGE represents biomass as a graded resource. Two grades of biomass are included, with fixed costs for each. The total supply from the first, less-expensive grade is limited, but the second, more-expensive grade is allowed to compete unhindered in the market.

Technology and Technological Change

Technology is the broad set of processes covering know-how, experience, and equipment used by humans to produce services and transform resources. In the three models participating in this scenario, the relationship between things that are produced and things that are used in the production process are represented mathematically. In the jargon of the models, the relationship between things that are produced and things that are used in the production process is referred to as a production function.

The three modeling groups differed substantially in their representation of technology depending on their overall design objectives. Differences also resulted from data limitations and computational feasibility, which force tradeoffs between the inclusion of engineering detail and the representation of the interaction among the segments of a modern economy that determines supply, demand, and prices (see Box 2.1).

All three of the models applied here follow a hybrid approach to the representation of energy technology, involving substantial detail in some areas and more aggregate representations in others, and some of the choices that flow from the distinct design of each can be seen in Table 2.1. They represent energy demand, as described in Section 2.2.3, with the application of an autonomous energy efficiency improvement (AEEI) factor to represent non-price-induced trends in energy use. However, AEEI parameter values are not directly comparable across the models because each has a unique representation of the processes that together explain the multiple forces that have contributed historically to changes in the energy intensity of economic activity. In IGSM and MERGE, the AEEI captures non-price changes (including structural change not accounted for in the models) that can be energy using rather than energy saving. MERGE represents the AEEI as a function of GDP growth in each region. MiniCAM captures shifts among fuels through differing income elasticities, which change over time, and separately represents AEEI efficiency gains.

Box 2.1. Top-Down, Bottom-Up, and Hybrid Modeling

The models used in energy and environmental assessments are sometimes classified as either top- down or bottom-up in structure, a distinction that refers to the way they represent technological options. A top-down model uses an aggregate representation of how producers and consumers can substitute non-energy inputs for energy inputs or relatively energy-intensive goods for less energy-intensive goods. Often, these tradeoffs are represented by aggregate production functions or by utility functions that describe consumers' willingness and technical ability to substitute among goods.

The bottom-up approach begins with explicit technological options, and fuel substitution or changes in efficiency occur as a result of discrete changes from one specific technology to another. The bottom-up approach has the advantage of being able to represent explicitly the combination of outputs, inputs, and emissions of types of capital equipment used to provide consumer services (e.g., a vehicle model or building design) or to perform a particular step in energy supply (e.g., a coal-fired powerplant or wind turbine). However, a limited number of technologies are often included, which may not well represent the full set of possible options that exist in practice. Also, in a pure bottom- up approach, the demands for particular energy services are often characterized as fixed (unresponsive to price), and the prices of inputs such as capital, labor, energy, and materials are exogenous.

On the other hand, the top-down approach explicitly models demand responsiveness and input prices, which usually require the use of continuous functions to model at least some parts of the available technology set. The disadvantage of the latter approach is that production functions of this form will poorly represent switch points from one technology to another – as from one form of electric generation to another or from gasoline to biomass blends as vehicle fuel. In practice, the vast majority of models in use today, including those applied in this scenario, are hybrids in that they include substantial technological detail in some sectors and more aggregate representations in others.

Other areas shown in Table 2.1 where there are significant differences among the models are in energy conversion – from fossil fuels or renewable sources to electricity and from solid fossil fuels or biomass to liquid fuels or gas. In IGSM, discrete energy technologies are represented as energy supply sectors contained within the input-output structure of the economy. Those sources of fuels and electricity that now dominate supply are represented as production functions with the same basic structure as the other sectors of the economy. Technologies that may play a large role in the future (e.g., power plants with CCS or oil from shale) are introduced as discrete technologies using a production function structure similar to that for existing production sectors and technologies. They are subject to economy-wide productivity improvements (e.g., labor, land, and energy productivity), with the effect on cost depending on the share of each factor in the technology production function. MERGE and MiniCAM also characterize energy-supply technologies in terms of discrete technologies. In the MERGE scenarios in this research, technological improvements are captured by allowing for the introduction of more advanced technologies in future periods. In the MiniCAM scenarios, the cost and performance of technologies are assumed to improve over time, and new technologies become available in the future. Similar differences among the models hold for other conversion technologies, such as coal gasification, coal liquefaction, or liquids from biomass.

The entry into the market of new sources and their levels of production by region are determined endogenously in all three models and depend on the relative costs of supply. It should be emphasized that the versions of the models used in this research do not explicitly represent the processes of technological change, for example, public and private R&D, spillovers from innovation in other economic sectors, and learning-by-doing. A number of recent efforts have been made to incorporate such processes and their effects as an endogenous component of modeling exercises. In most cases, these studies have not been applied to models of the complexity needed to meet the requirements of this scenario product.

Because of the differences in structure among these models, there is no simple technology-bytechnology comparison of performance and cost across particular sources of supply or technological options. This situation exists for a variety of reasons. First, cost is an output of the three models and not an input. In the three models here technologies are defined in many cases not in terms of some exogenously specified cost, but rather as a function of inputs whose prices change across simulations and over time. The three models differ in many regards. Each model defines the scope of a technology differently. Sectoral definitions, technology definitions, and data sources all vary across the three models. For example, one model has a service sector while another has a buildings sector. There is then, no common definition for technologies, technology descriptors and hence for a set of comparable costs. The detailed scenario documentation for each of the three modeling groups provides more information about the technology assumptions employed by three modeling groups. These are documented in Paltsev et al. (2005) for IGSM and in Clarke et al. (2007) for MiniCAM. Assumptions for MERGE are included in the version of the model posted at http://www.stanford.edu/group/MERGE.

The influence of differing technology specifications and assumptions is evident in the scenarios discussed in Chapters 3 and 4. For example, in the absence of efforts to control GHG emissions, motor fuel is drawn ever more heavily from high-emitting sources. Oil from shale comes in under the resource and technology assumptions used in the IGSM scenarios, whereas liquids from coal figure prominently in the MERGE scenarios, and the MiniCAM scenarios include an intermediate mix of both. Furthermore, because each model assumes market mechanisms operate efficiently, the marginal cost of reducing GHG emissions – that is the cost of reducing the last tonne of GHG – is equal to the price of carbon in every technology employed in every sector and in every country of the world. When stabilization conditions are imposed, CCS takes on a key role in all the scenarios over the time period considered in this research. Nuclear power contributes heavily in MERGE and in MiniCAM scenarios, whereas the potential role of this technology is overridden in the IGSM scenarios by an assumption of non- climate restraints on expansion due to concerns over issues such as safety, waste, and proliferation. Finally, although differences in emissions in the reference scenario contribute to variations in the difficulty of achieving stabilization, alternative assumptions about technological improvements also play a prominent role.

Land Use and Land-Use Change

The models used in this research were developed originally with a focus on energy and fossil carbon emissions. The integration of the terrestrial biosphere, including human activity, into the climate system is less highly developed. Each model represents the global carbon cycle, including exchanges among the atmosphere, natural vegetation, and soils; the effects of human land use and responses to carbon policy; and feedbacks to the global climate. No model represents all of these possible responses and interactions, and the level of detail varies substantially among the models. For example, the models differ in their handling of natural vegetation and soils and in their responses to change CO_2 concentrations and climate. Furthermore, land-use practices (e.g., low- or no-till agriculture and biomass production) and changes in land use (e.g., afforestation, reforestation, or deforestation) that influence GHG emissions and the sequestration of carbon in terrestrial systems are handled at different levels of detail. Indeed, improved two-way linking of global economic and climate analysis with models of physical land use (land use responding to climate and economic pressures and climate responding to changes in the terrestrial biosphere) is the subject of ongoing research in these modeling groups.

In IGSM, land is an input to agriculture, biomass production, and wind and/or solar energy production. Agriculture is a single sector that aggregates crops, livestock, and forestry. Biomass energy production is modeled as a separate sector, which competes with agriculture for land. Markets for agricultural goods and biomass energy are international, and demand for these products determines the price of land in each region and its allocation among uses. In other sectors, returns to capital include returns to land, but the land component is not explicitly identified. Anthropogenic emissions of GHGs (importantly, CH_4 and N_2O) are estimated within IGSM as functions of agricultural activity and assumed levels of deforestation. The response of terrestrial vegetation and soils to climate change and CO_2 increase is captured in the Earth system component of the model, which provides a detailed treatment of biogeochemical and land-surface properties of terrestrial systems. However, the biogeography of natural ecosystems and human uses remains unchanged over the simulation period, with the area of cropland fixed to the pattern of the early 1990s. Balance in the carbon cycle between ocean uptake, land-use and land-use change, and anthropogenic emissions is achieved in IGSM with an adjustment factor to ensure that the recent trend in atmospheric CO_2 increase is replicated. This adjustment factor is best interpreted as what carbon uptake due to forest regrowth must have been, given the representation of terrestrial and ocean systems in IGSM. The need for such an adjustment factor reflects the continuing scientific uncertainty in the carbon cycle. In other words, with fossil emissions and concentrations relatively well known, the total uptake is known but the partitioning of the uptake between terrestrial and ocean systems is uncertain (Sabine et al. 2004). IGSM does not simulate carbon price-induced changes in carbon sequestration (e.g., reforestation and tillage), and change among land-use types in the EPPA component of IGSM is not fed to the terrestrial biosphere component of the model.

The MERGE modeling group assumed a neutral terrestrial biosphere across all scenarios. That is, it is assumed that the net CO_2 exchange with the atmosphere by natural ecosystems and managed systems – the latter including agriculture, deforestation, afforestation, reforestation, and other land-use change – sums to zero.

MiniCAM includes a model that allocates the land area in a region among various components of human use and unmanaged land – with changes in allocation over time in relation to income, technology, and prices – and estimates the CO_2emissions (or sinks) that result. Land conditions and associated emissions are parameterized for a set of regional sub-aggregates. The supply of primary agricultural production (four food crop types, pasture, wood, and commercial biomass) is simulated regionally with competition for a finite land resource based on the average profit rate for each good potentially produced in a region. In stabilization scenarios, the value of carbon stored in the land is added to this profit, based on the average carbon content of different land uses in each region. This allows carbon mitigation policies to explicitly extend into land and agricultural markets. The model is solved by clearing a global market for primary agricultural goods and regional markets for pasture. The biomass market is cleared with demand for biomass from the energy component of the model. Exogenous assumptions are made for the rate of intrinsic increase in agricultural productivity, although net productivity can decrease in the case of expansion of agricultural lands into less productive areas (Sands and Leimbach 2003). Unmanaged land can be converted to agro-forestry, which in general leads to net CO_2 emissions from tropical regions in the early decades. Emissions of non-CO_2 GHGs are tied to relevant drivers, for example, with CH_4 from ruminant animals related to beef production. MiniCAM thus treats the effects on carbon emissions of gross changes in land use (e.g., from forests to biomass production) using an average emission factor for such conversion. The pricing of carbon stocks in the model provides a counterbalance to increasing demand for biomass crops in stabilization scenarios.

Emissions of CO_2 and Non-CO_2 Greenhouse Gases

In all three models, the main source of CO_2 emissions is fossil fuel combustion, which is computed on the basis of the carbon content of each of the underlying resources: oil, natural gas, and coal. Special adjustments are made to account for emissions associated with the additional processing required to convert coal, tar sands, and shale sources into products equivalent to those from conventional oil. Other industrial CO_2 emissions also are included, primarily from cement production.

As required for this research, all three models include representations of emissions and abatement of CH_4, N_2O, HFCs, PFCs, and SF_6 (plus aerosols and other substances not considered in this scenario). The models use somewhat different approaches to represent abatement of non-CO_2 GHGs. IGSM includes the emissions and abatement possibilities directly in the production functions of the sectors that are responsible for emissions of the different gases. Abatement possibilities are represented by substitution elasticities in a nested structure that encompasses GHG emissions and other inputs, benchmarked to reflect bottom-up studies of abatement potential. This construction is parallel to the representation of fossil fuels in production functions, where abatement potential is similarly represented by the substitution elasticity between fossil fuels and other inputs, with the specific set of substitutions governed by the nest structure. Abatement opportunities vary by sector and region.

In MERGE, CH4 emissions from natural gas use are tied directly to the level of natural gas consumption, with the emissions rate decreasing over time to represent reduced leakage during the transportation process. Non-energy sources of CH_4, N_2O, HFCs, PFCs, and SF_6 are based largely on the guidelines provided by the EMF Study No. 21 on Multi-Gas Mitigation and Climate Change (de la Chesnaye and Weyant 2006). The EMF developed baseline projections from 2000 through 2020. For all gases but N_2O and CO_2 , the baseline for beyond 2020 was derived by extrapolation of these estimates. Abatement cost functions – the relationship between levels of emissions reductions and the costs of these reductions – for these two gases are also based on EMF 21, which provided estimates of the abatement potential for each gas in each of 11 cost categories in 2010. These abatement cost curves are directly incorporated in the model and extrapolated after 2010 following the baseline. There is also an allowance for technical advances in abatement over time.

MiniCAM calculates emissions of CH_4, N_2O, and seven categories of industrial sources for HFCs, PFCs, and SF_6. Emissions are determined for over 30 sectors, including fossil fuel production, transformation, and combustion; industrial processes; land use and land-use change; and urban emissions. For details, see Smith (2005) and Smith and Wigley (2006). Emissions are proportional to driving factors appropriate for each sector, with emissions factors in many sectors decreasing over time according to an income-driven logistic formulation. Marginal abatement cost (MAC) curves from the EMF-21 study are applied, including shifts in the curves for CH_4 due to changes in natural gas prices. Any below-zero reductions in MAC curves are assumed to apply in the reference scenario.

Earth Systems Components

The Earth system components of the models represent the response of the atmosphere, ocean, and terrestrial biosphere to emissions and increasing concentrations of GHGs and other substances. Representation of these processes, including the carbon cycle (Box 2.2), is necessary to determine emissions paths consistent with stabilization because these systems determine how long each of these substances remains in the atmosphere and how they interact in altering the Earth's radiation balance. Each model includes such physical-chemical-biological components, but incorporates different levels of detail. The most elaborated Earth system components are found in IGSM (Sokolov et al. 2005), which falls in a class of models referred to as Earth System Models of Intermediate Complexity (Claussen et al. 2002). These are models that fall between the full three-dimensional atmosphere-ocean general circulation models (AOGCMs) and energy balance models with a box model of the carbon cycle. The Earth system components of MERGE and MiniCAM fall in the class of energy balance-carbon cycle box models. Table 2.1 shows how each of the models treat different components of the Earth systems.

IGSM has explicit spatial detail, resolving the atmosphere into multiple layers and by latitude, and it includes a terrestrial vegetation model with multiple vegetation types that are also spatially resolved. A version of IGSM with a full three-dimensional ocean model was used for this scenario, and it includes temperature-dependent uptake of carbon. IGSM models atmospheric chemistry, resolved separately for urban (i.e., heavily polluted) and background conditions. Processes that move carbon into or out of the ocean and vegetation are modeled explicitly. IGSM also models natural emissions of CH_4 and N_2O, which are weather and/or climate-dependent. The model includes a radiation code that computes the net effect of atmospheric concentrations of the GHGs studied in this research. Also included in the global forcing is the effect of changing ozone and aerosol levels, which result from emissions of CH_4 and non-GHGs, such as NO_X and volatile organic hydrocarbons; SO_X; black carbon; and organic carbon from energy, industrial, agricultural, and natural sources.

Box 2.2. The Carbon Cycle

Although an approximate atmospheric lifetime is sometimes calculated for CO_2, the term is potentially misleading because it implies that CO_2 put into the atmosphere by human activity always declines over time by some stable removal process. In fact, the calculated concentration of CO_2 is not related to any mechanism of destruction, or even to the length of time an individual molecule spends in the atmosphere, because CO_2 is constantly exchanged between the atmosphere and the surface layer of the ocean and with vegetation. Instead, it is more appropriate to think about how the quantity of carbon that the Earth contains is partitioned between stocks of in-ground fossil resources, the atmosphere (mainly as CO_2), surface vegetation and soils, and the surface and deep layers of the ocean. When stored carbon is released into the atmosphere, either from fossil or terrestrial sources, atmospheric concentrations of CO_2 increase, leading to disequilibrium with the ocean, and more carbon is taken up than is cycled back. For land processes, vegetation growth may be enhanced by increases in atmospheric CO_2, and this change could augment the stock of carbon in vegetation and soils. As a result of the ocean and terrestrial uptake, only about half of the carbon currently emitted remains in the atmosphere. Over millennial time scales, oceans would continue to remove carbon until a large fraction, presently about 80%, would ultimately be removed to the oceans, leaving about 20% as a permanent increase in the atmospheric CO_2 concentration. But this large removal only occurs because current levels of emissions lead to substantial disequilibrium between atmosphere and ocean. Lower emissions would lead to less uptake, as atmospheric concentrations come into balance with the ocean and interact with the terrestrial system. Rising temperatures themselves will reduce uptake by the ocean, and will affect terrestrial vegetation uptake, processes that the models in this scenario variously represent.

An important policy implication of these carbon-cycle processes as they affect stabilization scenarios is that stabilization of emissions near the present level will not lead to stabilization of atmospheric concentrations. CO_2 concentrations were increasing in the 1 990s at just over 3 ppmv per year, an annual increase of 0.8%. Thus, even if societies were able to stabilize emissions at current levels, atmospheric concentrations of CO_2 would continue to rise. As long as emissions exceed the rate of uptake, even very stringent abatement will only slow the rate of increase.

The carbon cycle in MERGE relates emissions to concentrations using a convolution ocean carbon-cycle model and assuming a neutral biosphere (i.e., no net CO_2 exchange). It is a reduced-form carbon cycle model developed by Maier-Reimer and Hasselmann (1987). Carbon emissions are divided into five classes, each with different atmospheric lifetimes. The behavior of the model compares favorably with atmospheric concentrations provided in the IPCC's Third Assessment Report (TAR) (IPCC 2001) when the same SRES scenarios of emissions are simulated in the model (Nakicenovic et al. 2000). MERGE models the radiative effects of GHGs using relationships consistent with summaries by the IPCC, and applies the median aerosol forcing from Wigley and Raper (2001). The aggregate effect is obtained by summing the radiative forcing effect of each gas.

MERGE's physical Earth system component is embedded in the inter-temporal optimization framework, thus allowing solution of an optimal allocation of resources through time, accounting for damages related to climate change, or optimizing the allocation of resources with regard to other constraints such as concentrations, temperature, or radiative forcing. In this research, the second of these capabilities is applied, with a constraint on radiative forcing (see Chapter 4). In contrast, the IGSM and Mini- CAM Earth system models are driven by emissions as simulated by the economic components. In that regard, they are simulations rather than optimization models.

MiniCAM uses the MAGICC model (Wigley and Raper 2001, 2002) as its biophysical component. MAGICC is an energy-balance climate model that simulates the energy inputs and outputs of key components of the climate system (sun, atmosphere, land surface, and ocean) with parameterizations of dynamic processes such as ocean circulations. It operates by taking anthropogenic emissions from the other MiniCAM components, converting these to global average concentrations (for gaseous emissions), then determining anthropogenic radiative forcing relative to preindustrial conditions, and finally computing global mean temperature changes. The carbon cycle is modeled with both terrestrial and ocean components. The terrestrial component includes CO_2 fertilization and temperature feedbacks; the ocean component is a modified version of the Maier-Reimer and Hasselmann (1987) model that also includes temperature effects on the terrestrial biosphere. land-use change emissions from the MiniCAM's land-use change component are fed into MAGICC so that the global carbon cycle is consistent with the amount of natural vegetation. Reactive gases and their interactions are modeled on a global-mean basis using equations derived from results of global atmospheric chemistry models (Wigley et al. 2002).

In MiniCAM, global mean radiative forcing for CO_2, CH_4, and N_2O are determined from GHG concentrations using analytic approximations. Radiative forcing for other GHGs are taken to be proportional to concentrations. Radiative forcing for aerosols (for sulfur dioxide and for black and organic carbon) are taken to be proportional to emissions. Indirect forcing effects, such as the effect of CH_4 on stratospheric water vapor, are also included. Given radiative forcing, global mean temperature changes are determined by a multiple box model with an upwelling-diffusion ocean component. The climate sensitivity is specified as an exogenous parameter. MAGICC's ability to reproduce the global mean temperature change results of AOGCMs has been demonstrated (Cubasch et al. 2001, Raper and Gregory 2001).

Although aerosols and ozone are not included in the computation of the radiative forcing targets that are the focus of these scenarios, they are nonetheless included in these scenarios as noted above. That is, the radiative forcing stabilization levels identified in Table 1.2 and the radiative forcing levels reported in subsequent chapters account for only that part of radiative forcing due to those GHGs covered by the target. The models can simulate total radiative forcing including additional positive forcing from ozone and dark aerosols and negative forcing from sulfate aerosols. As shown by Prinn et al. (In Press), even for very large changes in emissions related to these substances, the temperature effect is small, in large part because aerosols and ozone have offsetting cooling and warming effects. To the extent temperature is affected by these substances, however, they have a small, indirect influence on the scenarios because trace gas cycles are climate-dependent. For example, climate affects vegetation and ocean temperature and, thus, carbon uptake, and natural emissions of CH_4 and N_2O, and the lifetime of CH_4 also depends on climate. Because the net effect of these substances on temperature is small, the feedback effect on trace gas cycles also is very small. However, to the extent these feedbacks are represented in the models as discussed above, they are included in the calculation of required emissions reduction because the temperature paths, while not reported here, are simulated in the models and affect the CO_2 and non-CO_2 GHG concentrations. By the same token, the gases included under the Montreal Protocol, which are being phased out, are nonetheless included in these models and exert some influence on temperature.

Note that although the models used in this research have capabilities to evaluate various climate change effects, with few exceptions, they do not include the consequences of such feedback effects as: temperature on home heating and cooling requirements; local climate change on agricultural productivity; CO_2 fertilization on agricultural productivity (though a CO_2 fertilization effect is included in the terrestrial carbon cycle models employed by IGSM and MiniCAM); climate on water availability for applications ranging from crop growing to power plant cooling. Such improvements are left to future research.

3. Reference Scenarios

In the reference scenarios, energy consumption grows significantly and the energy system continues to rely on fossil fuels, leading to an increase in CO_2 emissions of roughly 3 to 3½ times the present level by 2100. Combined with increases in the non-CO_2 GHGs and net uptake by the ocean and terrestrial biosphere, radiative forcing from the GHGs considered in this research reaches 6.4 W/m^2 to 8.6 W/m^2 from preindustrial by 2100.

Introduction

This chapter introduces the reference scenarios developed by the three modeling groups. These scenarios are plausible future paths, not predictions, for by the very nature of their construction they lack the features of predictions or best-judgment forecasts. For example, they assume that in the post-2012 period existing measures to address climate change expire and are never renewed or replaced, which is an unlikely occurrence. Rather, they have been developed as points of departure to highlight the implications for energy use and other human activities of the stabilization of radiative forcing. Each of the modeling groups could have created a range of other plausible reference scenarios by varying assumptions about rates of economic growth, the cost and availability of alternative energy options, assumptions about non-climate environmental regulations, and so forth.

Other than to standardize reporting conventions and GHG emissions mitigation policies (or lack thereof), the three modeling groups developed their reference scenarios independently as each judged appropriate. As noted in Chapter 2, the three models were developed with somewhat different original design objectives. They differ in (a) their inclusiveness, (b) their specifications of key aspects of economic structure, and (c) their choice of values for key parameters. These choices then lead to different characterizations of the underlying economic and physical systems that these models represent.

Moreover, even if the models were identical in structure, the independent choice of key assumptions by the modeling groups leads to differences among scenarios. For example, as will be discussed, the reference scenarios differ in their specification of the technical details of virtually every aspect of the future global energy system, ranging from the cost and availability of oil and natural gas to the prospects for nuclear power. These differences affect emissions in the reference scenarios and the nature and cost of stabilization regimes.

Finally, the modeling groups did not attempt to harmonize assumptions about non-climaterelated policies. Such differences matter both in the reference and stabilization scenarios. For example, the MiniCAM reference scenario assumes a larger effect of CH_4 emission-control technologies deployed for economic reasons, which leads to lower reference scenario CH_4 emissions than in the reference scenarios from the other modeling groups. Similarly, the IGSM modeling group assumed that non-climate concerns would limit the deployment of nuclear power, while the MERGE and MiniCAM modeling groups assumed that nuclear power would be allowed to participate in energy markets on the basis of energy cost alone.

This variation in modeling approaches and assumptions is one of the strengths of this research, for the resulting differences across scenarios can help shed light on the implications of differing assumptions about the way key forces may evolve over time. It also provides three independent starting points for consideration of stabilization goals.

Although there are many reasons to expect that the three reference scenarios would be different, it is worth noting that the modeling groups met periodically during the research process to review progress and to exchange information. Thus, while not adhering to any formal protocol of standardization, the three reference scenarios are not entirely independent either.

Development of a reference scenario involves the elaboration of one path from among a range of uncertain outcomes. Thus, it should be further emphasized that the three reference scenarios were not designed in an attempt to span the full range of potential future conditions or to shed light on the probability of the occurrence of future events. That is a much more ambitious undertaking than the one reported here.

The remainder of this chapter describes the ref- erence scenarios developed by the three modeling groups working forward from underlying drivers to implications for radiative forcing. (Chapter 4 proceeds in the other direction, imposing the stabilization levels on radiative forcing and exploring the implications.) The presentation begins with a summary of the underlying socioeconomic assumptions, most notably for population and economic growth. There follows a discussion of the evolution of the global energy system over the twenty-first century in the absence of additional GHG controls and discusses the associated prices of fuels. The energy sector is the largest but not the only source of anthropogenic GHG emissions. Also important is the net uptake or release of CO_2 by the oceans and the terrestrial biosphere. The next section shows how the three reference scenarios handle this aspect of the interaction of human activity with natural Earth systems. Finally, the anthropogenic emissions are described, taking into account both the energy sector and other sources, such as agriculture and various industrial activities. This last section draws together all these various components to present reference scenarios of the consequences of anthropogenic emissions and the processes of CO_2 uptake and non-CO_2 gas destruction for the ultimate focus of the research: atmospheric concentrations and global radiative forcing.

Socioeconomic Assumptions

GHGs are a product of modern life. Population increase and economic activity are major determinants of the scale of human activities and ultimately of anthropogenic GHG emissions. In the reference scenarios, the global population rises from 6 billion in the year 2000 to between 8.6 and 9.9 billion in 2100. Economic activity grows through 2100 across the globe. Developed nations continue to expand their economies at historical rates, and developing nations make significant progress toward improved standards of living.

Reference scenarios are grounded in a larger demographic and economic story. Each uses population as the basis for developing scenarios of the scale and composition of economic activity for each region. For population assumptions, the IGSM modeling group adopted a regionally detailed United Nations (U.N.) projection for the period 2000-2050 (UN 2001) and extended this scenario to 2100 using information from a longer-term U.N. study (UN 2000). The Mini- CAM assumptions are based on a median scenario by the U.N. (UN 2005) and a Millennium Assessment Techno-Garden Scenario from the International Institute for Applied Systems Analysis (O'Neill 2005). Near-term population assumptions for the MERGE scenarios come from the Energy Information Administration's International Energy Outlook.

Population increases substantially across the scenarios by the end of the century, but all of the scenarios portray the population growth rate as slowing to near zero, if not turning negative, by the end of the century (Table 3.1 and Figure 3.1). As a result, by 2050 more than 75% of all the change between the year 2000 and 2100 has occurred. A demographic transition from high birth and death rates to low death rates and eventually to low birth rates is a feature of most demographic scenarios, reflecting assumptions that birth rates will decline to replacement levels or below. For some countries, birth rates are already below replacement levels, and just maintaining these levels would result in population decline for these countries. A key uncertainty in all demographic scenarios is whether a transition to less-than-replacement levels is a more or less permanent feature of those countries where it has occurred and whether such a pattern will be repeated in other countries.

The differences among the scenarios lie in nuances of this pattern. The MiniCAM scenarios exhibit a peak in global population around the year 2070 at slightly more than 9 billion people, after which the population declines to 8.6 billion. The MERGE and IGSM scenarios, on the other hand, both employ demographic assumptions by which the global population stabilizes but does not decline during this century. By 2100, populations range from 8.6 to 9.9 billion across the scenarios, which is an increase of roughly 40% to 65% from the 6 billion on Earth in 2000. In total, the difference between the demographic scenarios is relatively small: they differ by only 3% in 2030 and by less than 10% until after 2080.

Table 3.1. Population(million) by Region Across Models,2000-2100

IGSM Population by Region (million)						
Region	**2000**	**2020**	**2040**	**2060**	**2080**	**2100**
U.S.	283	334	379	396	395	393
Western Europe	390	388	368	331	302	289
Japan	127	126	116	113	118	119
Former Soviet Union	291	278	260	243	234	230
Eastern Europe	97	91	83	74	67	64
China	1282	1454	1500	1429	1365	1334
India	1009	1291	1503	1610	1635	1643
Africa	793	1230	1749	2163	2390	2500
Latin America	419	538	627	678	701	713
Rest of the World	1366	1848	2269	2521	2614	2652
MERGE Population by Region (million)						
U.S.	276	335	335	335	335	335
Western Europe	390	397	397	397	397	397
Japan	127	126	126	126	126	126
Former Soviet Union/Eastern Europe	411	393	393	393	393	393
China	1275	1429	1478	1493	1498	1499
India	1017	1312	1427	1472	1489	1496
Africa/Latin America/ Rest of World	2566	3538	4209	4677	5003	5228
MiniCAM Population by Region (million)						
U.S.	283	334	371	396	412	426
Western Europe	457	486	481	456	421	399
Japan	127	127	121	113	103	95
Former Soviet Union	283	284	283	275	261	253
Eastern Europe	124	119	111	100	87	80
China	1385	1578	1591	1506	1407	1293
India	1010	1312	1472	1513	1443	1300
Africa	802	1197	1521	1763	1893	1881
Latin America	525	670	786	869	929	952
Rest of World	1055	1454	1779	1976	2012	1918

Regional aggregations are different in the three models. For example, MiniCAM includes Turkey in Western Europe, but IGSM an dMERGE do not.

The variation in population among the scenarios is greater for the U.S. than for the globe. The U.S. population (Figure 3.1) increases from about 280 million in the year 2000 to between 335 million and 425 million by 2100. Although the MiniCAM global population is the lowest of the three scenarios in 2100, it is the highest for the U.S. The higher U.S. population in MiniCAM reference scenarios compared to the scenarios from the other two modeling groups can be traced to different assumptions about net migration.

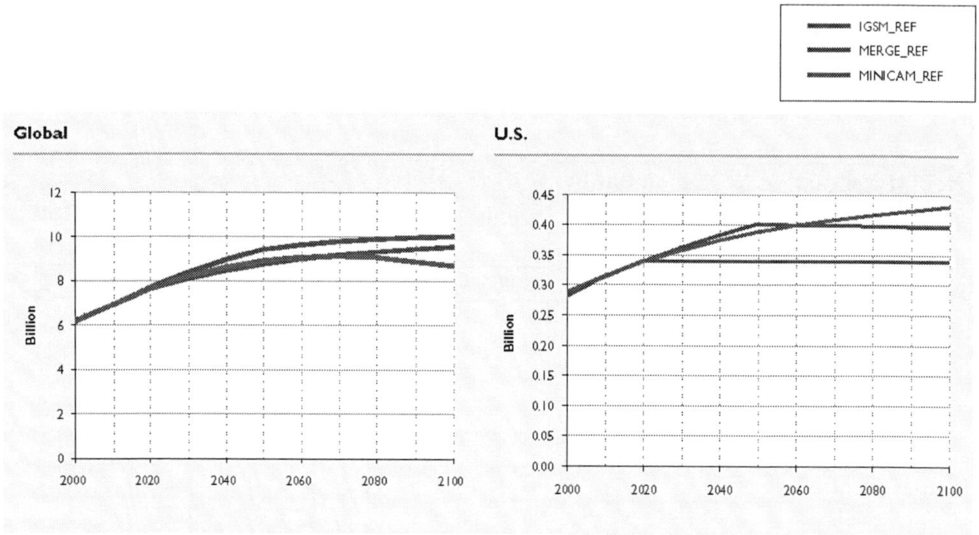

Assumed growth in global and U.S. population is similar among the three models. Global population in 2100 spans a range from about 8.5 to 10 billion. U.S. population in 2100 spans a range from less than .35 to over .45 billion.

Figure 3.1. World and U.S. Population Across Models

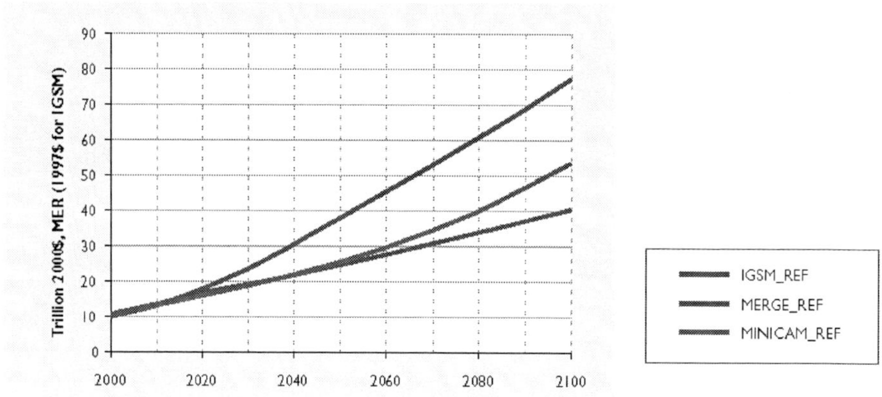

U.S. economic growth is driven, in part, by labor force growth and, in part, by assumptions about productivity growth of labor and, in part, by other factors such as by savings and investment. Annual average growth rates are 1.4% for the MERGE reference scenario, 1.7% for the MiniCAM reference scenario, and 2.2% for the IGSM reference scenario. By comparison, U.S. GDP grew at an annual average rate of 3.4% from 1959-2004 (CEA 2005).

Figure 3.2. U.S. GDP Across Reference Scenarios (1997$ for IGSM, 2000$ for MERGE and MiniCAM, MER)

As discussed in Chapter 2, GDP, while ostensibly an output of all three models, is in fact largely determined by assumptions about labor productivity and labor force growth, which are model inputs. None of the three modeling groups began with a GDP goal and derived sets of input factors that would generate that level of activity. Rather, each began with assessments about potential growth rates in labor productivity and labor force and used these, through differing mechanisms, to compute GDP. In MiniCAM, labor productivity and labor force growth are the main drivers of GDP growth. In MERGE and IGSM, savings and investment and productivity growth in other factors (e.g., materials, land, and energy) contribute as well. All three models derive labor force growth from the underlying assumptions about population.

The alternative scenarios of population and productivity growth lead to differences among the three reference scenarios in U.S. GDP growth (Figure 3.2). There is relatively little difference among the three trajectories through the year 2020. After 2020, however, the scenarios diverge, with the lowest scenario of U.S. GDP (MERGE) at roughly half of that of the highest scenario (IGSM) by the end of the century. The labor productivity growth assumptions for the U.S. in the IGSM scenario are the highest of the three, and the U.S. population assumptions are also relatively high in the IGSM scenarios. The relatively lower labor productivity growth assumptions used in the MERGE and MiniCAM scenarios lead to lower levels of GDP. The lower population growth assumptions employed in the MERGE scenarios give the MERGE reference scenario the lowest GDP in 2100.

Table 3.2 shows GDP across regions in the three reference scenarios. Differences in the absolute levels of GDP increase result from relatively small differences in rates of per capita growth. Although difficulties arise in comparisons of GDP across countries (see Box 3.1), the growth rates underlying these scenarios are usefully compared with historical experience. Long-term growth rates developed from reconstructed data (Table 3.3) show that consistent rapid growth is a phenomenon of industrialization, starting in the 1 800s in North America and Europe and gradually spreading to other areas of the world. By the end of the period 1950 to 1973, it appeared that the phenomenon of rapid growth had taken hold in all major regions of the world. Since 1973, it has been less clear to what degree that conclusion holds. Growth slowed in the 1 970s in most regions, the important exceptions being China, India, and several South and East Asian economies. In Africa, Latin America, Eastern Europe, and the former Soviet Union, growth slowed in this period to rates more associated with preindustrial times.

With this historical experience as background, the differences in GDP growth among the reference scenarios can be explained. Demographic trends, slowing population, and labor force growth all combine to influence overall GDP growth in the scenarios. With respect to the developed countries, the per capita income growth rate for the U.S. in the IGSM reference scenario is about the average for North America for the period 1950-2000. The MiniCAM reference scenario has lower growth, reflecting an assumption that an aging population will lead to lower labor force participation, and the result of this demographic maturation is a lower future rate of per capita GDP growth compared to history. U.S. growth rates in the MERGE reference scenario are similar to those of MiniCAM reference scenario.

GDP growth patterns for Western Europe and Japan are similar to one another within reference scenarios but vary across models. The IGSM reference scenario follows the post World War II trend in per capita GDP growth, but the MiniCAM and MERGE scenarios anticipate a break from the trend with lower per capita growth in GDP as a consequence of changes in underlying demographic trends

Table 3.2. GDP for Key Regions Across Reference Scenarios, 2000-2100

IGSM GDP by Region (trillions of $1997, MER)						
Region	**2000**	**2020**	**2040**	**2060**	**2080**	**2100**
U.S.	9.1	16.9	29.3	44.4	59.8	76.4
Western Europe	9.2	15.8	27.0	41.5	57.2	74.2
Japan	4.4	7.5	13.8	21.8	30.0	38.6
Former Soviet Union	0.6	1.4	2.9	4.8	7.2	10.2
Eastern Europe	0.3	0.6	1.2	2.1	3.3	4.9
China	1.2	3.3	6.9	12.8	19.9	28.9
India	0.5	1.1	2.0	3.3	5.2	8.0
Africa	0.6	1.3	2.0	3.3	5.0	7.4
Latin America	1.6	3.0	6.3	11.5	18.0	25.9
Rest of the World	4.4	8.6	14.9	23.9	35.3	49.9
MERGE GDP by Region (trillions of $2000, MER)						
U.S.	9.8	16.1	20.9	26.8	33.1	39.6
Western Europe	9.8	14.4	19.9	26.9	35.0	43.6
Japan	4.6	6.0	7.7	9.6	11.7	13.9
Former Soviet Union/Eastern Europe	1.0	1.9	3.6	6.6	11.9	20.4
China	1.2	3.1	7.4	17.3	38.5	78.6
India	0.5	1.5	3.6	8.3	18.5	39.2
Africa/Latin America/ Rest of World	6.5	14.6	27.5	49.3	85.1	141.9
MiniCAM GDP by Region (trillions of $2000, MER)						
U.S.	9.8	15.1	21.1	28.8	38.9	52.6
Western Europe	8.6	11.1	13.3	16.1	19.4	23.7
Japan	4.7	5.9	7.1	8.6	10.2	12.0
Former Soviet Union	0.4	0.8	1.4	2.3	3.6	5.7
Eastern Europe	0.4	0.7	1.4	2.4	4.0	6.6
China	1.2	4.8	11.6	20.8	34.1	49.3
India	0.5	1.6	4.8	10.7	19.5	32.0
Africa	0.6	1.2	2.1	3.9	7.7	13.8
Latin America	2.0	3.3	5.0	8.8	16.1	26.9
Rest of World	3.2	6.3	12.5	22.6	37.4	56.6

This table reports GDP for all regions of the globe, but accounts or inconsistency in regional aggregations across models. Note that while regions are generally comparable, slight differences exist in regional coverage, particularly in aggregate regions. Differences in 2000 arise from these differences as do differences in regional deflators and regional exchange rates.

(Note: IGSM is in $1997 and 1997exchange rates; MERGE uses$1997 and 1997 exchange rates restated to $2000 by the ratio of U.S. GDP for 2000in $1997 and $2000;MiniCAM is in $2000 and2000 exchange rates.)

. As with the U.S., the MiniCAM reference scenario exhibits a decline in average labor force participation in other developed regions as populations age, resulting in lower growth in per capita GDP compared to the IGSM reference scenario. The GDP growth pattern in the MERGE reference scenario is similar to that of MiniCAM reference scenario.

GDP growth patterns for developing regions show greater differences from historical experience. Notably, all three modeling groups chose assumptions leading to consistent growth in many non- Organization for Economic Cooperation and Development (OECD) regions at rates experienced by industrializing countries. However, growth rates are not homogeneous. Growth in China and India is generally higher than for regions such as Latin America and Africa, as it has been in recent decades. The IGSM reference scenario shows somewhat less growth for the non-OECD regions compared to the MiniCAM and MERGE reference scenarios. These are just one set of possible economic assumptions from each modeling group and are not intended to be expressions of what the groups view as desirable performance. Clearly, more rapid growth in developing countries, if gains spread to lower income groups within these regions, could be the basis for improving the outlook for people in these areas.

Energy Use, Prices, and Technology

The Evolving Structure of Energy Use

In the reference scenarios, global primary energy consumption expands dramatically over the century, growing to between 3 and 4 times its 2000 level of roughly 400 EJ. This growth results from a combination of forces, including rising economic activity, increasing efficiency of energy use, and changes in energy consumption patterns. Growth in per capita energy consumption occurs despite a continuous decline in the energy intensity of economic activity. The improvement in energy intensity reflects, in part, assumptions of substantial technological change in all three reference scenarios.

In all three reference scenarios a range of fossil resources is available to supply the bulk of the world's increasing demand for energy. Fossil fuels provided almost 90% of the energy supply in the year 2000 and remain the dominant energy source in all three reference scenarios throughout the twenty-first century despite a phasing out of conventional petroleum resources. Differing among the reference scenarios, however, is the mix of fossil fuels. The IGSM reference scenario has relatively more oil, derived from shale; the MERGE reference scenario has relatively more coal with a substantial amount of the increase used to produce liquid fuels; and the MiniCAM reference scenario has relatively more natural gas.

In all three reference scenarios, non-fossil fuel energy use grows substantially, reaching levels in 2100 that range from around half to levels that exceed total global energy consumption in 2000. The reference scenarios differ in terms of the mix of non-fossil resources. The substantial growth in non-fossil fuel energy use does not forestall substantial growth in fossil fuel consumption.

Energy production and consumption are closely associated with emissions of GHGs, particularly CO_2, because of the dominant role of fossil fuels in the energy sector. Figure 3.3 shows global primary energy consumption over the century and its composition by fuel type in the three reference scenarios. Not surprisingly, given the assumptions about economic growth, primary energy consumption grows substantially in all of the reference scenarios: from approximately 400 EJ/yr in the year 2000 to roughly between 1275 EJ/yr and 1500 EJ/yr by the end of this century. Combined with population growth, all three reference scenarios include a growing per capita use of energy for the world (Figure 3.4). The per capita growth in primary energy consumption for the world is very similar for Mini- CAM and IGSM reference scenarios, with trends diverging somewhat late in the century.

Box 3.1. Exchange Rates and Comparisons of Real Income among Countries

Models used in this type of research typically represent the economy in real terms, following the common assumption that inflation is a purely monetary phenomenon that does not have real effects, but issues occur in comparing income across regions in terms of what currency exchange ratesare most appropriate. The models do not represent the factors that govern exchange-rate determination and, therefore, cannot represent changes. However, modeling international trade in goods requires either an exchange rate or a common currency. Rather than separately model economies in native currencies and use a fixed exchange to convert currencies for trade, the equivalent and simpler approach is to convert all regions to a common currency at average market exchange rates(MER) for the base year of the model.

At the same time, it is widely recognized that using market exchange rates to compare countries can have peculiar implications. Country A might start with a larger GDP than country B when converted to a common currency using that year's exchange rates, and grow faster in real terms than B, yet could later have a lower GDP than B using exchange rates in that year. This paradoxical situation can occur if A's currency depreciates relative to B's. Depreciation and appreciation of currencies by 20% to 50% over just a few years is common, so the example is not extreme. Interest in making cross-country comparisons that are not subject to such peculiarities has led to development of indices of international purchasing power. A widely used index is purchasing power parity (PPP), whose development was sponsored by the World Bank. PPP-type indices have the advantage of being more stable over time and are thought to better reflect relative living standards among countries than MER. Thus, analysts drawing comparisons among countries have found it preferable to use PPP-type indices rather than MER. Although the empirical foundation for the indices has been improving, the theory for them remains incomplete, and thus there is a limited basis on which scenarios of future changes in PPP can be developed. Some hypothesize that differences close as real income gaps narrow, but the evidence for this outcome is weak, in part due to data limitations.

Controversy regarding the use of MER arose around the SRES produced by the IPCC (Nakicenovic et al. 2001) because they were reported to model economic convergence among countries, yet reported economic attributes of the scenarios in MER. Assessing convergence implies a cross-country comparison, but that would only be strictly meaningful if MER measures were corrected for acountry's real international purchasing power. In developing the scenarios for this research, no assumptions were made regarding convergence. Growth prospects and other parameters for theworld's economies were assessed relative to their own historical performance. The models used inthis research are simulated in MER, as this is consistent with modeling of trade in goods. To the extent GDP results are provided, international comparisons are to made with great caution; for example, even global GDP for an historical period will differ if exchange rates of different years are used.

The MERGE reference scenario has relatively slower growth in per capita primary energy consumption early in the century, with accelerated growth later. On the other hand, per capita primary energy consumption in the U.S. differs substantially among the reference scenarios. U.S. per capita primary energy consumption in MERGE and IGSM reference scenarios increases substantially, while it declines gradually over the century in the MiniCAM reference scenario.

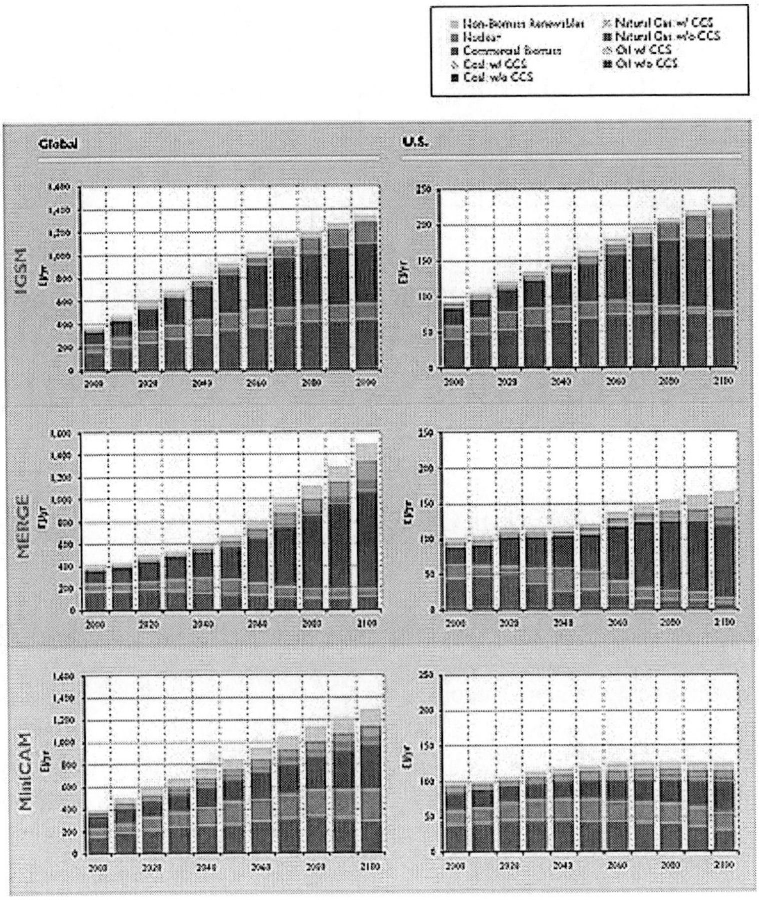

Global total primary energy consumption grows to between three and four times today's levels over the century in the reference scenarios, while U.S. primary energy consumption grows to between 1 and 21/2 times today's levels. Fossil fuels remain a major energy source, despite substantial increases in the consumption of non-fossil energy sources.

[Notes. i. Oil consumption includes that derived from tar sands and oil shales, and coal consumption includes that used to produce synthetic liquid and gaseous fuels. ii. Primary energy consumption from nuclear power and non-biomass renewable electricity are accounted for at the average efficiency of fossil-fired electric facilities, which vary over time and across scenarios. This long-standing convention means that, all other things being equal, increasing efficiency of fossil-electric energy lowers the contribution to primary energy from these sources.]

Figure 3.3. Global and U.S. Primary Energy Consumption by Fuel Across Reference Scenarios (EJ/yr).

The growth in total and per capita primary energy consumption arises despite substantial improvements in energy technology assumed in all three scenarios. The ratio of U.S. primary energy consumption to GDP (primary energy intensity) declines throughout the century in all three reference scenarios (Figure 3.5). These patterns represent a continuation of changes in primary energy intensity that have occurred in recent decades in the U.S. In 2100, each dollar of real GDP is produced with only 40% of the primary energy consumed in 2000 in the MERGE reference scenario, only 30% of the energy in the IGSM reference scenario, and only 25% in the MiniCAM reference scenario.

Table 3.3. Historical Annual Average per Capita GDP Growth Rates

Region	1500-1820	1820-1870	1870-1913	1913-1950	1950-1973	1973-2001
North America	0.34	1.41	1.81	1.56	2.45	1.84
Western Europe	0.14	0.98	1.33	0.76	4.05	1.88
Japan	0.09	0.19	1.48	0.88	8.06	2.14
Former U.S.SR	0.10	0.63	1.06	1.76	3.35	-0.96
Eastern Europe	0.10	0.63	1.39	0.60	3.81	0.68
Africa	0.00	0.35	0.57	0.92	2.00	0.19
Latin America	0.16	-0.03	1.82	1.43	2.58	0.91
China	0.00	-0.25	0.10	-0.62	2.86	5.32
India	-0.01	0.00	0.54	-0.22	1.40	3.01
Other Asia	0.01	0.19	0.74	0.13	3.51	2.42

Source. Maddison, 2001

Globally and in the U.S., primary energy consumption over the century remains dominated by fossil fuels. In this sense, the three reference scenarios tell a consistent story about future global energy, and all three run counter to the view that the world is running out of fossil fuels. Although reserves and resources of conventional oil and gas are limited in all three reference scenarios, the same cannot be said of coal and unconventional liquids and gases. In all three reference scenarios, the world economy moves from current conventional fossil resources to increased exploitation of some combination of the extensive (if more costly) global resources of heavy oils, tar sands, and shale oil, and to synfuels derived from coal. The three reference scenarios exhibit a different mix of these sources. The IGSM reference scenario exhibits a relatively higher share of oil production (including unconventional oil); the MERGE reference scenario exhibits a relatively higher coal share; and the MiniCAM reference scenario exhibits a higher share for natural gas.

The relative contribution of oil to primary energy supply differs across the reference scenarios, but all three include a decline in the share of conventional oil. Thus, these scenarios represent three variations on a theme of energy transition precipitated by limited availability of conventional oil and continued expansion of final demands for liquid fuels, mainly for passenger and freight transport.

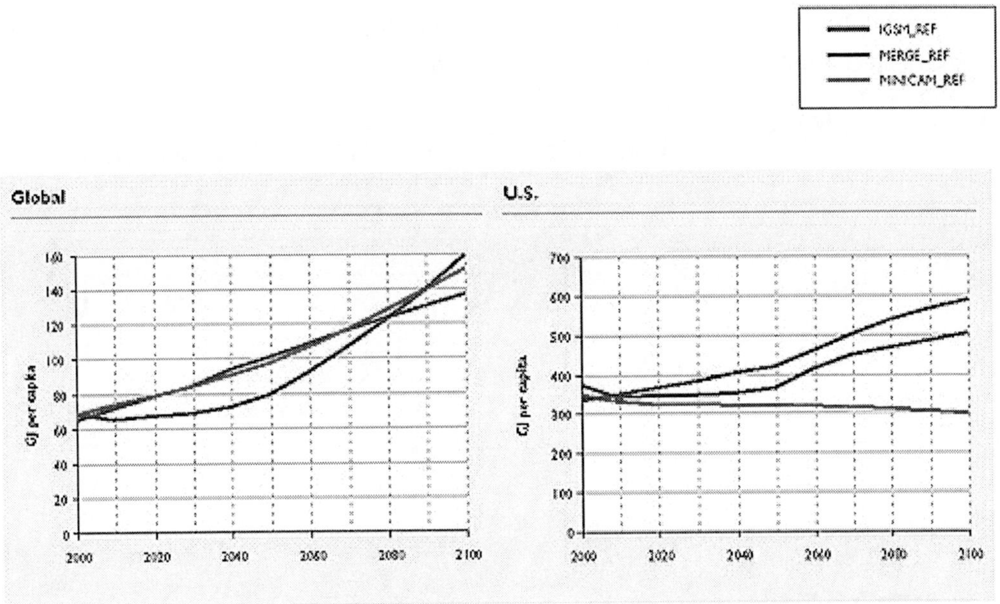

All three reference scenarios include growing global per capita primary energy consumption. However, even after 100 years of growth, global per capita primary energy consumption is about 1/2 of the current U.S. level. U.S. per capita primary energy consumption varies more substantially among the reference scenarios.

[Note. Primary energy consumption from nuclear power and non- biomass renewable electricity are accounted for at the average efficiency of fossil-fired electric facilities, which vary over time and across scenarios.]

Figure 3.4. Global and U.S. Primary Energy Consumption per Capita Across Reference Scenarios (GJ per capita)

In the IGSM reference scenario, limits on the availability of conventional oil resources lead to the development of technologies to exploit unconventional oil, such as oil sands, heavy oils, and shale oil. These resources are large and impose no meaningful constraint on production during the twenty-first century. Thus, despite the fact that production costs are higher than for conventional oil, total oil production (conventional plus shale) expands throughout the century, although oil as a primary energy source declines as a share of total energy with the passage of time.

The transition plays out differently in the MERGE reference scenario. Although it begins the same way (that is, the transition is initiated by limits on conventional oil resources), declining production of conventional oil leads to higher oil prices and makes alternative fuels, especially those derived from coal liquefaction, economically competitive. Thus, there is a transition away from conventional oil (and gas) and a corresponding expansion of coal production. The large difference between the MERGE and IGSM scenarios regarding primary oil thus reflects the role of coal liquefaction rather than a fundamentally different scenario of the need for liquid fuels.

The MiniCAM reference scenario depicts yet a third possible transition. Again, it begins with limited conventional oil resources leading to higher oil prices. Higher oil prices then lead to the development and deployment of technologies that access unconventional oil, such as oil sands, heavy oils, and shale oils. However, it also leads to expanded production of natural gas and to expanded production of coal to produce synthetic liquids, as in the MERGE reference scenario.

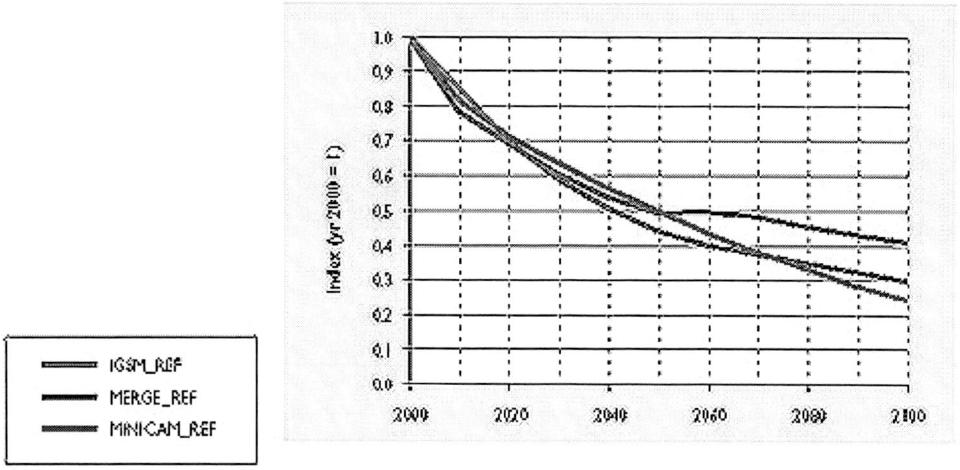

U.S. total primary energy intensity – primary energy consumption per dollar of GDP – continues to decline in the reference scenarios. In recent decades, the rate of decline has been about 14% per decade. U.S. primary energy intensity declines about 12% per decade in the IGSM reference scenario, about 13% per decade in the MiniCAM reference scenario, and about 9% per decade in the MERGE reference scenario.

[Note. Primary energy consumption from nuclear power and non-biomass renewable electricity are accounted for at the average efficiency of fossil-fired electric facilities, which vary over time and across scenarios.]

Figure 3.5. U.S. Primary Energy Intensity: Consumption per Dollar of GDP Across Reference Scenarios (Index, yr 2000 = 1.0)

Primary energy consumption patterns also reflect assumptions about the availability of low- cost alternatives to conventional fossil fuels. In all three reference scenarios, non-fossil sources increase both their absolute and relative roles in providing energy to the global economy, with their share growing to roughly 20% to 30% of total supply by 2100. In the IGSM reference scenario, which has the lowest consumption from non-fossil resources, the magnitude of total consumption from these resources in 2100 is 65% the size of the total global primary energy consumption in 2000, which is more than a 500% increase in the level of production of non-fossil energy. In the MERGE reference scenario, which has the highest contribution from non-fossil resources, total primary energy consumption from these sources in 2100 exceeds total primary energy consumption in 2000. Despite this growth, the continued availability of relatively low-cost fossil energy supplies, combined with continued improvements in the efficiency with which they are used, allows fossil energy forms to remain competitive throughout the century.

The three reference scenarios tell different stories about non-fossil energy (much of which is covered below in the discussion of electricity generation). The IGSM reference scenario assumes political limits on the expansion of nuclear power, so it grows only to about 50% above the 2000 level by 2100. However, growing demands for energy and for liquid fuels in particular lead to the development and expansion of bioenergy, both absolutely and as percentage of total primary energy consumption.

In contrast, the MERGE reference scenario assumes that a new generation of nuclear technology becomes available and that societies do not limit its market penetration, so the share of nuclear power in the economy grows with time. In addition, renewable energy forms, both commercial biomass and other forms such as wind and solar, expand production during the century.

The MiniCAM reference scenario also assumes the availability of a new generation of nuclear energy technology that is both cost competitive and unrestrained by public policy. Nuclear power, therefore, increases market share although not to the extent found in the MERGE reference scenario. Non-biomass renewable energy supplies become increasingly competitive as well. In the MiniCAM reference scenario, bioenergy production is predominantly recycled wastes, with a modest contribution from commercial biomass farming toward the end of the century.

The three reference scenarios for the U.S. are similar in character to the global ones (Figure 3.3). The transition from conventional oil to alternative sources of liquid fuels and changes in electricity production affect energy markets and patterns in the U.S. However, primary energy consumption grows somewhat more slowly in the U.S. than in the world in general. As with the world total, the U.S. energy system remains dominated by fossil fuels in all three reference scenarios. The MERGE and IGSM reference scenarios have similar contributions from non-fossil energy, but the sources in the MERGE reference scenario are predominantly nuclear and other renewables, whereas it is biomass in the IGSM reference scenario. The MiniCAM reference scenario has the smallest overall contribution from non-fossil sources split relatively evenly between nuclear, biomass, and other renewables.

Trends in Fuel Prices

Historically, oil prices have been highly variable, with the volatility often related to political events (Figure 3.6). Prices were in the $15 to $20 range (in the constant 2006 dollars shown in the figure) until the increases in the 1 970s and early 1980s that resulted from disruptions in the Middle East. In inflation-adjusted terms, prices declined from peaks in the late 1970s to vary around the $20 level in the latter half of the 1980s and 1990s. The period 2000 to 2005 has again seen rising prices of oil and other fossil energy sources, which suggests the possibility of a long-term trend toward rising prices. Depletion alone would suggest rising prices because of a combination of rents associated with a limited resource and the exhaustion of easily recoverable grades of oil. Global demand continues to grow, putting increasing pressure on supply. Improvements in technology that reduce the cost of recovering known deposits and facilitate discovery of new ones are opposing these forces toward higher prices.

The three models used for these scenarios employ time steps of 5 to 15 years (see Chapter 2) and, thus, are not set up to analyze short-term variability in prices. Their long-term trends are best interpreted as multi-year averages.

The three reference scenarios paint similar, but by no means identical, pictures of future energy prices. The price paths in the three reference scenarios reflect assumptions regarding both energy resources and energy technologies, and they shed light on these assumptions. For example, the price of oil is related to the marginal cost of bio-fuels, and the prices of both reflect, among other things, the technology options assumed to be available for their production.

Figure 3.7 shows mine-mouth coal prices, electricity producer prices, natural gas producer prices for the U.S., and the world oil (producer) price. All four energy markets – oil, natural gas, coal, and electricity – are shaped by the supply of, and demand for, these commodities. These fuels also are interconnected because users can substitute one fuel for another, thus higher prices in one fuel market will tend to increase demand for and the price of other fuels. Oil markets are driven by the rising cost of conventional oil and the transition to more expensive unconventional sources to supply a growing demand for liquid fuels, mainly for transportation. Thus, the oil prices in the scenarios result from the interplay between increasing demands for liquid fuels, the available technology, and the availability of liquids derived from these other sources.

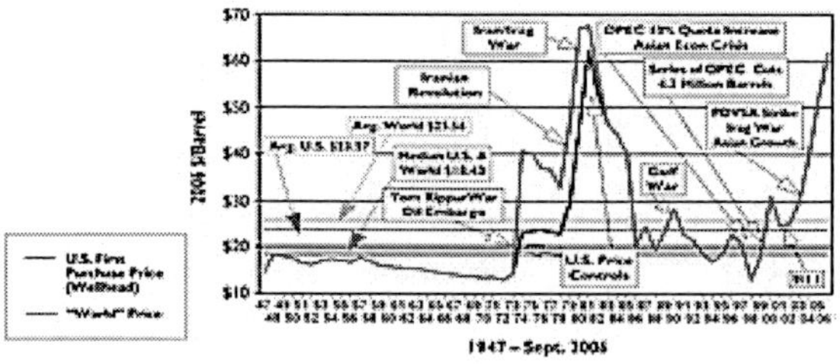

Crude oil prices have historically been highly variable. (Figure courtesy of James Williams, WTRG Economics)

Figure 3.6. Long-Term Historical Crude Oil Prices

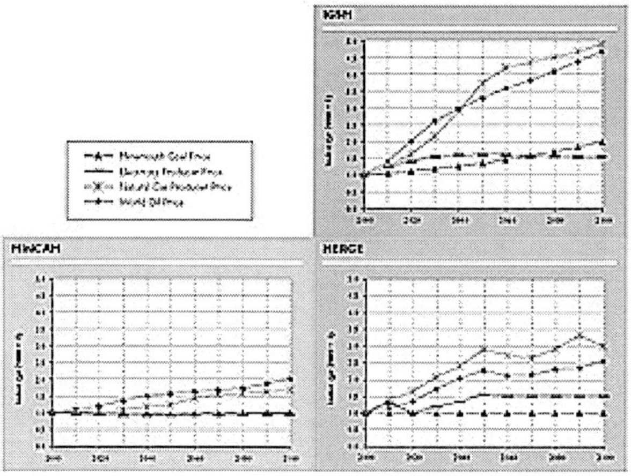

Energy prices through 2100 cover a wide range among the reference scenarios, but generally show a rising trend relative to recent decadal averages. Prices in the MERGE reference scenario are intermediate; by 2100 the crude oil price is about that observed in 2005 (3 times the 2000 level). The MiniCAM reference scenario has the lowest prices, with crude oil price about twice 2000 levels in 2100, somewhat below the level reached in 2005. The IGSM reference scenario has the highest prices, which for crude oil would be about 50% to 60% higher in 2100 than the price level of 2005.

Figure 3.7. Indices of Energy Prices Across Reference Scenarios (Indexed to yr 2000 = 1.0)

Natural gas prices tell a similar story. Assumptions regarding the ultimately recoverable natural gas resource vary, as does the cost structure of the resource, leading to differences among the models. Like the demand for oil, the demand for natural gas grows, driven by increasing population and per capita incomes. As is the case for oil, the price of natural gas tends to be driven higher in the transition from inexpensive conventional resources to less easily accessible grades of the resource and to substitutes, such as gas derived from coal or biological sources. The different degrees and rates of price escalation reflect different technology assumptions in the three reference scenarios.

Coal prices do not rise as fast as oil and natural gas prices in any of the three reference scenarios. The reason is the abundance of the coal resource base. The different patterns of coal price movement with time in the three scenarios reflect differences in assumptions about the rate of resource depletion, its grade structure, and improvements in extraction technology.

The stability of electricity prices compared with oil and natural gas prices is a reflection of the variety of technologies and fuels available to produce electricity, their improvement over time, and the fact that fuel is just one component of the cost of electricity. The details underlying this electric sector development are reported next.

Electricity Production and Technology

Electricity production steadily increases in both the U.S. and the world in the reference scenarios, although the scale and generation mix differ among the three reference scenarios (Figure 3.8). All the reference scenarios depict a continued role for coal. The IGSM reference scenario is dominated by coal, which accounts for more than half of all power production by the end of the twenty-first century. This characteristic of the IGSM reference scenario is consistent with its limited growth in nuclear power. In contrast, nuclear power penetrates the market based on economic performance, and non-biomass renewable energy gains market share in the MERGE reference scenario. Limited natural gas resources lead to a peak and decline in gas use in the first half of the century. In the Mini- CAM reference scenario, coal supplies the largest share of power, but natural gas is relatively abundant and provides a significant portion as well, as do nuclear and non-biomass renewable energy forms.

Non-Electric Energy Use

An important consideration in scenarios of the future energy system is conversion losses as relatively lower-grade resources are converted to higher-grade fuels for use in final applications such as space conditioning, lighting, and mechanical power. Figure 3.9 identifies the energy content of primary fuels for the U.S. in the year 2000 and where conversion losses occur. It shows the energy loss in the conversion from fuel to electricity to be 29.6 exajoules while the energy content of the electricity is 13.0 exajoules. Other losses occur when fuels are used to create the mechanical power to, for example, propel vehicles or when efficiency of conversion to heat, light, or mechanical energy is less that 100%. The potential for reducing such losses is one reason why energy intensity of the economy can continue to improve.

However, in the future other fuel transformation activities may become important and fundamentally change energy-flow patterns, as higher-grade resources are exhausted and lower- grade resources that require more conversion are used. As already discussed, the potential exists for coal and commercial biomass to be converted to liquids and gases – a technology thus far implemented only at a small scale. Furthermore, fuels and electricity may be transformed into hydrogen, creating fundamentally new branches of the energy system. Like electricity, these new branches will have conversion losses, and those losses can be important.

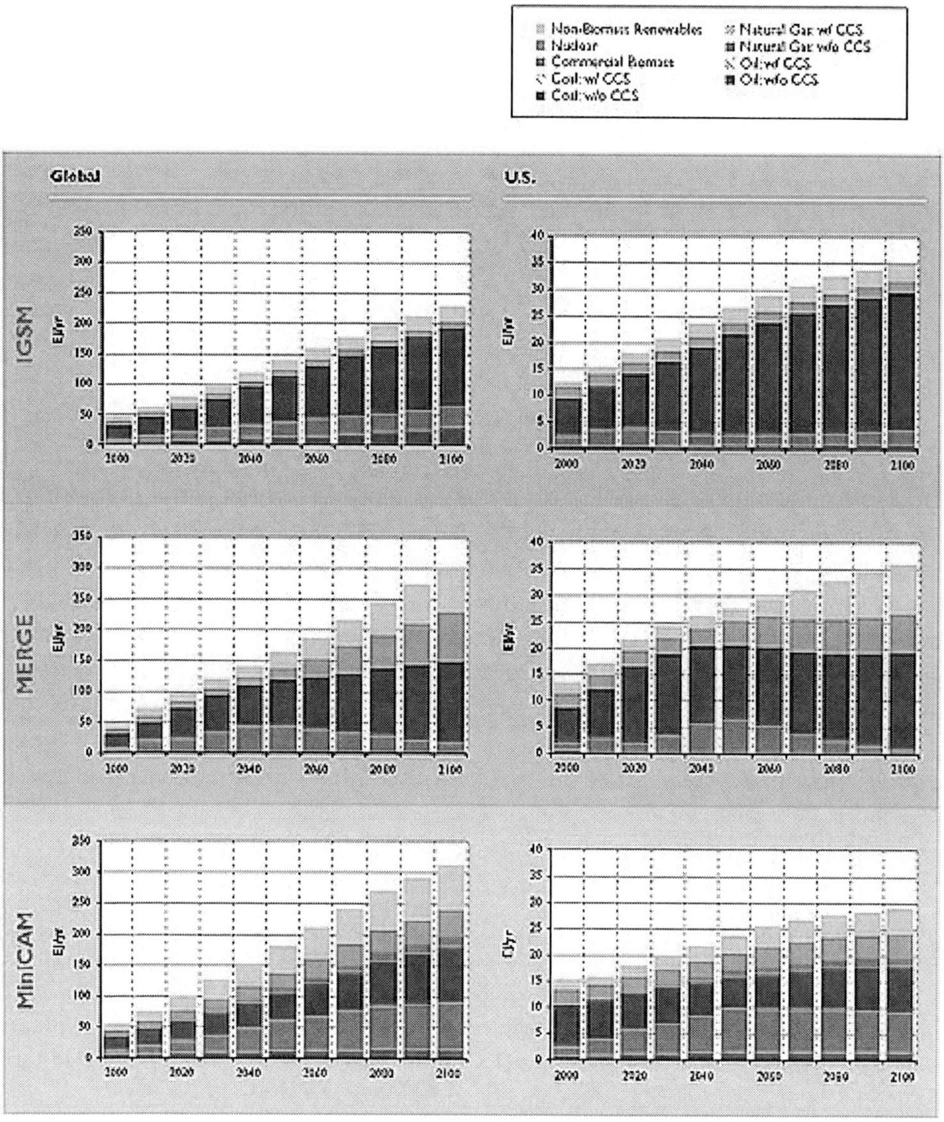

Global and U.S. electricity production in the reference scenarios show continued use of coal, especially the IGSM reference scenario, which assumes that nuclear energy expansion is limited by safety, waste, and proliferation concerns. The MERGE and MiniCAM reference scenarios are based on the assumption that nuclear energy is unconstrained by non-climate concerns, so these scenarios exhibit greater expansion. They also include greater contributions from renewable energy sources and somewhat greater use of electricity overall compared with IGSM reference scenario.

Figure 3.8. Global and U.S. Electricity Production by Source Across Reference Scenarios (EJ/yr of electricity).

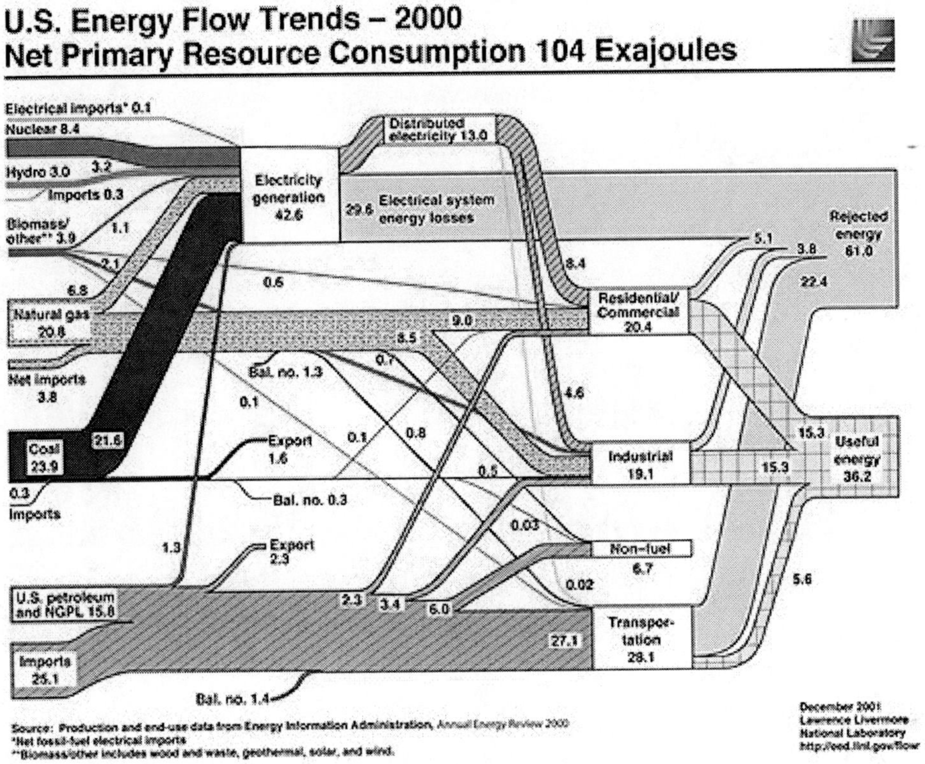

Primary energy is transformed into different energy carriers that can easily be used for specific applications (e.g., space conditioning, light, and mechanical energy), but in the process losses occur. Of the 104 exajoules, of primary energy consumed in the U.S. in the year 2000, only an estimated 36 exajoules, were actually useful. Each of the models used in this research represents such conversion processes. Assumptions about efficiency improvements in conversion and end use are one of the reasons why energy intensity per dollar of GDP falls in the reference scenarios.

Figure 3.9. U.S. Energy Flow Diagram and Non-Electrical Energy Use for the Year 2000

Figure 3.10 shows non-electric energy use in the reference scenario, and it is important to realize that these patterns of non-electric use also can imply significant conversion losses. This prospect plays a strong role in the MERGE reference scenario, in which coal and biomass go into liquefaction and gasification plants. To a lesser extent, these conversions are also present in the MiniCAM and IGSM scenarios. In addition, in the MiniCAM reference scenario some nuclear and renewable energy appears in non- electricity uses to produce hydrogen. In the IGSM and MiniCAM reference scenarios, oil use is the largest single non-electric energy source, reflecting continued growth in demand for liquids by the transportation sectors. In the MERGE reference scenario, increasingly expensive conventional oil is supplanted by coal- based liquids. This phenomenon also has implications for energy intensity in that improvements in end-use energy intensity can be offset, in part, by losses in converting primary fuels to end-use liquids or gases.

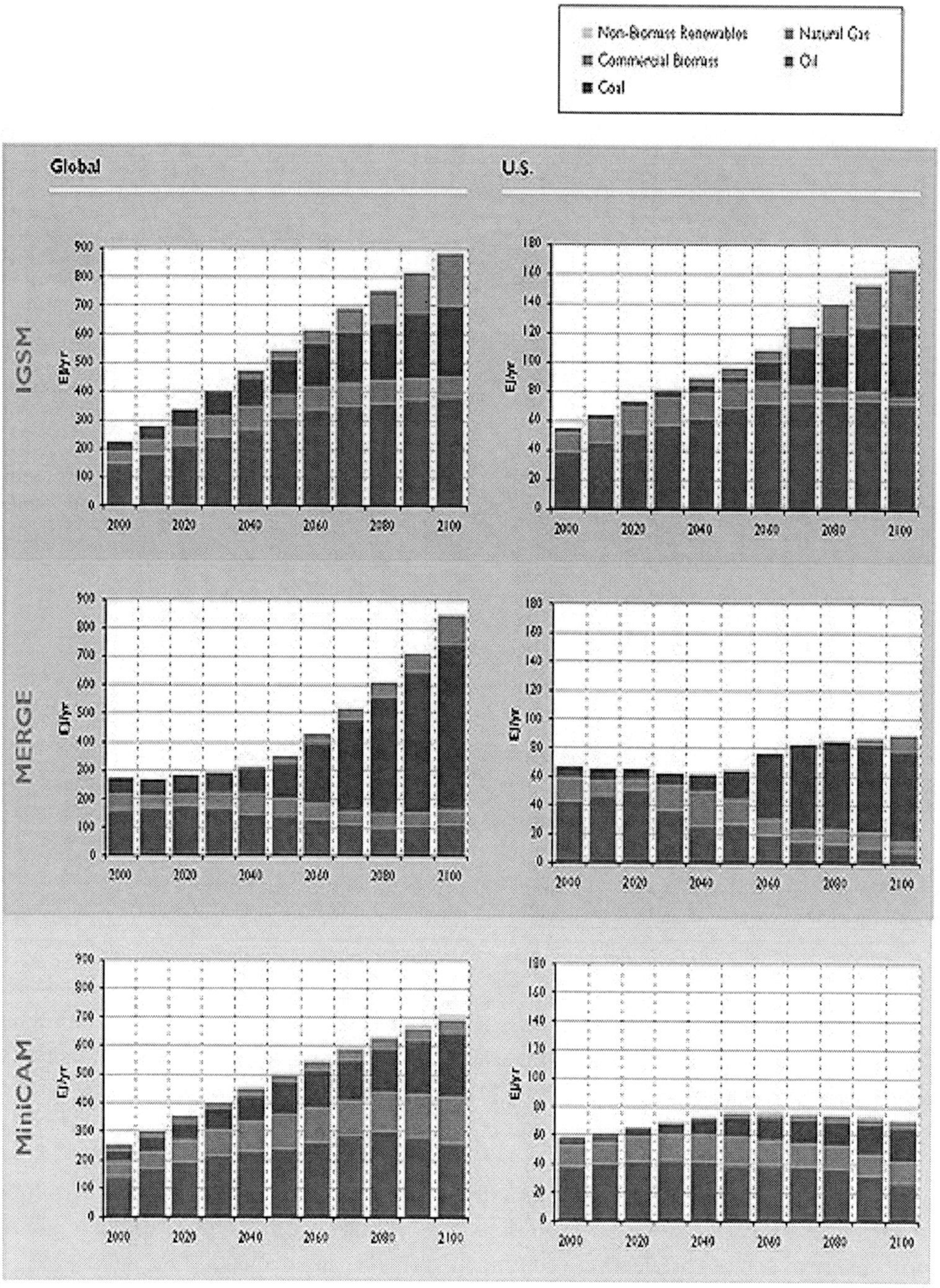

As with electricity production, non-electric energy consumption remains heavily dependent on fossil fuels with some penetration of biomass energy. Primary energy is reported here, and the resurgence of coal in the reference scenarios is due to its use to produce synthetic liquids or gas. [Notes. Oil consumption includes that derived from tar sands and oil shales, and coal consumption includes that used to produce synthetic liquid and gaseous fuels.]

Figure 3.10. Global and U.S. Primary Energy Consumption in Non-Electric Applications Across Reference Scenarios (EJ/yr)

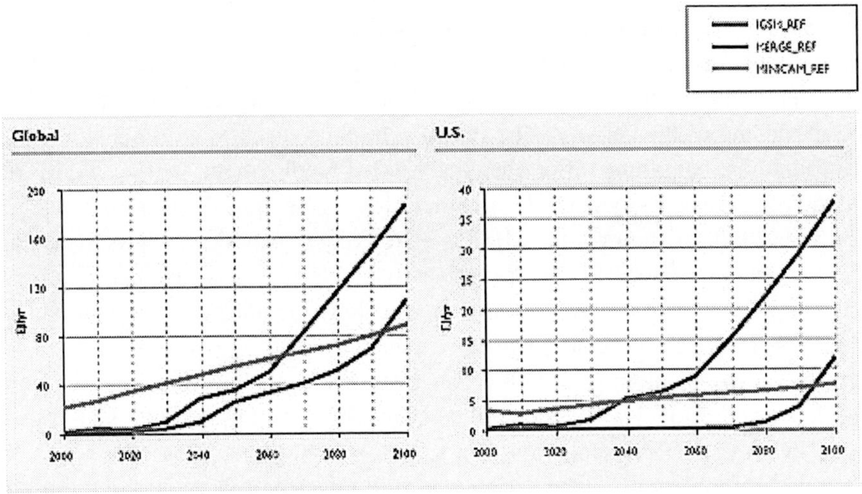

The MiniCAM scenarios include waste-derived biomass fuels as well as commercial biomass and, thus, show significant use in 2000. The IGSM and MERGE scenarios include only commercial biomass energy beyond that already used. Globally, the IGSM and MERGE reference scenarios include more biomass production than does the MiniCAM reference scenario toward the end of the century.

Figure 3.11. Global and U.S. Production of Biomass Energy Across Reference Scenarios (EJ/yr).

Land Use and Land-Use Change

The three reference scenarios take different approaches to emissions from land use and land-use change. The MERGE reference scenario assumes that the biosphere makes no net contribution to the carbon cycle. In the IGSM and MiniCAM reference scenarios, the net contribution of the terrestrial biosphere is to remove carbon from the atmosphere, which results from the countervailing forces of land-use change emissions from deforestation and other human activities and the net uptake from unmanaged systems.

An important aspect of land use and land-use change in the scenarios from all three modeling groups is the production of bio-fuels for energy. Both IGSM and MiniCAM take account of the competition for scarce land resources in developing scenarios of bioenergy production and consumption. MERGE takes the availability of bio-fuels as an exogenous input based on extra- model analysis. Global and U.S. biomass production is displayed in Figure 3.11. The IGSM and MiniCAM scenarios use somewhat different definitions, which account for the difference in 2000. The numbers presented for the IGSM scenarios account only the production of biomass energy beyond that now used and do not include traditional use of biomass or, for example, the own-use of wood wastes for energy in the forest products industry. The MiniCAM scenarios explicitly account for some current uses of biomass energy, such as that used in the pulp and paper industry, and separately consider the future potential for bio-fuels derived from wastes and residue along with energy crops grown explicitly for their energy content.

Apparent differences among the models need to be considered in light of this differential accounting. In the MiniCAM reference scenario, biomass production tends to be higher in early years because it is accounting waste and residue-derived bio-fuels explicitly. These waste and residue-derived bio-fuels account for all of the biomass production in the MiniCAM reference scenario in the early part of the century and the majority of all biomass production at the end of the century. The IGSM reference scenario exhibits strong growth in bio-fuels production beginning after the year 2020. Deployment in the IGSM reference scenario is driven primarily by a world oil price that in the year 2100 is over 4.5 times the price in the year 2000. In contrast, the MiniCAM reference scenario, with its lower long-term world oil price, includes insufficient incentive to create a substantial market for biomass crops. However, the MiniCAM reference scenario does include an increasing share of the potentially recoverable bio-waste as a source of energy.

Land use has implications for the carbon cycle as well. IGSM applies its component Terrestrial Ecosystem Model with a prescribed scenario of land use, and this land-use pattern is employed in all the IGSM scenarios. Thus, in the IGSM scenarios commercial biomass production must compete with other agricultural activities for cultivated land, but the extent of cultivated land does not change from scenario to scenario. Because the land-use pattern is fixed in the IGSM scenarios, changes in the net flux of carbon to the atmosphere reflect the behavior of the terrestrial ecosystem in response to changes in CO_2 and climatic effects that are considered within the IGSM's Earth system component. Taken together, these effects lead to the negative net emissions from the terrestrial ecosystem (Figure 3.12), which contrasts with the neutral biosphere assumed in the MERGE reference scenario. (Note that one tonne C is equivalent to 3.67 tonnes CO_2. See Box 3.2 for more on converting between units of carbon and units of CO_2.)

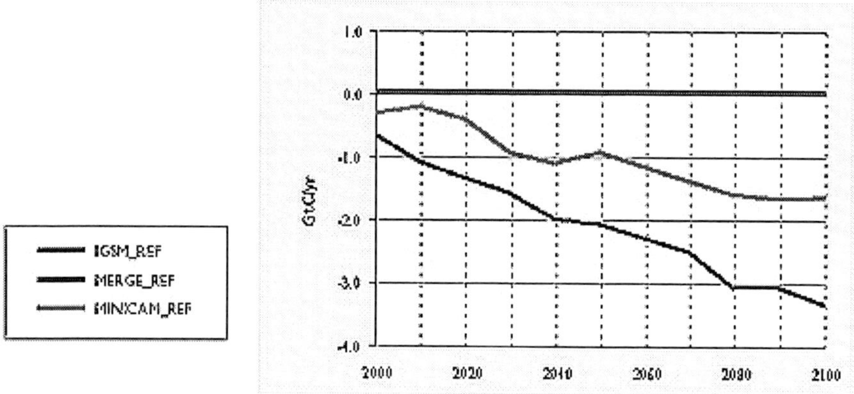

Global net emissions of CO_2 from terrestrial systems, including net deforestation, serve as a slight net sink in 2000 that grows over time in the IGSM and MiniCAM reference scenarios, mainly because of reduced deforestation and CO_2 fertilization of plants. The MERGE scenarios assume a neutral terrestrial system.

Figure 3.12. Global Net Emissions of CO_2 from Terrestrial Systems Including Net Deforestation Across Reference Scenarios (GtC/yr).

MiniCAM uses the terrestrial carbon cycle model of MAGICC (Wigley and Raper 2001, Wigley and Raper 2002) to determine the aggregate net carbon flux to the atmosphere. However, unlike either IGSM or MERGE, MiniCAM determines the level of terrestrial emissions as an output from an integrated agriculture-land-use module rather than as the product of a terrestrial model with fixed land use. Thus, the MiniCAM scenarios exhibit the same types of CO_2 fertilization effects as the IGSM scenarios, but they also represent interactions between the agriculture sector and the distribution of natural terrestrial carbon stocks.

Emissions, Concentrations, and Radiative Forcing

The growth in the global economy in the reference scenarios and the changes in the composition of the global energy system lead to growing emissions of GHGs over the century. Emissions from fossil fuel burning and cement production more than triple from 2000 to 2100 in all three reference scenarios. With growing emissions, GHG concentrations rise substantially over the twenty-first century, with CO_2 concentrations increasing by 2½ to over 3 times preindustrial levels. Increases in non-CO_2 GHG concentrations vary more widely across the reference scenarios. Radiative forcing from the GHGs considered in this research reaches 6.4 W/m^2 to 8.6 W/m^2 from preindustrial by 2100, with the non-CO_2 GHGs accounting for 20% to 25% of the instantaneous forcing in 2100.

Moderating the effect on the atmosphere of anthropogenic CO_2 emissions is the net uptake by the ocean and the terrestrial biosphere. As atmospheric CO_2 grows in the reference scenarios, the rate of net uptake by the ocean increases as well. Also, mainly through the effects of CO_2 fertilization, increasing atmospheric levels of CO_2 spur plant growth and net carbon uptake by the terrestrial biosphere. Differences among scenarios of these effects are, in part, a reflection of variation in their sub-models of the carbon cycle.

Greenhouse Gas Emissions

Calculating Greenhouse Gas Emissions

Emissions of CO_2 from fossil fuels are the sum of emissions from each of the different fuel types, and for each type, emissions are the product of a fuel-specific emissions coefficient and the total combustion of that fuel. Exceptions to this treatment occur if a fossil fuel is used in a non-energy application (e.g., as a feedstock for plastic) or if the carbon is captured and stored in isolation from the atmosphere. All three of the modeling groups assumed the availability of CCS technologies and treated the leakage from such storage as zero over the time period considered in this research, although they assumed that carbon capture technologies capture and store less than 100% of the CO_2. CCS increases the costs of electricity production with no attendant benefits, absent actions to constrain carbon emissions, so CCS is not deployed in any of the reference scenarios.

Although bioenergy such as wood, organic waste, and straw are hydrocarbons like the fossil fuels (only much younger), they are treated as if their use had no net carbon release to the atmosphere. Any fossil fuels used in their cultivation, processing, transport, and refining are accounted for. Nuclear and non-biomass renewables, such as wind, solar, and hydroelectric power, have no direct CO_2 emissions and therefore have a zero carbon coefficient. Like bioenergy, emissions associated with the construction and operation of conversion facilities are accounted with the associated emitting source.

The calculation of net emissions from terrestrial ecosystems, including land-use change, is more complicated, and each model employs its own technique. IGSM employs the Terrestrial Ecosystem Model, which is a state-of-the-art terrestrial carbon-cycle model with a detailed, geographically disaggregated representation of terrestrial ecosystems and associated stocks and flows of carbon on the land. The IGSM scenarios, therefore, incorporate fluxes to the atmosphere as a dynamic response of managed and unmanaged terrestrial systems to the changes in the climate and atmospheric composition.

MiniCAM builds its net terrestrial carbon flux by summing both emissions from changes in the stocks of carbon from human-induced land-use change and the natural system response, represented in the reduced-form terrestrial carbon module of MAGICC. As noted above, Mini- CAM employs a simpler reduced-form representation of terrestrial carbon reservoirs and fluxes; however, its scenario is fully integrated with its agriculture and land-use module, which in turn is directly linked to energy and economic activity in the energy portion of the model. As noted above, the MERGE modeling group assumed no net emissions from the terrestrial biosphere.

Differing approaches among the modeling groups are used to account for the non-CO_2 GHGs. They begin with a current inventory of these gases and link growth in emissions to relevant activity levels. Because emissions are associated with very narrow activities, in some cases below the sectoral resolution of the models, emissions growth may be benchmarked to more detailed forecasts of activities.

Reference Scenarios of Fossil Fuel CO_2 Emissions

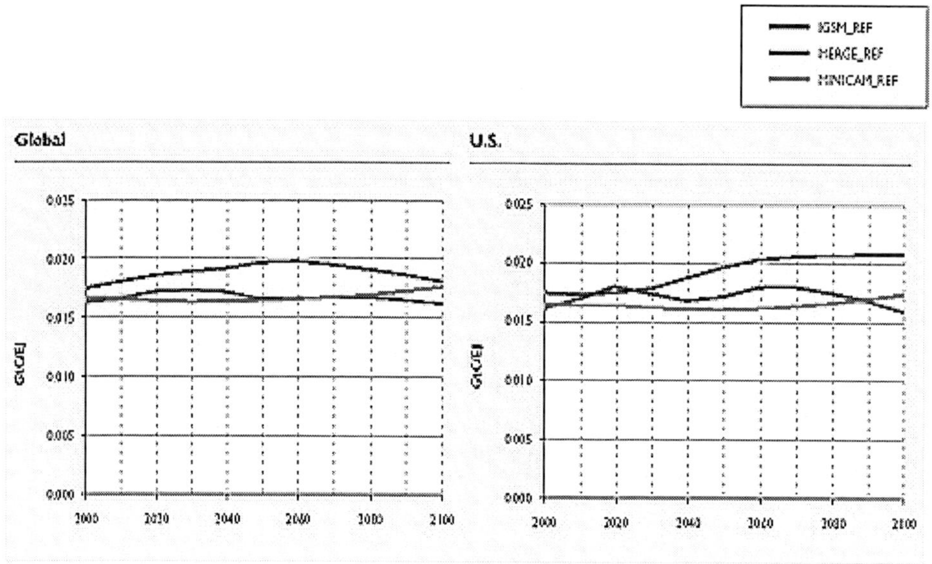

The CO_2 intensity of energy use changes little over the century in the three reference scenarios, reflecting the fact that fossil fuels remain important sources of energy. Potential reductions in the CO_2 intensity of energy from more carbon-free or low-carbon energy sources is offset by a move to more carbon-intensive shale oil or synthetic fuels from coal.

Figure 3.13. Global and U.S CO_2 Emissions from Fossil Fuel Combustion and Industrial Sources Relative to Primary Energy Consumption (GtC/EJ)

Box 3.2. Reporting Conventions for Carbon Emissions and Prices

Two different conventions have been used to report emissions and prices in past studies of CO_2emissions and concentrations. One convention is based on the total mass of emitted CO_2. Emissions are commonly expressed in tonnes of CO_2 and prices in terms of dollars per tonne of CO_2. The second convention is based on the carbon component of the emitted CO_2. Emissions are expressed in tonnes of carbon and prices in terms of dollars per tonne of carbon. This chapter uses the second approach throughout. In contrast, emissions of non-CO_2 GHGs, such asCH_4 or N_2O, are reported in terms of their full mass.

It is important to be clear on which convention is used, but it is easy to convert between the two based on the molecular com-position of CO_2. One molecule of CO_2 includes one carbon atom,with a molecular weight of 12, and two oxygen atoms, each witha molecular weight of 16. The total molecular weight of CO_2istherefore 44, and carbon represents 12/44 of this weight. Emissions expressed in terms of $CO_2$are therefore larger than when expressed in terms of the carbon component of CO_2: one tone of CO_2 is equivalent to 44/12, or 3.67, tonnes of carbon. Conversely, emissions prices are lower when reported in units ofCO_2 because the price must be spread over a larger weight; $100 per tonne of carbon is equivalent to $27 per tonne of CO_2.

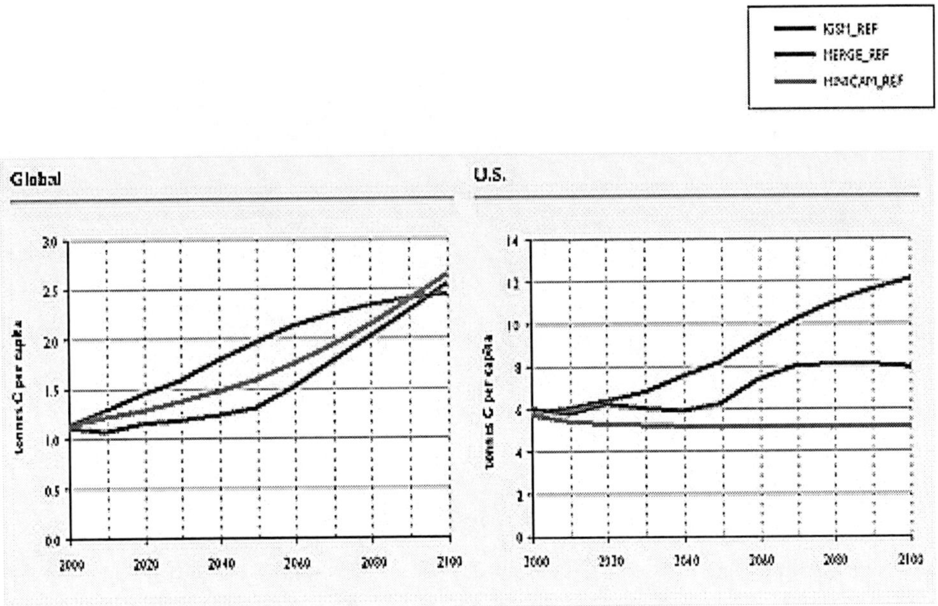

Global per capita fossil fuel and industrial CO_2 emissions grow in all three reference scenarios. However even after 100 years of growth, global per capita CO_2emissions are slightly less than 1/2 of the 2000 U.S. level in the three scenarios. There is greater divergence in U.S. CO_2 emissions per capita over the century among the reference scenarios.

Figure 3.14. Global and U.S. Emissions of $CO_2$from Fossil and Other Industrial Sources per Capita Across Reference Scenarios (tonnes per capita).

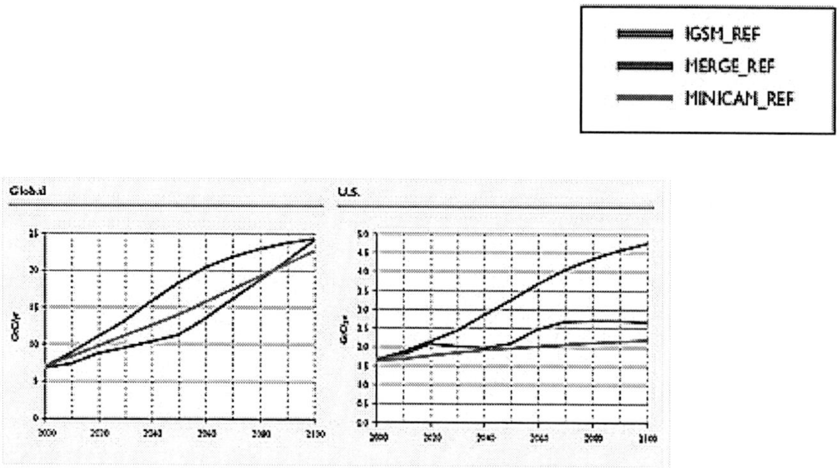

Global emissions of CO_2 from fossil fuel combustion and other industrial sources, mainly cement production, grow throughout the century in all three reference scenarios. By 2100, global emissions are between 22.5 GtC/yr and 24.0 GtC/yr. U.S. emissions are more varied across the reference scenarios. By 2 100, U.S. emissions are between 2 GtC/yr and 5 GtC/yr. Note that CO_2 from land-use change is excluded from this figure

Figure 3.15. Global and U.S. Emissions of CO_2 from Fossil Fuels and Industrial Sources (CO_2 from land-use change excluded) Across Reference Scenarios (GtC/yr)

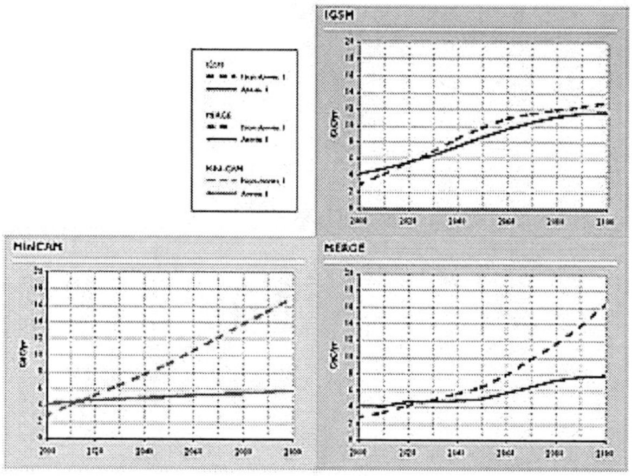

Emissions of fossil fuel and industrial CO_2 in the Non-Annex I countries exceed Annex I emissions for all three reference scenarios by 2030 or earlier. The MERGE and MiniCAM reference scenarios exhibit continued relative rapid growth in emissions in Non-Annex I regions after that, so that emissions are on the order of twice the level of Annex I by 2100. The IGSM reference scenario does not show continued divergence, due in part to assumptions of relatively slower economic growth in Non- Annex I regions and faster growth in Annex I than the scenarios from the other modeling groups. The IGSM reference scenario also shows increased emissions in Annex I as those nations become producers and exporters of shale oil, tar sands, and synthetic fuels from coal

Figure 3.16. Global Emissions of Fossil Fuel and Industrial CO_2 by Annex I and Non-Annex I Countries Across Reference Scenarios (GtC/yr).

All three reference scenarios include a transition from conventional oil production to some other source of liquid fuels based primarily on other fossil sources, either unconventional liquids or coal. As a consequence, carbon-toenergy ratios cease their historic pattern of decline (Figure 3.13). While the particulars of the reference scenarios differ, no reference scenario shows a dramatic reduction in carbon intensity over this century.

Substantial increases in total energy use with no or little decline in carbon intensity lead to substantial increases in CO_2 emissions per capita (Figure 3.14) and in global totals (Figure 3.15). Emissions of CO_2 from fossil fuel use and industrial processes increase from less than 7 GtC/yr in 2000 to between 22.5 and 24.0 GtC/yr by 2100. These global emissions are higher than in many earlier studies such as IS92a, where emissions were 20 GtC/yr in 2100 (Leggett et al. 1992). Global emissions from these reference scenarios are closer to those from the higher scenarios in the IPCC SRES (Nakicenovic et al. 2000); particularly those included under the headings A1FI and A2. U.S. emissions trajectories are more varied than the global trajectories. By 2100, U.S. emissions are between 2 GtC/yr and 5 GtC/yr.

The three reference scenarios display a larger share of emissions growth outside of the Annex I nations – the developed nations of the OECD as well as Eastern Europe and the former Soviet Union[5] (Figure 3.16). Annex I emissions are highest and Non-Annex I emissions lowest in the IGSM reference scenarios. At least in part, this is because of two factors underlying the IGSM scenarios. First, the demand for liquids is satisfied by expanding production of unconventional oil, which has relatively high carbon emissions at the point of production. The U.S., with major resources of shale oil, switches from being an oil importer to an exporter but is responsible for CO_2 emissions associated with shale oil production. Second, assumed rates of productivity growth in Non-Annex I nations are lower in the IGSM scenarios than in those of the other two models.

In contrast, the MERGE reference scenario assumes that liquids come primarily from coal, a fuel that is more broadly distributed around the world than unconventional oils. The MERGE scenarios also exhibit higher rates of labor productivity in the Non-Annex I nations than the IGSM scenarios. Finally, the MERGE reference scenario has a greater deployment of nuclear power, leading to a lower carbon-to-energy ratio. These three features combine to produce lower Annex I emissions and higher Non-Annex I emissions than in the IGSM reference scenario. The MiniCAM reference scenario has Annex I emissions similar to those of the MERGE reference scenario, but higher Non- Annex I emissions.

The range of global fossil fuel and industrial CO_2 emissions across the three reference scenarios is relatively narrow compared with the uncertainty inherent in these developments over a century. While it is beyond the scope of this research to conduct a formal uncertainty or error analysis, both higher and lower emissions trajectories could be constructed.

There are at least two approaches to developing a sensible context in which to view these scenarios. One is to compare them with others produced by analysts who have taken on the same or a largely similar task. The literature on emissions scenarios is populated by hundreds of scenarios of future fossil fuel and industrial CO_2 emissions. Figure 3.17 gives some sense of what earlier efforts have produced, although they should be used with care. Many were developed at earlier times and may be significantly at variance with events as they have already unfolded. Also, no effort was undertaken in constructing the collection in the figure to weight scenarios for the quality of underlying analysis. Scenarios for which no underlying trajectories of population or GDP are available are mixed in with efforts that incorporate the combined wisdom of a large team of interdisciplinary researchers working over the course of years. Moreover, it is not clear that the observations are independent.

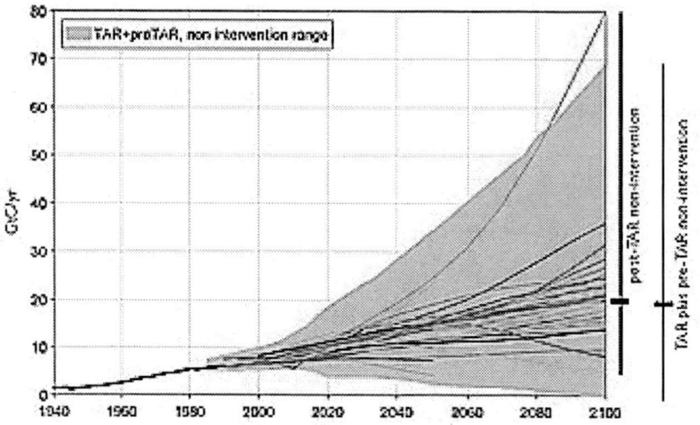

The 284 nonintervention, or reference, scenarios published before 2001 are included in the figure as the blue-shaded range. The thin lines are an additional 55 non-intervention scenarios published since 2001 . Two vertical bars on the right-hand side indicate the ranges for scenarios since 2001 (post-TAR non-intervention) and for those published up to 2001 (TAR plus pre-TAR non-intervention).

Source: Figure 4, Nakicenovic et al. 2006, with kind permission of Springer Science and Business Media.

Figure 3.17. Global Emissions of CO_2 from Fossil Fuel and Industrial Sources: Historical Development and Scenarios (GtC/yr).

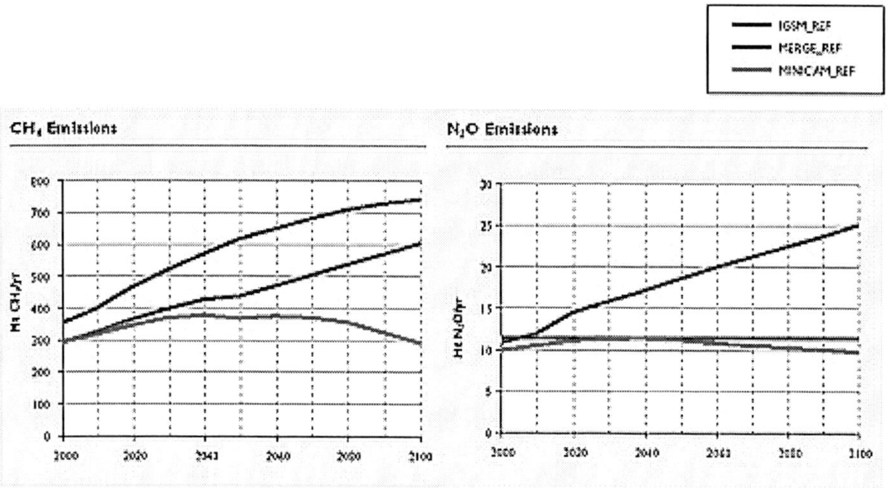

Global anthropogenic emissions of CH_4 and N_2O vary widely among the reference scenarios. There is uncertainty in year 2000 CH_4 emissions, with the IGSM reference scenario ascribing more of the emissions to human activity and less to natural sources. Differences in the scenarios reflect, to a large extent, different assumptions about whether current emissions rates will be reduced significantly for other reasons, for example, whether higher natural gas prices will stimulate capture of CH_4 for use as a fuel.

Figure 3.18. Global CH_4 and N_2O Emissions Across Reference Scenarios (Mt CH_4/yr and Mt N_2O/yr).

The clustering of year 2100 fossil fuel and industrial CO_2 emissions around 20 GtC/yr in both the pre- and post-IPCC TAR time frames coincides closely with the IPCC IS92a scenario. Many later scenarios were simply tuned to it, so are not independent assessments. For these reasons and others, looking to the open literature can provide some information, but caution in interpreting literature compilations is warranted.

Another approach to provide a context is systematic uncertainty analysis. There have now been several such analyses, including efforts by Nordhaus and Yohe (1983), Reilly et al. (1987), Manne and Richels (1994), Scott et al. (2000), and Webster et al. (2002). These studies contain many valuable lessons and insights. For the purposes of this research, one useful product of these uncertainty studies is an impression of the position of any one scenario within the window of futures that might pass a test of plausibility. Also useful is the way that the distribution of outcomes is skewed upward – an expected outcome when one considers that many model inputs, and indeed emissions themselves, are constrained to be greater than zero. Naturally, these uncertainty calculations present their own problems (Webster 2003).

Future Scenarios of Anthropogenic CH_4 and N_2O Emissions

The range of emissions for CH_4 and N_2O is wider than for CO_2 (Figure 3.18). Base-year emissions in the MERGE and MiniCAM reference scenarios are similar for N2O but diverge for CH4. In the IGSM reference scenario, CH_4 emissions are higher in the year 2000 than in the other scenarios, reflecting an independent assessment of historical emissions and uncertainty in the scientific literature regarding even historic emissions. Note that the IGSM reference scenario has a correspondingly lower natural CH_4 source (from wetlands and termites) that is not shown in Figure 3.18, balancing the observed concentration change, rate of oxidation, and natural and anthropogenic sources.

Both the IGSM and MERGE reference scenarios exhibit steadily growing CH_4 emissions throughout the twenty-first century as a consequence of the growth of CH_4-producing activities such as ruminant livestock herds, natural gas use, and landfills. Unlike CO_2, for which the combustion of fossil fuels without CCS leads inevitably to emissions, slight changes in activities can substantially reduce emissions of the non-CO_2 gases (Reilly et al. 2003). The MiniCAM reference scenario assumes that despite the expansion of human activities traditionally associated with CH_4 production, emissions control technologies will be deployed in response to local environmental regulations and in response to the economic value of CH_4. For this reason, CH_4 emissions peak and decline in the MiniCAM reference scenario.

Future Scenarios of Anthropogenic F-Gas Emissions

A set of industrial products that act as GHGs are combined under the term, F-gases, which refers to an element that is common to them, fluorine. Several are replacements for the CFCs that have been phased out under the Montreal Protocol. They are usefully divided into two groups: (1) a group of HFCs, most of which are short-lived, and (2) the long-lived PFCs and SF_6. Figure 3.19 presents the reference scenarios for these GHGs. The IGSM and MERGE reference scenarios exhibit strong growth in the short-lived species, while the MiniCAM reference scenario exhibits about half as much growth over the century. Emissions of the long-lived gases are very similar among the reference scenarios. PFCs are used in semiconductor production and are emitted as a byproduct of aluminum smelting; they can be avoided relatively cheaply. Emissions from the main use of SF_6 in electric switchgear can easily be abated by recycling to minimize venting to the atmosphere. Many of the abatement activities have already been undertaken, and the modeling groups assumed they will continue to be used.

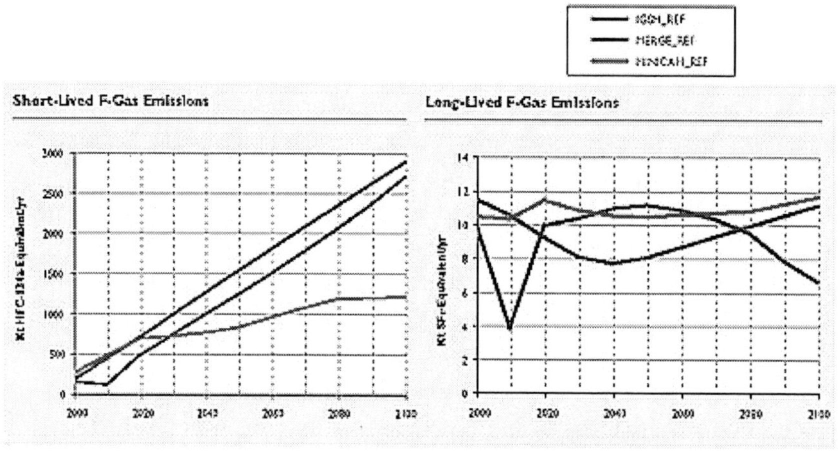

Figure 3.19. Global Emissions of Short-Lived and Long-Lived F-Gases (Kt HFC- 1 34a-Equivalent/yr and Kt SF6-Equivalent/yr).

The Carbon Cycle: Net Ocean and Terrestrial CO_2 Uptake

The stock of carbon in the atmosphere at any time is determined from an initial concentration of CO_2 to which is added anthropogenic emissions from fossil fuel and industrial sources and from which is subtracted net CO_2 transfer from the atmosphere to the ocean and terrestrial systems. Each of the three participating models represents these processes differently.

The three reference scenarios display strong increases in ocean uptake of CO_2 (Figure 3.20), reflecting modeled mechanisms that become increasingly active as CO_2 accumulates in the atmosphere. The IGSM reference scenario has the least active ocean, which results from its three- dimensional ocean representation that shows less uptake, in part, as a result of rising water temperatures and CO_2 levels in the surface layer and, in part, as a result of a slowing of mixing into the deep ocean. The MERGE reference scenario has the most active ocean, and uptake rates continue to increase over the century. As will be discussed in Chapter 4, the three ocean models produce more similar behavior in the stabilization scenarios; for example, the MERGE and MiniCAM Level 1 and Level 2 scenarios have almost identical ocean uptake.

As discussed above, the net transfer of CO_2 from the atmosphere to terrestrial systems includes many processes, such as deforestation (which transfers carbon from the land to the atmosphere), uptake from forest regrowth, and the net effects of atmospheric CO_2 and climate conditions on vegetation. As noted earlier, MERGE employs a neutral biosphere: by assumption, its net uptake is zero with processes that store carbon assumed to just offset those that release it. Taken together with its more active ocean system in the reference scenario, the behavior of the carbon cycle in total is similar to the other two models, especially MiniCAM. IGSM and MiniCAM employ active terrestrial biospheres, which on balance remove carbon from the atmosphere (Figure 3.12). Both the MiniCAM and the IGSM reference scenarios display the net effects of deforestation, which declines in the second half of the century, combined with terrestrial processes that accumulate carbon in existing terrestrial reservoirs. The IGSM and MiniCAM reference scenarios also include feedback effects of a changing climate.

Greenhouse Gas Concentrations

Radiative forcing is related to the concentrations of GHGs in the atmosphere. The relationship between emissions and concentrations of GHGs is discussed in Box 2.2. The concentration of gases that reside in the atmosphere for long periods of time – decades to millennia – is more closely related to cumulative emissions than to annual emissions. In particular, this is true for CO_2, the gas responsible for the largest contribution to radiative forcing. This relationship can be seen for CO_2 in Figure 3.21, where cumulative emissions over the period 2000 to 2100, from the three reference scenarios and the twelve stabilization scenarios, are plotted against the CO_2 concentration in the year 2100. The plots for all three models lie on essentially the same line, indicating that despite considerable differences in representation of the processes that govern CO_2 uptake, the aggregate response to increased emissions is very similar. This basic linear relationship also holds for other long-lived gases, such as N_2O, SF_2, and the other long-lived F-gases.

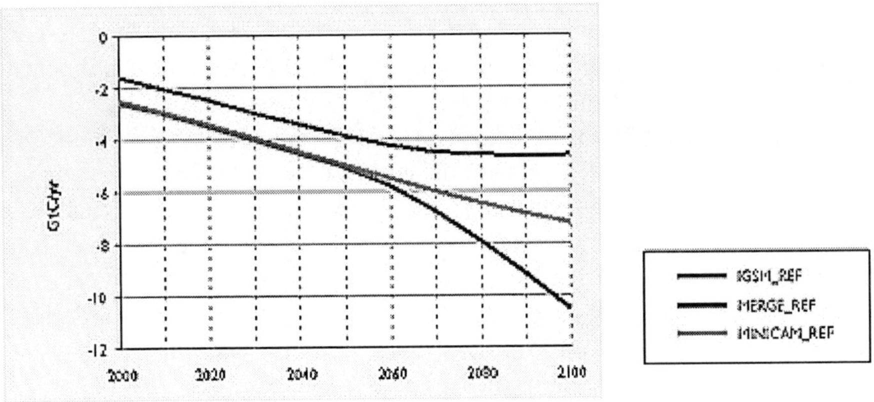

The IGSM reference scenario, which is based on the IGSM's three-dimensional ocean model, exhibits less CO_2 uptake than the other two reference scenarios and, after some point, little additional increase in uptake even though concentrations are rising. The MiniCAM reference scenario exhibits some slowing of ocean uptake, although not as pronounced as in the IGSM reference scenario. There is no slowing of uptake in the MERGE reference scenario. Although the MERGE reference scenario has higher ocean uptake in the latter half of the century, the effects of this increase are offset by the assumption of a neutral biosphere. Hence the aggregate behavior of its carbon cycle tends to be more similar to thatin the other two reference scenarios, especially the MiniCAM reference scenario (Figure 3.22). The three ocean models produce more similar behavior in the stabilization scenarios.

Figure 3.20. CO_2 Uptake from Oceans Across Reference Scenarios (GtC/yr, expressed in terms of net emissions).

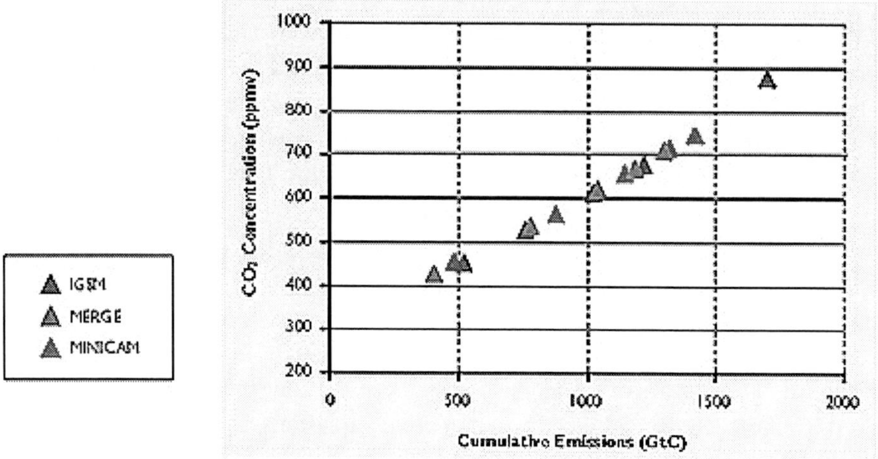

Despite differences in how the carbon cycle is handled in each of the three models, the scenarios exhibit a very similar response in terms of concentration level for a given level of cumulative emissions. [Note. The cumulative emissions do not include emissions from land use and land-use change].

Figure 3.21. Relationship Between Cumulative CO_2 Emissions from Fossil Fuel Combustion and Industrial Sources, 2000-2100, and Atmospheric CO_2 Concentration in 2100 Across All Scenarios.

GHG concentrations rise in all three reference scenarios. CO_2 concentrations increase from 370 ppmv in year 2000 to somewhere in the range of 700 to 875 ppmv in 2100 (Figure 3.22). The preindustrial concentration of CO_2 was approximately 280 ppmv. While all three reference scenarios display the same increasing pattern, by the year 2100 there is a difference of approximately 175 ppmv among the three scenarios. This difference has implications for radiative forcing and emissions mitigation (discussed in Chapter 4).

Increases in the concentrations of the non-CO_2 GHGs vary across the reference scenarios. The concentrations of CH_4 and N_2O in the Mini- CAM reference scenario are on the low end of the range, reflecting assumptions discussed above about use of CH_4 for energy and emissions control for non-climate reasons. The IGSM reference scenario has the highest concentrations for all of the substances. The differences mainly reflect differences in anthropogenic emissions, but they also are influenced by the way each model treats natural emissions and sinks for the gases. The IGSM scenarios include climate and atmospheric feedbacks to natural systems, which tend to result in an increase in natural emissions of CH_4 and N_2O. Also, increases in other pollutants generally lengthen the lifetime of CH_4 in the IGSM scenarios because the other pollutants deplete the atmosphere of the hydroxyl radical (OH), which is the removal mechanism for CH_4. These feedbacks tend to amplify the difference in anthropogenic emissions among the reference scenarios. The concentrations of the short-lived and long-lived F-gases are also presented in Figure 3.22.

Radiative Forcing from Greenhouse Gases

Contributions to radiative forcing are a combination of the abundance of the gas in the atmosphere and its heat-trapping potential (radiative efficiency). Of the directly released anthropogenic gases, CO_2 is the most abundant, measured in parts per million; the others are measured in parts per billion. However, the other GHGs are about 24 times (CH_4), to 200 times (N_2O), to thousands of times (SF_6 and PFCs) more radiatively efficient than CO_2. Thus, what they lack in abundance they make up for, in part, with radiative efficiency. However, CO_2 is still the main contributor to radiative forcing among these substances, and all three reference scenarios exhibit an increasing relative contribution from CO_2.

The three models display essentially the samerelationship between GHG concentrations andradiative forcing, so the three reference scenarios also all exhibit higher radiative forcing,growing from roughly 2.1 W/m^{2}from preindustrial in 1998 to between 6.4 W/m^{2}and 8.6W/m^2 in 2100. The differences among radiative forcing in 2100 imply differences in the amount of emissions reductions required to stabilize as the four radiative forcing levels in this research. For example, the emissions reductions required for stabilization in the IGSM stabilization scenarios are larger than those required in the Mini-CAM stabilization scenarios, because the radiative forcing reaches 8.6 W/m^{2}in 2100 in the IGSM reference scenario and 6.4 W/m^{2}in he MiniCAM reference scenario.

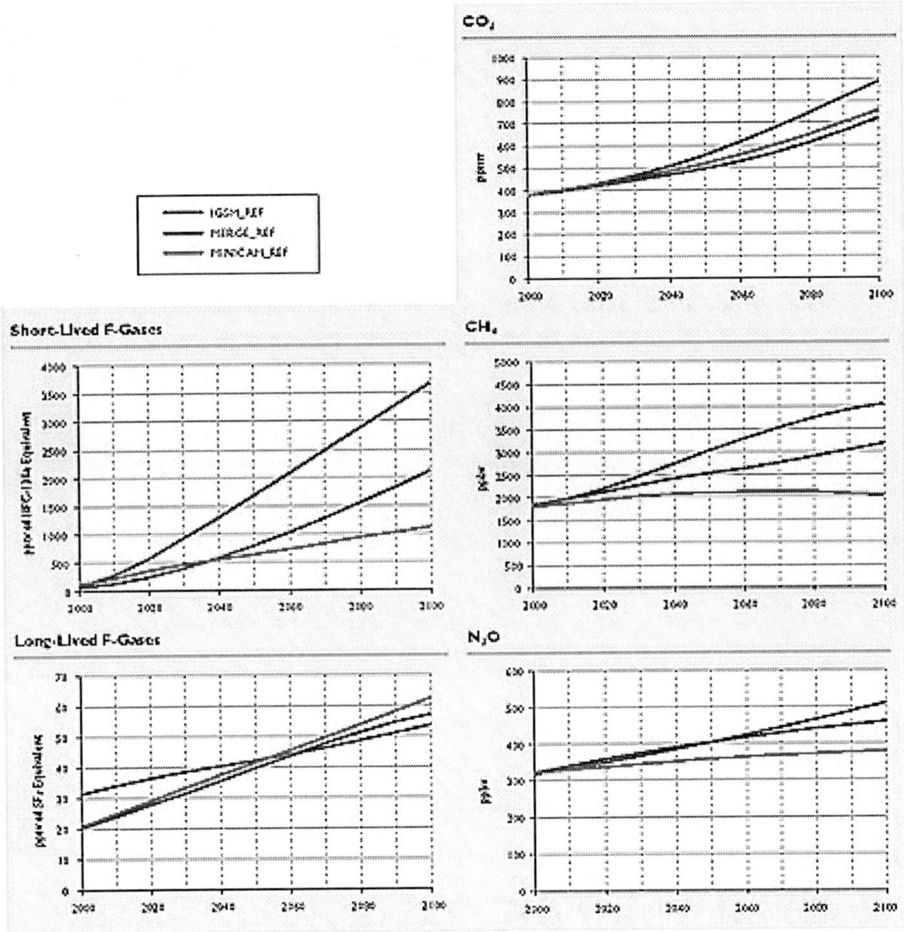

Differences in concentrations for CO_2, CH_4, and N_2O across the reference scenarios reflect differences in emissions and treatment of removal processes. By 2100, CO_2 concentrations range from about 700 ppmv to 900 ppmv, CH_4 concentrations range from 2000 ppbv to 4000 ppbv, and N_2O concentrations range from about 380 ppbv to 500 ppbv.

Figure 3.22. Atmospheric Concentrations of CO_2, CH_4, N_2O, and F-gases Across the Reference Scenarios (units vary).

The relative contribution of CO_2 to radiative forcing increases over the century in all three reference scenarios (Figure 3.23). In 2000, the non-CO_2 GHGs examined in this research contributed slightly above 30% of the estimated radiative forcing from preindustrial. In the IGSM reference scenario, the contribution of the nonCO_2 GHGs to radiative forcing falls slightly to about 26% by 2100. The MiniCAM reference scenario includes little additional increase in radiative forcing for non-CO_2 GHGs, largely as a result of assumptions regarding the control of CH_4 emissions for non-climate reasons, and thus has their share falling to about 18% by 2100. The MERGE reference scenario is intermediate, with the non-CO_2 GHG contribution falling to about 24%.

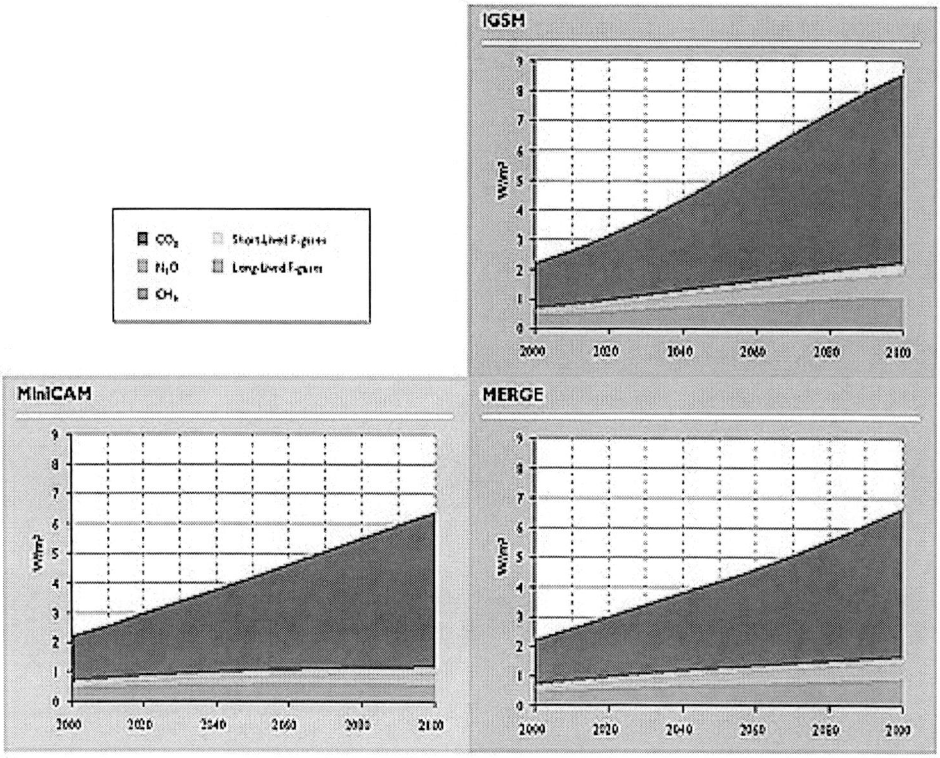

CO_2 accounts for more than 80% of the radiative forcing from the GHGs considered in this research by 2100 in all three reference scenarios. Radiative forcing in 21 00 ranges from about 6.4 W/m^2 to 8.6 W/m^2 from preindustrial in the reference scenarios.

Figure 3.23. Radiative Forcing by Gas Across Reference Scenarios (W/m^2 from preindustrial).

From the discussion above, it can be seen that the three reference scenarios contain many large-scale similarities. All have expanding global energy systems, all remain dominated by fossil fuel use throughout the twenty-first century, all generate increasing concentrations of GHGs, and all produce substantial increases in radiative forcing. Yet the reference scenarios differ in many details, ranging from demographics to labor productivity growth rates to the composition of energy supply to treatment of the carbon cycle. These differences shed light on important points of uncertainty that arise for the future. In Chapter 4, they will also be seen to have important implications for efforts to limit radiative forcing.

4. STABILIZATION SCENARIOS

In these scenarios, stabilizing radiative forcing at levels ranging from 3.4 W/m² to 6.7 W/m² above preindustrial levels (Level 1 to Level 4) implies significant changes to the world's energy and agriculture systems and leads to lower global economic output. Although all the stabilization scenarios require changes in the world's energy and agricultural systems, the three modeling groups produced scenarios with differing conceptions of how these changes might occur. The economic implications vary considerably among the scenarios, depending on the amount that emissions must be reduced and the evolution of technology, particularly in the post-2050 period.

Introduction

In Chapter 3, each modeling group developed scenarios of long-term GHG emissions associated with changes in key characteristics, such as demographics, economic growth, and technology. This chapter describes how such developments might affect or be affected by limits on radiative forcing. It illustrates that society's response to a limit on radiative forcing can take many paths, reflecting factors shaping the reference scenario and the availability and performance of emissions-reducing technologies. Control of GHG emissions requires changes in the global energy, economic, agriculture, and land-use systems.

It should be emphasized that the four radiative forcing stabilization levels considered in this research and detailed in Table 1.2 were chosen for illustrative purposes only. They reflect neither a preference nor a recommendation. In all the stabilization scenarios, it was assumed that radiative forcing would not be allowed to overshoot the radiative forcing levels along the path to long-term stabilization. Given this assumption, each modeling group had to make further decisions regarding the means of meeting these radiative forcing limits. Section 4.2 compares the approaches of the three modeling groups. Section 4.3 shows the effect of the three strategies on GHG emissions, concentrations, and radiative forcing. The implications for global and U.S. energy and industrial systems are explored in Section 4.4 and for agriculture and land-use change in Section 4.5. Section 4.6 discusses economic consequences of the measures to achieve the various radiative forcing stabilization levels in these scenarios.

Stabilizing Radiative Forcing: Model Implementations

Some features of scenario construction were coordinated among the three modeling groups, and others were left to their discretion. In three areas, a common set of approaches was adopted:

- Climate policies in the stabilization scenarios
- The timing of participation in stabilization scenarios
- Policy instrument assumptions in stabilization scenarios.

In two areas, the groups employed different approaches:

- The timing of CO_2 emissions mitigation
- Non-CO_2 emissions mitigation.

Climate Policies in the Stabilization Scenarios

For the stabilization scenarios, each modeling group assumed that, as in the reference scenarios, the U.S. will achieve its goal of reducing GHG emissions intensity (the ratio of GHG emissions to GDP) by 18% by 2012, although implementation of this goal was left to the judgment of each modeling group. Also, the Kyoto Protocol participants were assumed to achieve their commitments through the first commitment period, 2008 to 2012. In the reference scenarios, these policies were modeled as not continuing after 2012. In the stabilization scenarios, these initial period policies were superseded by the long-term control strategies imposed by each group.

Participation in Stabilization Scenarios

For the stabilization scenarios, it was assumed that policies to limit the change in radiative forcing would be applied globally after 2012, as directed by the Prospectus. Although it seems unlikely that all countries would simultaneously join such a global agreement, and the economic costs of stabilization would be greater with less-than-universal participation, the assumption that all countries participate does provide a useful benchmark.

Policy Instrument Assumptions in Stabilization Scenarios

Note that the issue of economic efficiency applies across both space and time. All of the scenarios assume an economically efficient allocation of reductions among nations in each time period, that is, across space. Thus, in these scenarios, GHG emissions in all regions and across all sectors of the economy were controlled by imposing a single price for each GHG at any point in time. As will be discussed in detail in Section 4.5, the prices of emissions for individual GHGs differ across the models. The implied ability to access emissions reduction opportunities wherever they are cheapest is sometimes referred to as *where* flexibility (Richels et al. 1996).

Timing of CO_2 Emissions Mitigation

The cost of stabilizing radiative forcing to any given level depends on the timing of the associated emissions reductions. There is a strong economic argument that costs will be lower if emissions reductions start slowly and then progressively ramp up, particularly for CO_2. Distributing emissions mitigation over time, such that larger efforts are undertaken later, reduces the current cost as a consequence of such effects as discounting, the preservation of energy-using capital stock over its natural lifetime, and the potential for the development of increasingly cost-effective technologies (Wigley et al. 1996).

Although 100 years is a very long time horizon for economic scenarios, it is not long enough to fully evaluate stabilization goals. For several of the radiative forcing stabilization levels, the scenarios are only approaching stabilization in 2100; radiative forcing is below the long-term stabilization levels and still rising, but the rate of increase is slowing. Stabilizing radiative forcing and associated atmospheric GHG concentrations requires that any emissions be completely offset by uptake or destruction processes. Because ocean and terrestrial uptake of $CO_2$is subject to saturation and system inertia, at least for the approximate CO_2 concentration levels considered in this research, emissions need to peak and subsequently decline during the twenty-first century or soon thereafter. In the very long term (many hundreds to thousands of years), emissions must decline to virtually zero for any CO_2 concentration to be maintained. Although there is some flexibility in the inter-temporal allocation of emissions, this allocation is inherently constrained by the carbon cycle. Given that anthropogenic CO_2 emissions rise with time in all three of the reference scenarios, the degree of CO_2 emissions reduction also increases steadily with time in the stabilization scenarios.

Different approaches were used by the modeling groups to determine the profile of emissions reductions over time and how the different GHGs contribute to meeting the radiative forcing stabilization levels. A major reason for the difference is the structure of the models. MERGE is an inter-temporal optimization model and is able to solve for the cost-minimizing allocation of emissions reductions across GHGs and over time to meet a given radiative forcing stabilization level. It thus offers insights regarding the optimal path of emissions reductions. A positive discount rate will lead to a gradual phase-in of emissions reductions, and the tradeoff among GHGs is endogenously calculated based on the contribution each makes toward the long-term goal (Manne and Richels 2001). The changing relative prices of GHGs over time can be interpreted as an optimal trading index for the GHGs that combines economic considerations with modeled physical considerations (lifetime and radiative forcing). The resulting relative weights are different from those derived using Global Warming Potential (GWP) indices, which are based purely on physical considerations (IPCC 2001). Furthermore, economically efficient indices for the relative importance of GHG emissions reductions will vary over time and across policy regimes.

IGSM and MiniCAM are simulation models and do not endogenously solve for optimal allocations over time and by GHG. However, the choice of price paths over time used in the stabilization scenarios for the IGSM and MiniCAM modeling groups take account of insights from economic principles that lead to a pattern similar to that computed by MERGE. The pattern was anticipated by Peck and Wan (1996) using a simple optimizing model with a carbon cycle and by Hotelling (1931) in a simpler context.

In the MiniCAM stabilization scenarios, the rate of increase in the carbon price was set equal to the rate of interest plus the average rate of carbon removal from the atmosphere by natural systems. This approach follows Peck and Wan (1996) and yields a resulting carbon price path similar in structure to that obtained in the MERGE scenarios. This carbon price path ensures that the present discounted marginal cost of having one tonne of carbon less in the atmosphere during one period in the future is exactly the same regardless of whether the removal takes place today or one period later. When marginal costs are equal over time, total costs cannot be reduced by making emissions mitigation either earlier or later.

As is the case in the MERGE scenarios, the exponential increase in the price of CO_2 continues until such time as radiative forcing is stabilized in the MiniCAM stabilization scenarios. Thereafter, the price is set by the carbon cycle. That is, once radiative forcing has risen to its stabilization level, additional CO_2 can only enter the atmosphere to the extent that natural processes remove it, otherwise CO_2 radiative forcing would be increasing. This is relevant in the Level 1 stabilization scenario and, to a lesser extent, in the Level 2 stabilization scenario. However, it is not relevant in the Level 3 or Level 4 scenarios because stabilization is not reached until after the end of the twenty-first century.

The IGSM scenarios are based on a carbon price path that rises 4% per year. The initial carbon price is set to achieve the required concentrations and radiative forcing. Thus, the rate of increase in the CO_2 price paths is identical for all stabilization scenarios, but the initial value of the carbon price is different. The lower the concentration of CO_2allowed, the higher the initial price. The insight behind this approach is that an entity faced with a carbon constraint and a decision to reduce emissions now or later would compare the expected return on that emissions reduction investment with the rate of return elsewhere in the economy. The 4% rate is taken to be this economy-wide rate of return. If the carbon price were rising more rapidly than the rate of return, investments in emissions reductions would yield a higher return than investments elsewhere in the economy, so that the entity would invest more in emissions reductions now (and possibly bank emissions permits to use them later). By the same logic, an increase in the carbon price lower than the rate of return would lead to a decision to postpone emissions reductions. It would lead to a tighter carbon constraint and a higher carbon price in the future. Thus, this approach is intended to be consistent with a market solution that would allocate emissions reductions through time.

Timing of Non-CO_2 Emissions Mitigation

Like CO_2, the contribution of non-CO_2 GHGs to radiative forcing depends on their concentrations. However, these gases are dissociated in the atmosphere over time so that the relationship between emissions and concentrations is different from that for CO_2, as are the sources of emissions and opportunities for emissions reductions. Each of the three modeling groups used its own approach to model control of nonCO_2 GHGs. As noted above, MERGE employs an inter-temporal optimization approach. The price of each GHG was determined so as to minimize the cost of stabilizing radiative forcing at each level. Thus, the price of each GHG was constant across regions at any point in time, but varied over time so as to minimize the cost of achieving each stabilization level.

In the MiniCAM stabilization scenarios, nonCO_2GHG prices were tied to the price of CO_2 using the GWPs of the gases. This procedure has been adopted by parties to the Kyoto Protocol and applied in the definition of the U.S. emissions intensity goal. The IGSM stabilization scenarios are based on the same approach as MiniCAM stabilization scenarios for determining the prices for HFCs, PFCs, and SF6, pegging the prices to that of CO_2 using GWP coefficients. For CH_4 and N_2O, however, independent emission stabilization levels were set for each gas in the IGSM scenarios because GWPs poorly represent the full effects of CH_4, and emissions trading at GWP rates leads to problems in defining what stabilization means when CH_4 and N_2O are involved (Sarofim et al. 2005). The relatively near-term stabilization for CH_4 in the IGSM scenarios implies that near- term emissions reductions result in economic benefit, an approach consistent with a view that there are risks associated with levels of radiative forcing below the long-term stabilization levels. This approach is different than that followed in the MERGE scenarios, where any value of CH_4 emissions reductions is derived only from the extent to which it contributes to meeting the long-term stabilization level. In the MERGE stabilization scenarios, reductions of emissions of short-lived species like CH_4 have very little consequence for a radiative forcing stabilization level that will not be reached for many decades, so the optimized result places little value on reducing emissions of short-lived species until the stabilization level is approached. A full analysis of the resulting climate change and its effects would be required to select between the approaches used in the MERGE and IGSM scenarios. The different stabilization paths in the scenarios from these two models provide a range of plausible scenarios for non-CO_2 GHG stabilization. The MiniCAM scenarios yield an intermediate result.

Implications for Radiative Forcing, Greenhouse Gas Concentrations, and Emissions

Despite significantly different radiative forcing levels in the reference scenarios, radiative forcing relative to preindustrial levels in 2100 is similar across models in all four stabilization scenarios. CO_2 concentrations are also similar in 2100 across the models. Scenarios with higher CO_2 concentrations for a given stabilization level generally have lower concentrations and emissions of nonCO_2 GHGs, trading off reductions in these substances to make up for higher forcing from CO_2.

All three modeling groups produced scenarios in which emissions reductions below levels in the reference scenarios were much smaller between 2000 and 2050 than between 2050 and 21 00. With one exception at the least stringent stabilization level, the stabilization scenarios were characterized by a peak and decline in global CO_2 emissions in the twenty-first century. In the most stringent scenarios, CO_2 emissions begin to decline immediately or within a matter of decades.

Implications for Radiative Forcing

Given that all the models were constrained to the same radiative forcing stabilization levels, radiative forcing from preindustrial for the year 2100 is similar across the models (Table 4. 1).[6] The differences across the models between the long-term stabilization levels and the radiative forcing levels in 2100 are smaller for Levels 1 and 2 than for Levels 3 and 4 because the latter allow a greater accumulation of GHGs in the atmosphere. For Levels 3 and 4, each modeling group required radiative forcing to be below the long-term limits in 2100 to allow for subsequent emissions to fall gradually toward levels required for stabilization.

The radiative forcing stabilization paths are shown in Figure 4.1. Even though they reflect different criteria used to allocate emissions reductions over time, the paths are very similar across models. The radiative forcing paths are dominated by radiative forcing associated with CO_2 concentrations, which in turn are driven by cumulative emissions. Thus, even fairly different time profiles of CO_2 emissions can yield relatively little difference in concentrations and radiative forcing.

Table 4.1. Radiative Forcing in the Year 2100 Across Scenarios

		Radiative Forcing in 2100 (W/m^{-2} from preindustrial)		
Stabilization Level	**Long-Term Radiative Forcing Limit (W/m^2 from preindustrial)**	**IGSM**	**MERGE**	**MiniCAM**
Reference	No Constraint	8.6	6.6	6.4
Level 4	6.7	6.1	6.2	6.1
Level 3	5.8	5.4	5.7	5.5
Level 2	4.7	4.4	4.7	4.5
Level 1	3.4	3.5	3.4	3.4

Although their totals are similar, the GHG composition of radiative forcing differs among the models. Figure 4.2 shows the breakdown among gases in 2100 for the reference scenario along with all four stabilization levels. Forcing is dominated by CO_2 in all scenarios at all stabilization levels, but there are variations among models. For example, the MiniCAM stabilization scenarios have larger contributions from CO_2 and lower contributions from the non-CO_2 gases than the scenarios from the other two models. Conversely, the MERGE stabilization scenarios have higher contributions from the non-CO_2 gases and lower contributions from CO_2 relative to the IGSM and MiniCAM stabilization scenarios.

Implications for Greenhouse Gas Concentrations

The relative GHG composition of radiative forcing across models in any scenario reflects differences in concentrations of the GHGs. The CO_2 concentration paths are presented in Figure 4.3, and the year 2100 atmospheric levels are shown in Table 4.3. Because the stabilization levels were specified in terms of total radiative forcing from the multiple GHGs, it is possible to meet those levels while varying from the approximate CO_2 concentration levels used to construct them (Table 1.2). That means CO_2 concentrations in 2100 differ across models for any stabilization level. For example, the CO_2 concentrations in the MiniCAM stabilization scenarios are generally higher than in IGSM and MERGE stabilization scenarios. Consequently, CH_4 and N_2O concentrations are systematically lower as can be seen in Figure 4.4 and Figure 4.5.

Differences in the GHG concentrations among the scenarios from the three models reflect differences in the way that tradeoffs were made among gases and differences in assumed emissions reduction opportunities for non-CO_2 GHGs compared to CO_2.

Approximate stabilization of CO_2 concentrations occurs by 2100 in all the Level 1 and Level 2 scenarios, but concentrations are still increasing in 2100 for the Level 3 and Level 4 scenarios, although at a slowing rate. An important implication of the less stringent stabilization levels is that substantial emissions reductions would be required after 2100. Sometime within the next century, all the stabilization paths would require emissions levels nearly as low as that for Level 1. Higher stabilization levels do not change the nature of long-term changes in emissions required in the global economy; they only delay when the emissions reductions must be achieved.

Table 4.2. CO_2Concentrations in the Year 2100 Across Scenarios (ppmv).

		CO_2 Concentration in 2100 (ppmv)		
Level	**Approximate Long-Term CO_2 Concentration Limit (ppmv)**	**IGSM**	**MERGE**	**MiniCAM**
Reference	—	875	711	746
Level 4	750	677	670	716
Level 3	650	614	619	656
Level 2	550	526	535	562
Level 1	450	451	426	456

The approximate CO_2 concentrations were used as a guide to develop the radiative forcing stabilization levels. The scenarios were required to meet the total radiative forcing limits. The CO_2 concentrations in the scenarios do not exactly match these approximations and differ among the modeling groups because of differences in the treatment of the forces that influence emissions of GHGs, possibilities for emissions reductions, and tradeoffs between reductions among GHGs.

In all the stabilization scenarios, as the rise in atmospheric concentrations slows, ocean uptake slows and even begins to decline. These natural removal processes are uncertain, and to some extent this uncertainty is reflected in differences in the scenarios from the three modeling groups, as shown in Figure 4.6. Ocean uptake is smallest in the IGSM scenarios. The MERGE scenarios have the highest uptake for the least stringent stabilization levels, and the MiniCAM and MERGE scenarios are almost identical under the most stringent stabilization levels.

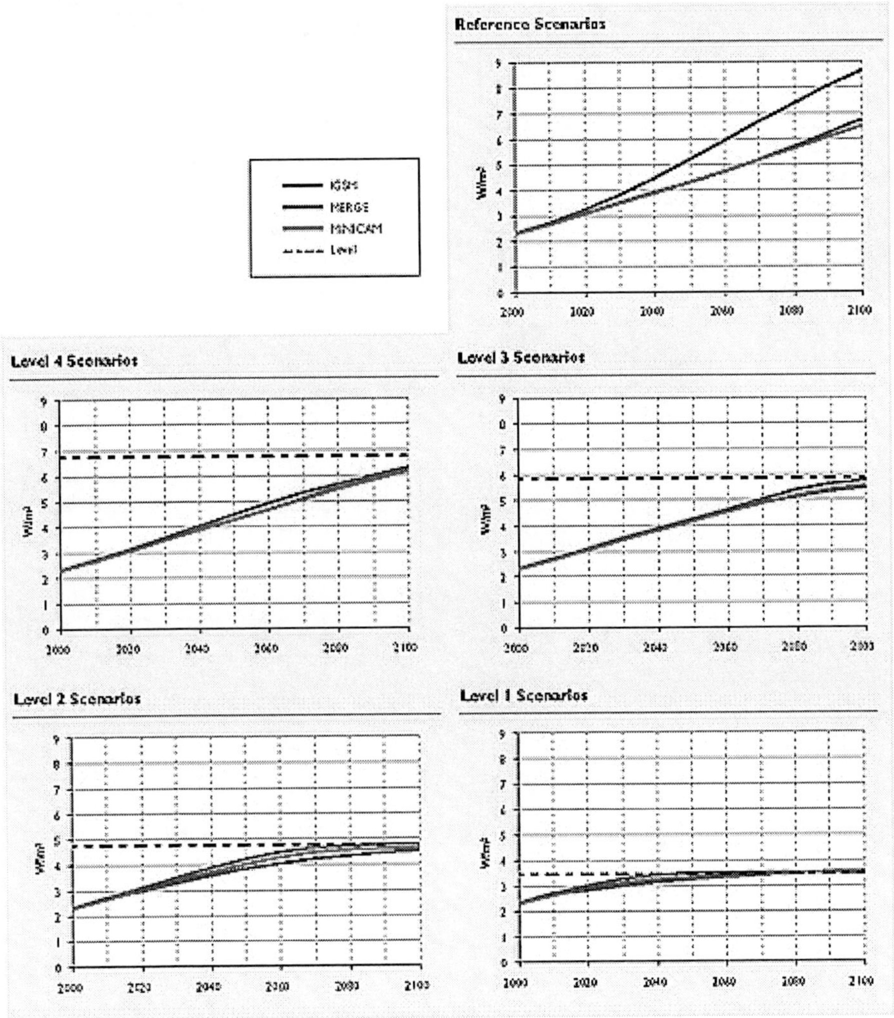

Radiative forcing trajectories differ across the stabilization levels but are similar among models for each stabilization level. The similarity across models reflects the design of the scenarios. Radiative forcing is stabilized or close to being stabilized this century in the Level 1and Level 2 scenarios. Radiative forcing remains below the long-term radiative forcing stabilization level in 2100 in the Level 3 and Level 4 stabilization scenarios, allowing for a gradual approach to stabilization in the following century.

Figure 4.1. Total Radiative Forcing by Year Across Scenarios (W/m^2 from preindustrial).

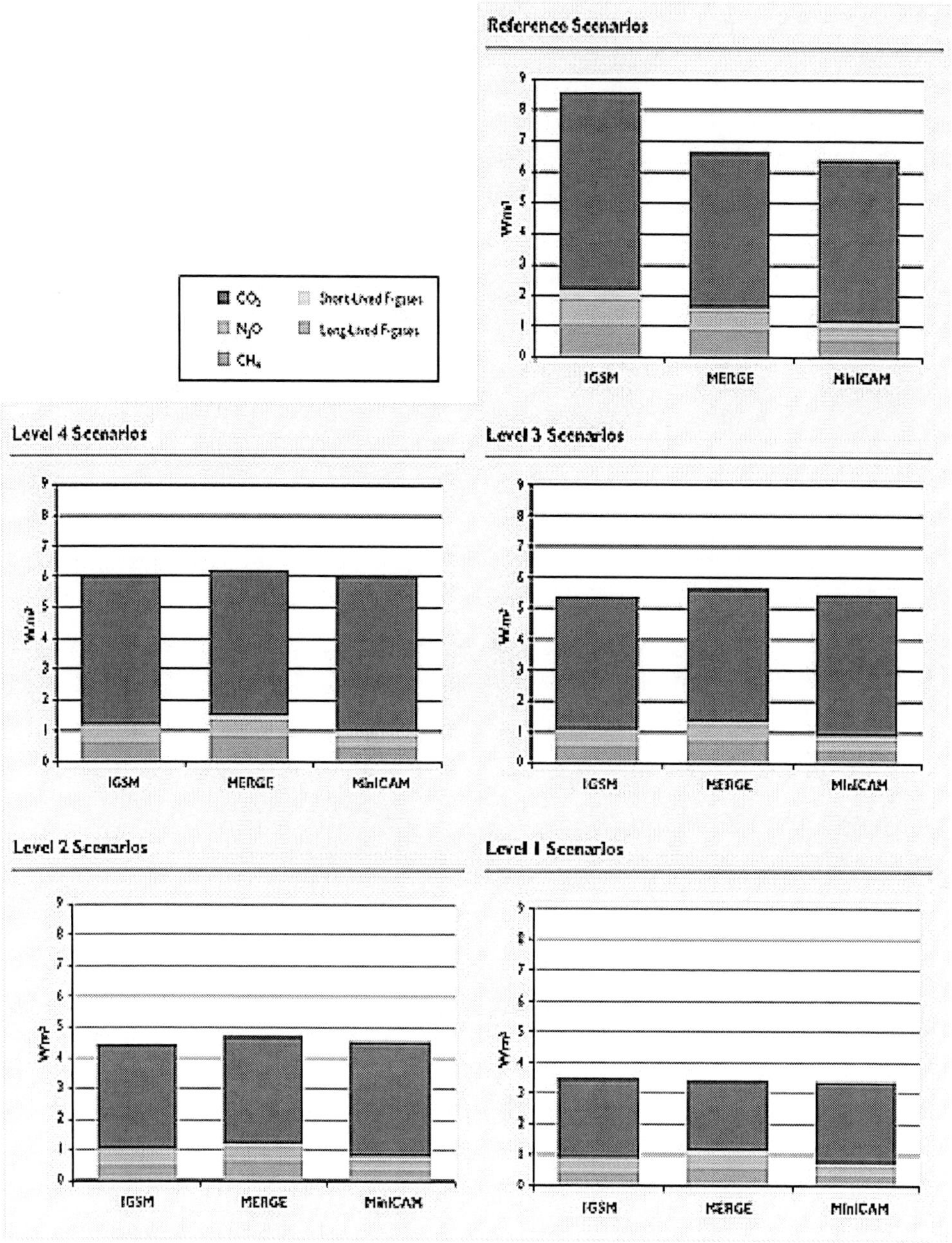

CO_2 is the main contributor to radiative forcing by the end of the century in the scenarios from all three modeling groups. The IGSM reference scenario has the highest contribution from non-CO_2GHGs among the three models. The MERGE stabilization scenarios have the highest contribution from non-CO_2GHGsamong the three models, implying greater non-CO_2controlefforts in the IGSM scenarios than in the MERGE scenarios. Contributions from non-CO_2GHGs are lowest in the MiniCAM scenarios, reflecting, inpart, assumptions about control of these substances for non-climate reasons.

Figure 4.2. Total Radiative Forcing by Gas in 2100Across Scenarios ($W/m^2$from preindustrial).

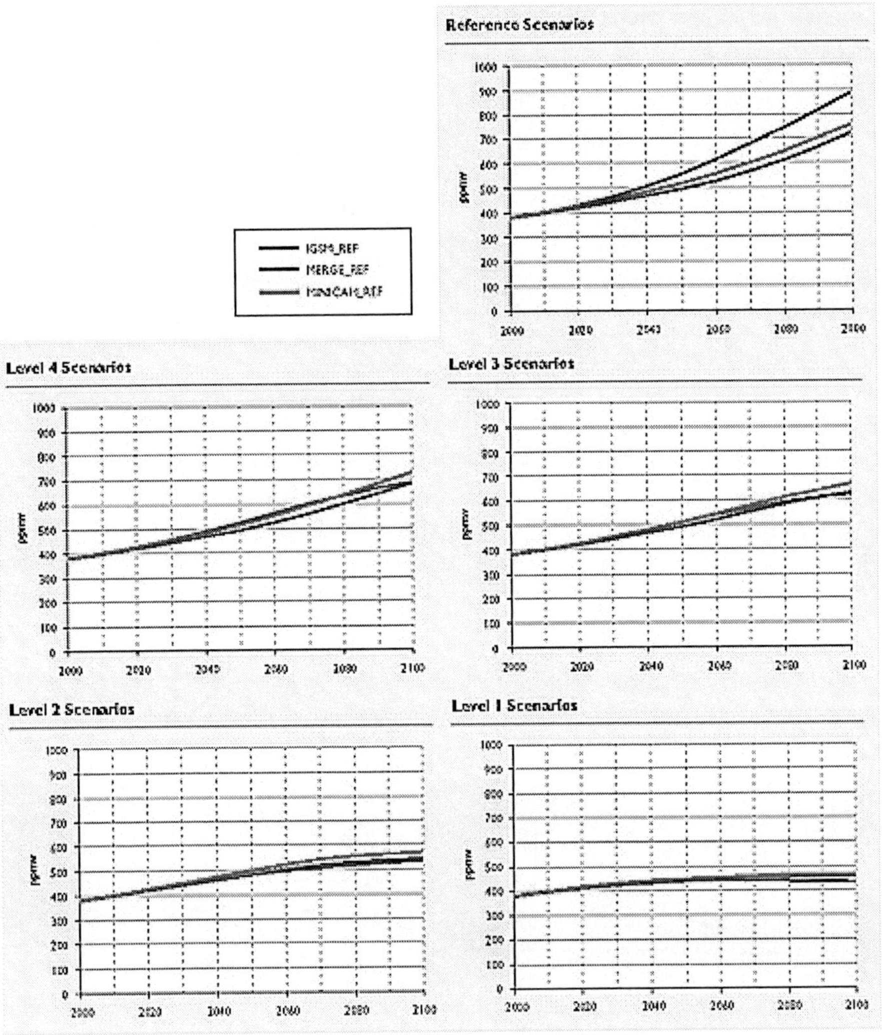

Atmospheric concentrations of CO_2 range from about700 ppmv to 900 ppmv in 2100 in the reference scenarios, with no sign of slowing. In the stabilization scenarios, differences in CO_2concentrations among models occur because of the relative contribution of other GHGs to meeting the radiative forcing stabilization levels, and because for Levels 3 and 4, the scenarios are based on a gradual approach to the stabilization level that will not be reached until the following century.

Figure 4.3. CO_2Concentrations Across Scenarios (ppmv)

Implications for Greenhouse Gas Emissions

Implications for Global CO_2 Emissions

Global CO_2emissions begin declining immediately after 2010 or in a matter of decades in all three Level 1 stabilization scenarios (Figure 4.7). The constraint is so tight that there is relatively little room for variation among models.

All three modeling groups show continuedemissions growth throughout the first half of the twenty-first century for Level 4, the least stringent stabilization levels, and the MiniCAM

Level 4 scenario exhibits increasing emissionsthroughout the century, although emissions areapproaching a peak by 2100.

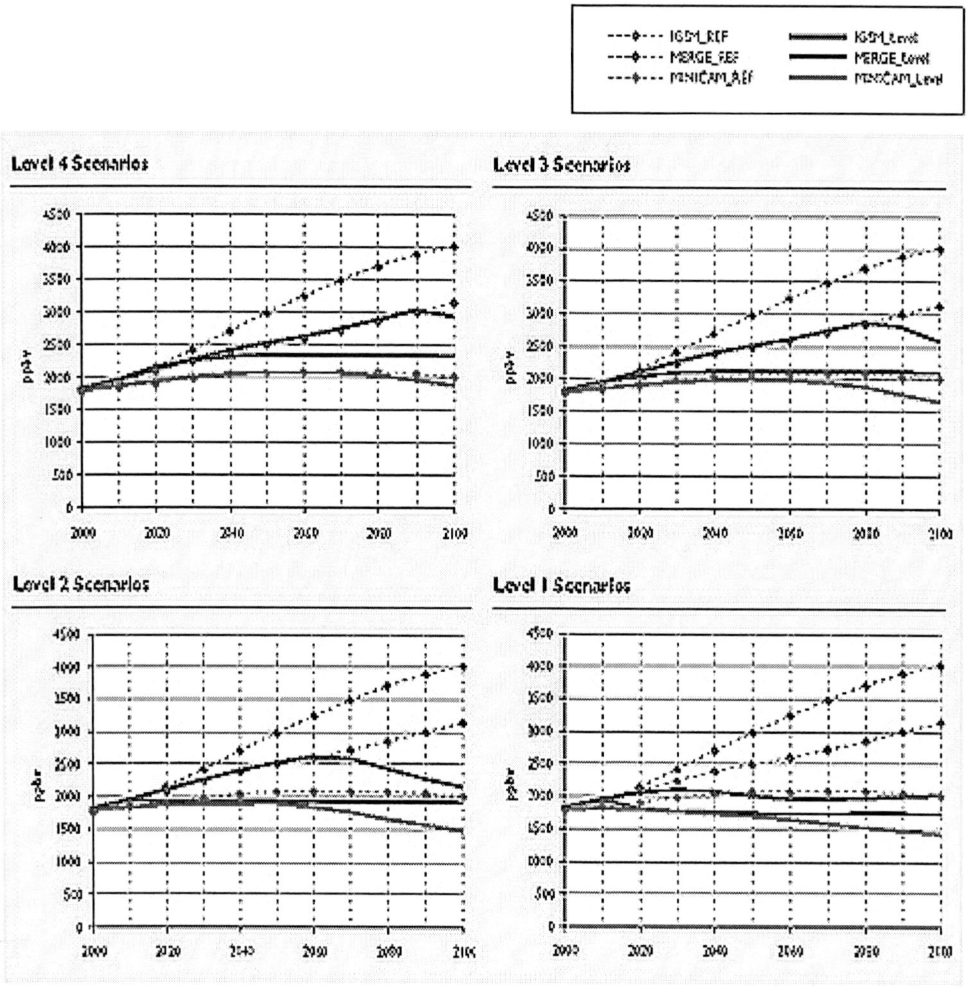

Differences among the models in CH_4 concentrations are larger than differences in CO_2 concentrations. These differences stem from differences in reference scenarios, assumptions about options for emissions reductions, and the methods used by the modeling groups for determining the relative emissions reductions among different GHGs. Reductions in non-CO_2 GHG emissions in the MiniCAM stabilization scenarios are based on 100-year GWPs. The MERGE stabilization scenarios are based on intertemporal optimization, leading to relatively little value for controlling CH_4 emissions until the stabilization level is approached due to the relatively short lifetime of CH_4. The IGSM stabilization scenarios are based on independent stabilization of CH_4.

Figure 4.4. CH_4 Concentrations Across Scenarios (ppbv).

The stabilization scenarios from all three modeling groups exhibit more emissions reduction in the second half of the twenty-first century than in the first half, as noted earlier, so the mitigation challenge grows with time.

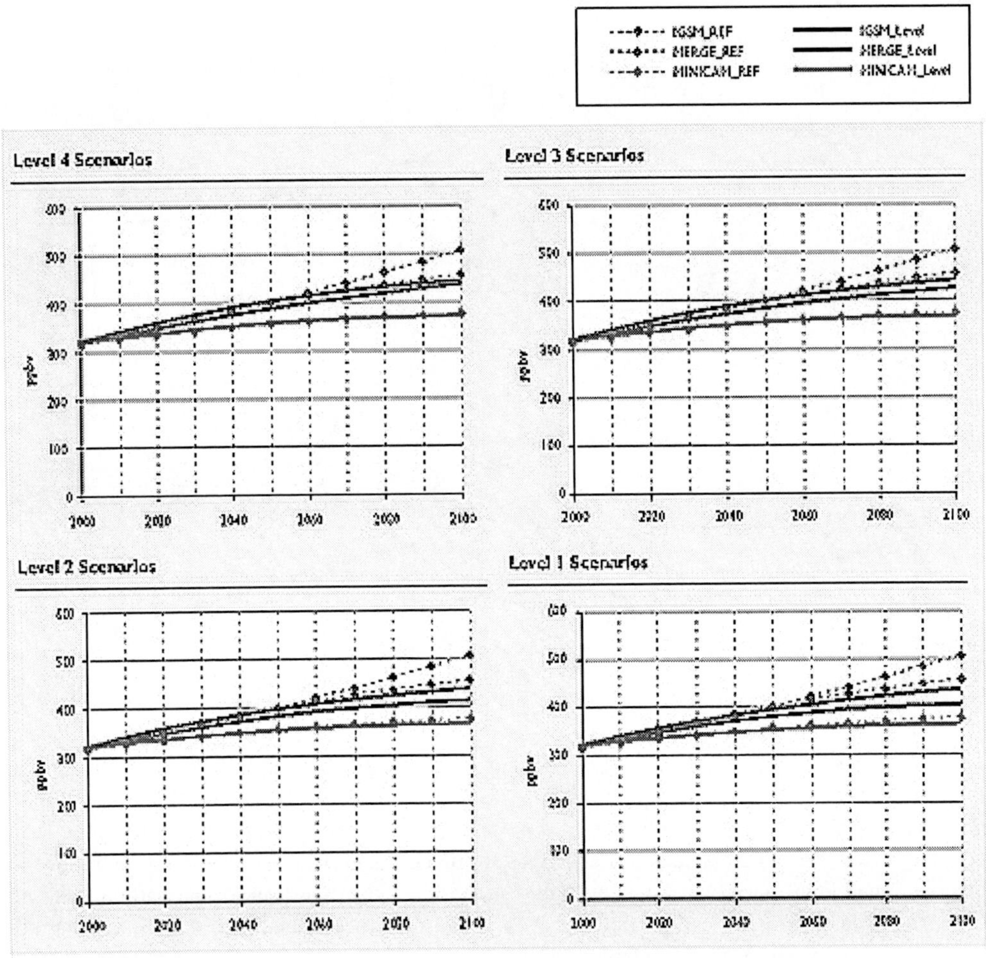

Atmospheric concentrations of N_2O range from about 375 ppbv to 500 ppbv in 2100 across the scenarios, with concentrations continuing to rise in the reference scenarios. Different approaches were used by the different modeling groups to develop emissions reductions, leading to differences in concentrations between the reference and stabilization scenarios.

Figure 4.5. N_2O Concentrations Across Scenarios (ppbv).

The precise timing and degree of departure from the reference scenario depend on many aspects of the scenarios and on each model's representation of Earth system properties, including the radiative forcing stabilization level, the carbon cycle, atmospheric chemistry, the character of technology options over time, the reference scenario CO_2 emissions path, the non-climate policy environment, the rate of discount, and the climate policy environment. For Level 4, 85% or more of emissions mitigation occurs in the second half of the twenty-first century in the scenarios from all three modeling groups. Even for Level 1, where the limit on radiative forcing is the tightest and near-term mitigation most urgent, 75% or more of the emissions reduction below reference scenario occurs in the second half of the century. While this is partly a result of the *when* flexibility assumption, continuing emissions growth in the reference scenarios means that the percentage reduction increases over time.

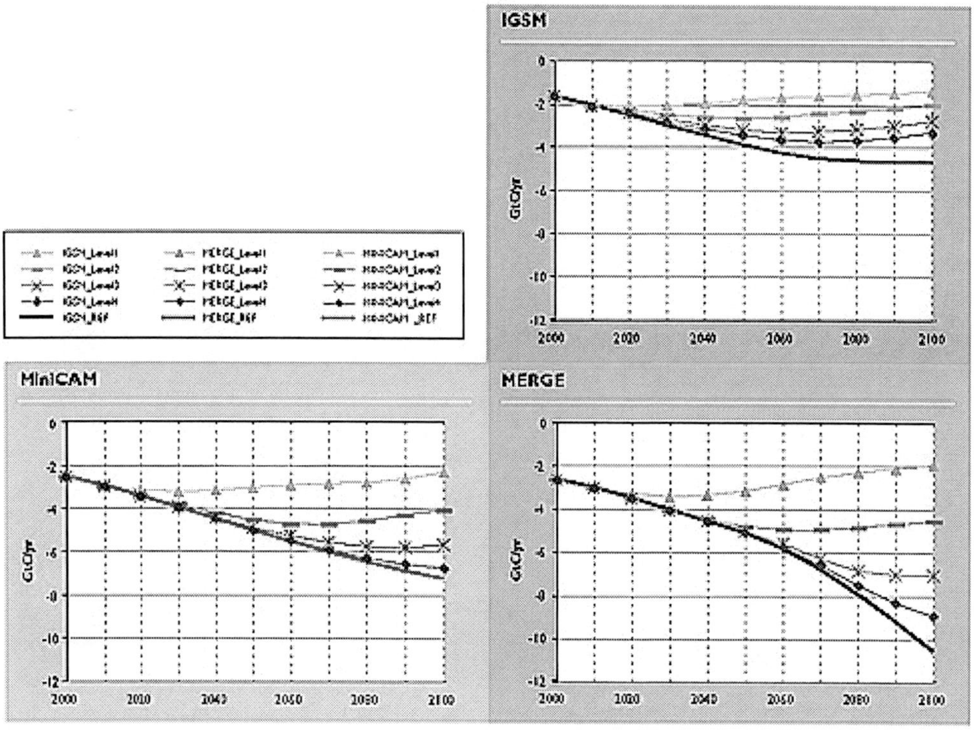

Oceans have taken up approximately one half of anthropogenic emissions of CO_2 since preindustrial times, and future ocean behavior is an important determinant of atmospheric concentrations. The three-dimensional ocean used for the IGSM scenarios shows the least ocean carbon uptake and considerable slowing of carbon up take even in the reference scenario as carbon concentrations continue to rise. The MERGE reference scenario shows thelargest uptake among the three models, and the MERGE stabilization scenarios have the greatest reductions from the reference scenario among the models. The MiniCAM scenarios are intermediate at most stabilization levels. At the more stringent stabilization levels, the MERGE and MiniCAM scenarios exhibit similar ocean uptake behavior.

Figure 4.6. Ocean CO_2Uptake Across Scenarios (GtC/yr, expressed in terms of net emissions).

All three of the modeling groups constructed reference scenarios in which Non-Annex 1 emissions were a larger fraction of the global total in the future than at present (Figure 3.16).

Because the stabilization scenarios are based on the assumption that all regions of the world face the same price of GHG emissions and have access to the same general set of technologies for mitigation, the resulting distribution of emissions mitigation between Annex I and Non- Annex I regions generally reflects the distribution of reference scenario emissions among them. So, when radiative forcing is restricted to Level 1, all three models find that more than half of the emissions mitigation occurs in Non-Annex I regions by 2050 because more than half of reference scenario emissions occur in Non-Annex I regions. Note that with the global policy specified so that a common carbon price occurs in all regions at any one time, emissions reductions occur separately from and mostly independent of the distribution of the economic burdens of reduction.

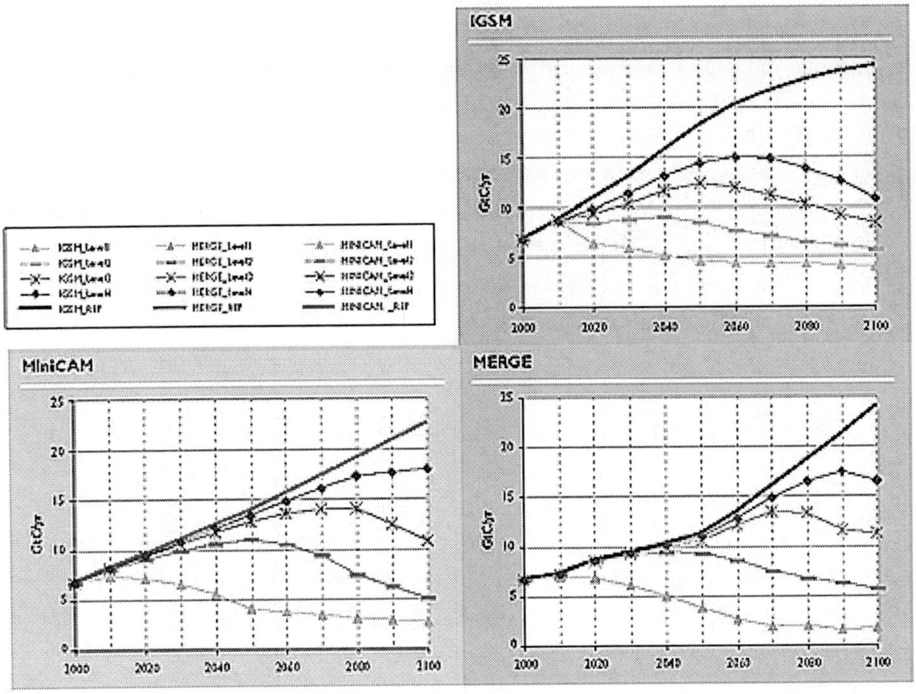

Fossil fuel CO_2 emissions vary among the reference scenarios, but the three differing emissions trajectories lead to emissions in 2100 in the range of 22.5GtC/yr to 24.0 GtC/yr. The timing of emissions reductions varies substantially across the stabilization levels. In the Level 1scenarios, global emissions begin to decline soon after the stabilization policy is put in place (as the scenarios were designed, after 2012), and emissions are below current levels by2100 in all of the Level 1 and Level 2 scenarios. Emissions peak sometime around the mid-century to early in the next century in the Level 3 and Level 4 scenarios and then begin a decline that would continue beyond 2100.

Figure 4.7. Fossil Fuel and Industrial CO_2Emissions Across Scenarios (GtC/yr).

Implications for Non-CO_2 Greenhouse Gas Emissions

The stabilization properties of the non-CO_2 GHGs differ due to their lifetimes (as determined by chemical reactions in the atmosphere), technologies for reducing emissions, and natural sources. CH_4 has a relatively short lifetime, and anthropogenic sources are a big part of CH_4 emissions. If anthropogenic emissions are kept constant, an approximate equilibrium between oxidation and net emissions will be established relatively quickly and concentrations will stabilize. The same is true for the relatively short-lived HFCs.

CH_4 emissions under stabilization are systematically lower the more stringent the stabilization level, as can be seen in Figure 4.8. The MiniCAM scenarios have the lowest CH_4 emissions among the models in the reference scenario and the stabilization scenarios. The assumed policy environment for CH_4 control is also important. Despite the fact that the IGSM reference scenario has higher reference CH_4 emissions than the MERGE reference scenario, the MERGE stabilization scenarios have higher emissions under stabilization in several instances. The reason is that the MERGE inter- temporal optimization approach leads to a low relative price for CH_4 emissions in the near term, which grows rapidly relative to CO_2, favoring strong reductions of CH_4 emissions only toward the end of the century, whereas CH_4

emissions were controlled based on quantitative limits in the IGSM stabilization scenarios, and these limits lead to substantial reduction early in the century. Thus, emissions in the MERGE stabilization scenarios sometimes exceed those in the IGSM stabilization scenarios until the relative CH_4 price rises sufficiently to induce substantial emissions reductions.

The very long-lived gases are nearly indestructible, thus for stabilization their emissions must be very near zero. Based on the assumptions used by all three modeling groups, it is possible, at reasonable cost, to achieve substantial reductions in long-lived gas emissions. While these substances are important, their emissions are not as difficult to reduce as those from fossil energy.

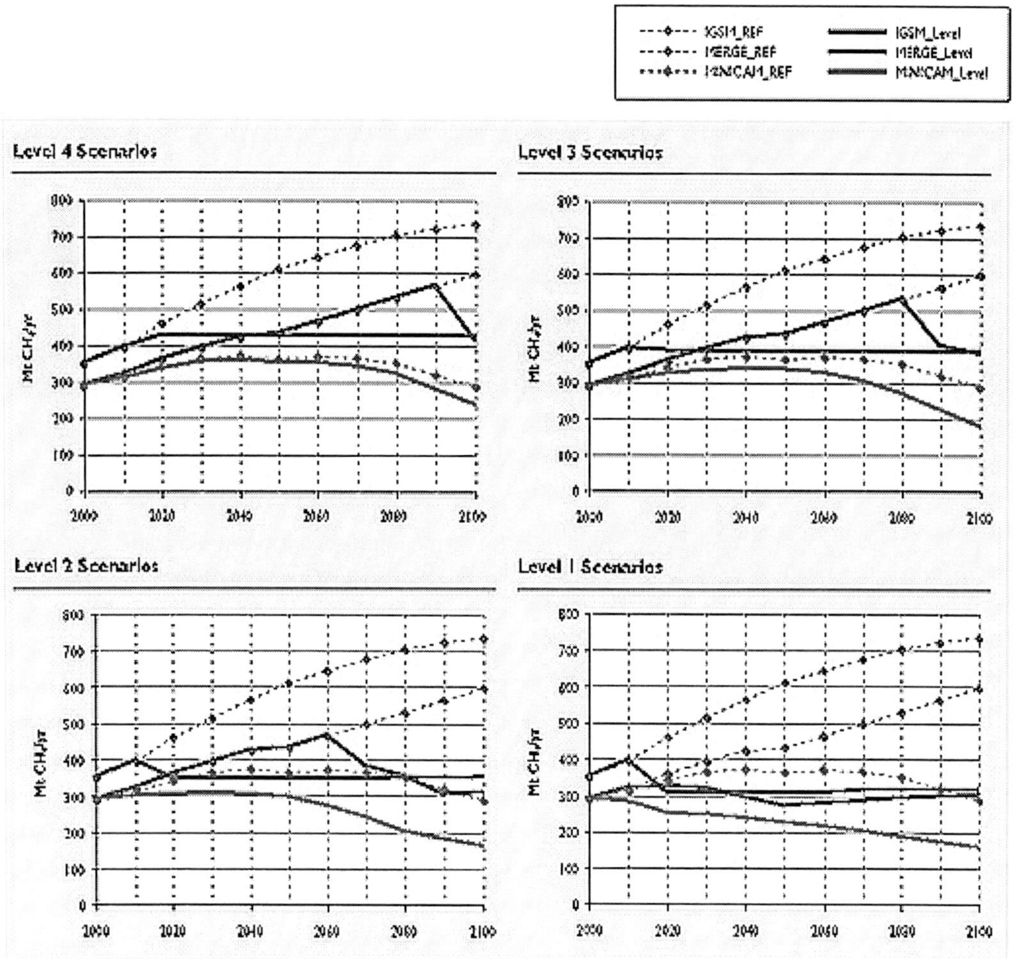

Emissions of anthropogenic CH_4 vary widely across the models, including differences in year 2000 emissions that reflect uncertainty about these emissions. With current concentrations and destruction rates relatively well known, the difference in current levels means that IGSM scenarios ascribe relatively more to anthropogenic sources and relatively less to natural sources than do the MERGE and MiniCAM scenarios. Wide differences in scenarios for the future reflect differing modeling approaches, outlooks for activity levels that lead to emission reductions, and assumptions about whether emissions will be reduced for non-climate reasons.

Figure 4.8. CH_4 Emissions Across Scenarios (Mt CH_4/yr).

N_2O is more problematic. A major anthropogenic source is from use of fertilizer for agricultural crops – an essential use. Moreover, its natural sources are important, and they are augmented by terrestrial changes associated with climate change. It is fortunate that N_2O is not a major contributor to radiative forcing because the technologies and strategies needed to achieve its stabilization are not obvious at this time. Nevertheless, differences in the control of N_2O are observed across models, as shown in Figure 4.9, although these differences are smaller than those for CH_4.

Implications for Energy Use, Industry, and Technology

In these scenarios, stabilizing radiative forcing requires a transformation of the global energy system, including reductions in the demand for energy and changes in the mix of energy technologies and fuels. This transformation is more substantial and takes place more quickly at the more stringent stabilization levels. Fossil fuel use and energy consumption are reduced in all the stabilization scenarios due to increased consumer prices for fossil fuels. CO_2 emissions from electricity production are reduced at relatively lower prices than CO_2 emissions from othersectors, such as transport, industry, and buildings. Emissions are reduced from electric power by increased use of technologiessuch as CCS, nuclear energy, and renewable energy. Other sectors respond to rising greenhouse gas prices by reducing demands for fossil fuels; substituting low- or non-emitting energy sources such as bioenergy and low-carbon electricity or hydrogen; and applying CCS where possible.

Changes in Global Energy Use

The degree and timing of change in the global energy system depends on the level at which radiative forcing is stabilized. Although differences in the reference scenarios developed by each of the three modeling groups led to different patterns of response, some important similarities emerge. The more stringent the radiative forcing stabilization level, the larger the change in the global energy system relative to the reference scenario; moreover, the scale of this change is increasing over time. Also, significant fossil fuel use continues at all four stabilization levels. This pattern can be seen in Figure 4.10, which shows the global primary energy consumption across the scenarios, and Figure 4.11, which shows the reference scenario from Chapter 3 with an additional plot of the net changes in the various sources of primary energy for each stabilization level.

Although atmospheric stabilization would take away much of the growth potential of coal over the century, its usage expands above today's levels by the end of the century in all the stabilization scenarios. In several of the Level 1 and Level 2 scenarios, the global coal industry declines in the first half of the century before recovering by 2100 to levels of production somewhat larger than today. Oil and natural gas also continue as contributors to total energy over the century although, as with coal, they are increasingly pushed from the energy mix as the stabilization level is tightened.

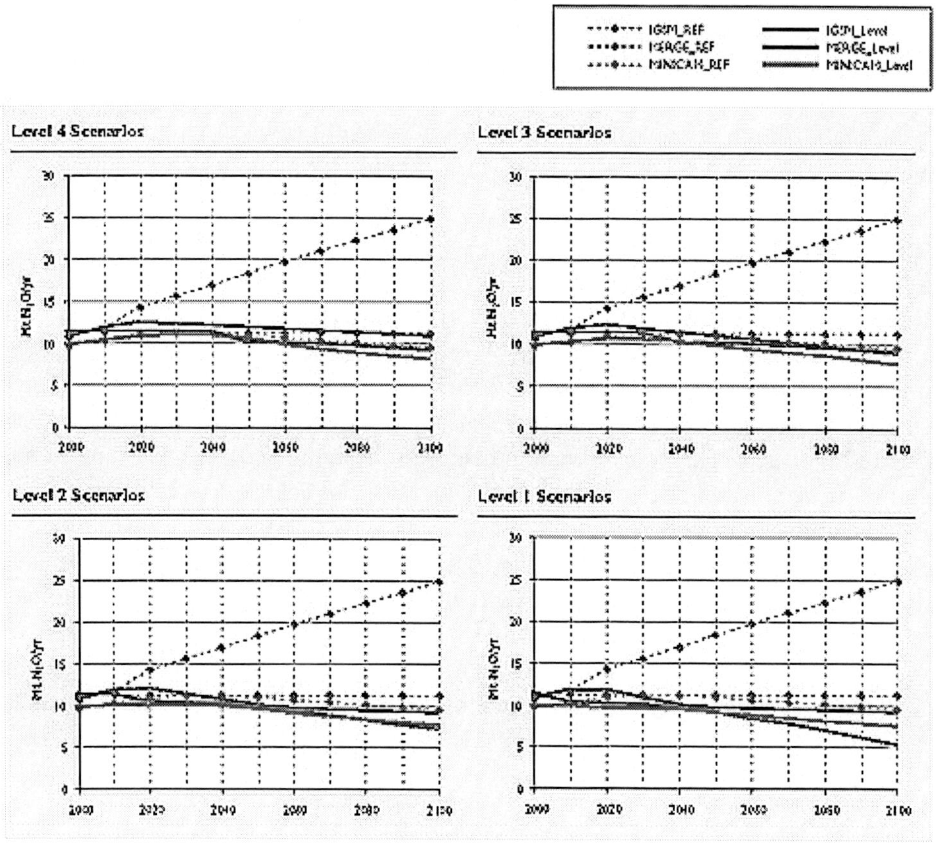

Anthropogenic emissions of N_2O are similar across models in the stabilization scenarios despite large differences in the reference scenarios.

Figure 4.9. N_2O Emissions Across Scenarios (Mt N_2O/yr).

One reason that fossil fuels continue to be utilized despite constraints on GHG emissions is that CCS technologies are available in the scenarios from all three modeling groups. Figure 4.10 shows that as the carbon price rises, CCS technology takes on an increasing market share. Section 4.4.2 addresses this pattern as well as the contribution of non-biomass renewable energy forms in greater detail.

Changes in the global energy system in response to constraints on radiative forcing reflect an interplay between technology options and the other assumptions that shaped the reference scenarios. For example, the MERGE reference scenario assumes relatively limited ability to access unconventional oil and gas resources and the evolution of a system that increasingly employs coal as a feedstock for the production of liquids, gases, and electricity. Against this background, a constraint on radiative forcing leads to reductions in coal use and end-use energy consumption. As the carbon price rises, nuclear and non-biomass renewable energy forms and CCS augment the response.

The IGSM scenarios assume greater availability of unconventional oil than the MERGE scenarios. Thus, the IGSM stabilization scenarios, in general, involve less reduction in coal use by the end of the century, but a larger decline in oil than in the MERGE stabilization scenarios. To produce liquid fuels for the transportation sector, the IGSM scenarios respond to a constraint on radiative forcing by growing biomass energy crops both earlier and more extensively than in the reference scenario. Also, reductions in energy demand are larger in the IGSM stabilization scenarios than in the scenarios from the other two models.

The MiniCAM stabilization scenarios include the smallest reductions in energy consumption among the models. The imposition of constraints on radiative forcing leads to reductions in oil, gas, and coal, as is the case with the IGSM and MERGE stabilization scenarios, but also leads to considerable expansion of nuclear power and renewable energy supplies. The largest supply response is in commercial bioderived fuels. These fuels are largely limited to bio-waste recycling in the MiniCAM reference scenario. As the price of CO_2 rises, commercial bioenergy becomes increasingly attractive. As will be discussed in Section 4.5, the expansion of the commercial biomass industry to produce hundreds of EJ/yr of energy has implications for crop prices, land use, land-use emissions, and unmanaged ecosystems.

The relative role of nuclear energy differs among the scenarios from the three modeling groups. The MERGE reference scenario deploys the largest amount of nuclear power, contributing 170 EJ/yr of primary energy in the year 2100. In the Level 1 stabilization scenario, deployment expands to 240 EJ/yr of primary energy in 2100. Nuclear power in the MiniCAM reference scenario produces 90 EJ/yr in the year 2100, which in the Level 1 stabilization scenario expands to more than 180 EJ/yr of primary energy in the year 2100. The IGSM stabilization scenarios show little change in nuclear power generation among the stabilization scenarios or compared with the reference, reflecting the assumption that nuclear levels are limited by policy decisions regarding safety, waste, and proliferation that are unaffected by climate policy.

Reductions in total primary energy consumption play an important role in all of the stabilization scenarios. In the IGSM stabilization scenarios, this is the largest single change in the global energy system. While not as dramatic as the IGSM stabilization scenarios, the MERGE and MiniCAM stabilization scenarios also exhibit reductions in energy demand. As will be discussed in Section 4.6, differences in primary energy reductions among the models reflect differences in the carbon prices required for stabilization, which are substantially higher in the IGSM stabilization scenarios than in the MERGE and MiniCAM stabilization scenarios. In all the stabilization scenarios, carbon price differences are reflected in the user prices of energy. Carbon prices, in turn, reflect technological assumptions that influence both the supply of alternative energy and the responsiveness of users to changing prices. The fuel and GHG prices discussed later in this chapter, therefore, can be instructive in understanding the character of technology assumptions employed in the models.

As noted throughout the preceding and following discussions, the economic equilibrium nature of these three models implies that technology deployments are a reflection of prices. Technologies are deployed up to the point where marginal cost is equal to price. For example, the prices of oil and carbon set the price at which bio-fuels compete. It is therefore possible to infer the marginal costs of bio-fuels when they first enter the market and how the marginal cost changes as the market expands.

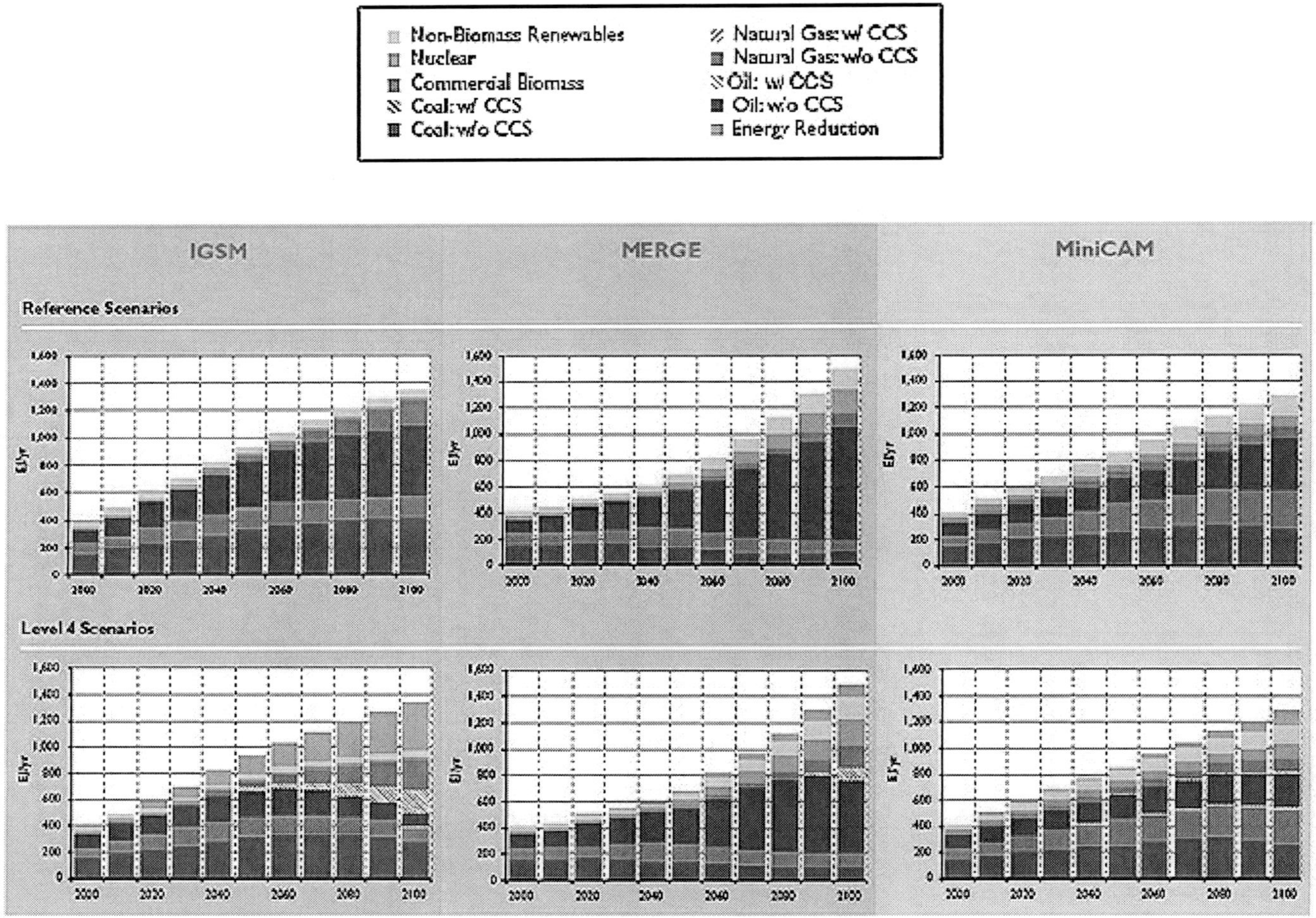

Figure 4.10 continued

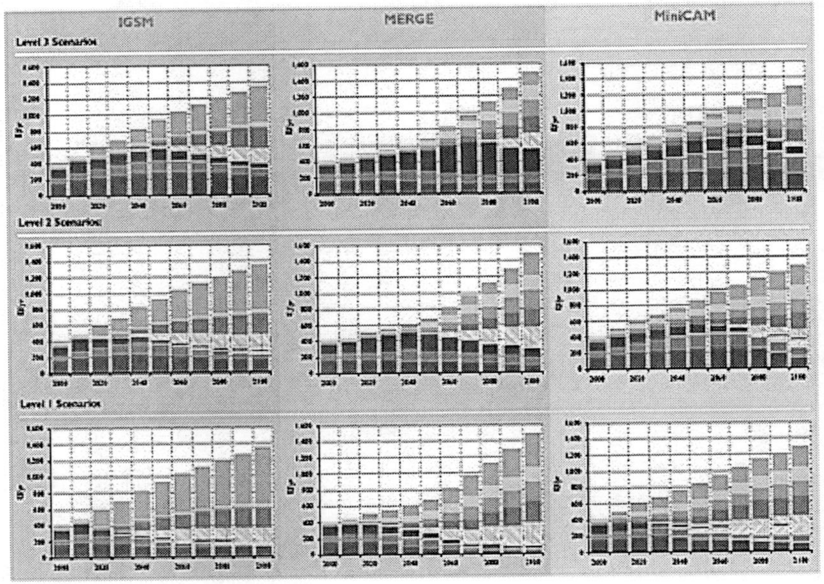

The transition to stabilization, reflected most fully in the Level 1 stabilization scenarios, means an eventual phase-out of fossil fuel use unless CCS is employed. Consumption of non-fossil energy sources increases 6-fold to 14-fold over the century in the Level 1 stabilization scenarios. In the IGSM stabilization scenarios, more of the emissions reductions are met through demand reductions than in the scenarios from the other two modeling groups, with 2100 energy use cut by up to one-half relative to the reference scenario in 2100. In the MiniCAM Level 1 scenario, in contrast, total energy is reduced by less than 20%. Levels 2, 3, and 4 require progressively less transformation compared with the reference scenarios in the coming century, delaying these changes until beyond 2100.

[Notes. i. Oil consumption includes that derived from tar sands and oil shales, and coal consumption includes that used to produce synthetic liquid and gaseous fuels. ii. Primary energy consumption from nuclear power and non-biomass renewable electricity are accounted for at the average efficiency of fossil-fired electric facilities, which vary over time and across scenarios. This long-standing convention means that, all other things being equal, increasing efficiency of fossil-electric energy lowers the contribution to primary energy from these sources.]

Figure 4.10. Global Primary Energy Consumption by Fuel Across Scenarios (EJ/yr).

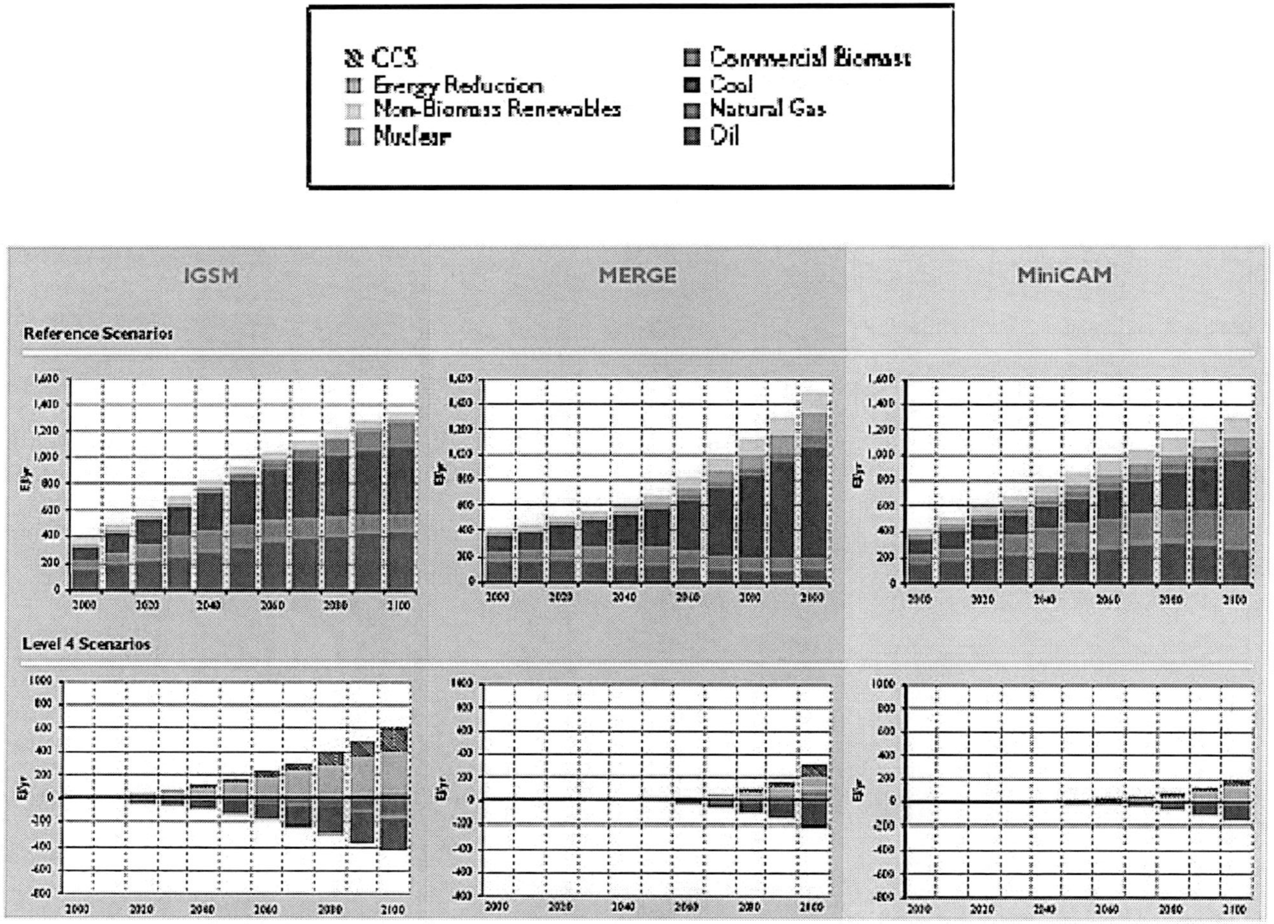

Figure 4.11 continued

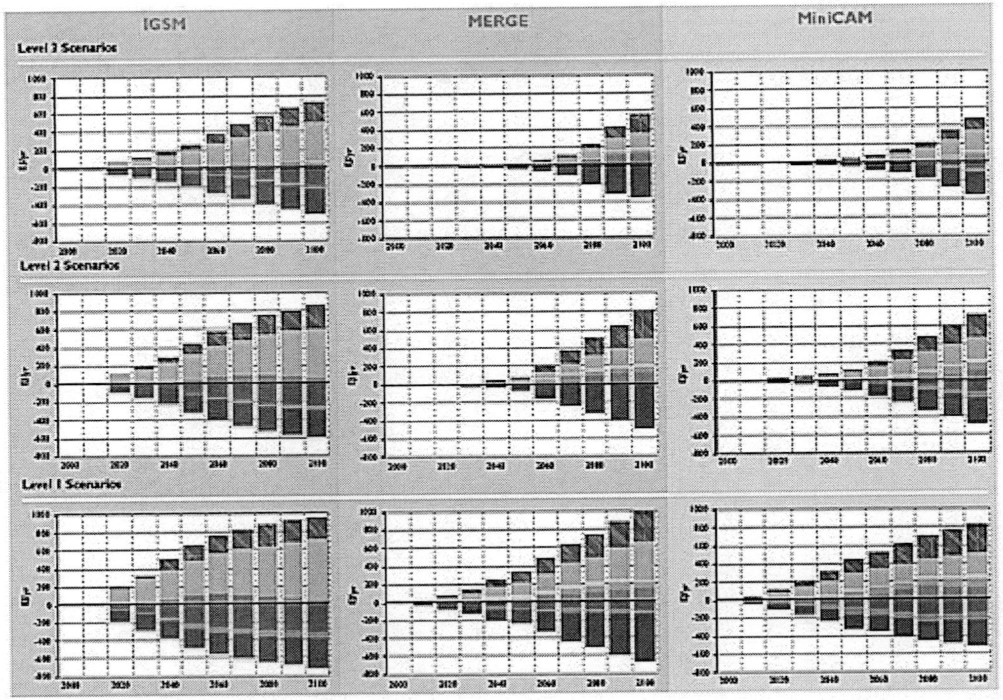

The energy system is significantly transformed from the reference scenarios in the stabilization scenarios from all three modeling groups. The transformation begins later in the Level 3 and 4 stabilization scenarios, but would need to continue into the following century. The transformation includes reductions in energy consumption, increased use of carbon-free sources of energy (biomass, other renewables, and nuclear power), and the addition of CCS. The contribution of each of these varies among the models, reflecting different assumptions about economic viability, non-climate policy, and resource limits.

[Notes. i. Oil consumption includes that derived from tar sands and oil shales, and coal consumption includes that used to produce synthetic liquid and gaseous fuels. ii. Primary energy consumption from nuclear power and non-biomass renewable electricity are accounted for at the average efficiency of fossil-fired electric facilities, which vary over time and across scenarios. This long-standing convention means that, all other things being equal, increasing efficiency of fossil-electric energy lowers the contribution to primary energy from these sources.]

Figure 4.11. Change in Global Primary Energy Consumption by Fuel Across Stabilization Scenarios, Relative to Reference Scenarios (EJ/yr).

It is worth reemphasizing that reductions in energy consumption are an important component of response at all stabilization levels. These reductions reflect a mix of three factors:

- Substitution of technologies that produce the same energy service with lower direct-plusindirect carbon emissions
- Changes in the composition of final goods and services, shifting toward consumption of goods and services with lower directplus-indirect carbon emissions
- Reductions in the consumption of energy services.

This chapter does not attempt to quantify the relative contribution of each of these responses. Each of the models has a different set of technology options, different technology performance assumptions, and different model structures. Furthermore, no well defined protocol exists that can provide a unique attribution among these three general processes.

Changes in Global Electric Power Generation

Across the scenarios, stabilization leads to substantial changes in electricity-production technologies, although the MERGE and MiniCAM stabilization scenarios exhibit relatively little change in electricity production. Indeed, across the models, the relative reductions in electricity production under stabilization are lower than relative reductions in total primary energy consumption. One reason for this is that electricity price increases are smaller relative to those for direct fuel use because the fuel input, while important, is only part of the consumer cost of electricity. Also, the long-term cost of the transition to low and non-carbon-emitting sources is relatively smaller in electricity production than in the remaining sectors taken as an average.

Table 4.3. Global Annual CO_2Capture and Storage in 2030, 2050, and 2100 for Four Stabilization Levels

		Annual Global Carbon Capture and Storage (GtC/yr)		
Stabilization Level	**Year**	**IGSM**	**MERGE**	**MiniCAM**
Level 4	2030	0.01	0.00	0.09
	2050	0.44	0.00	0.15
	2100	4.12	2.31	0.72
Level 3	2030	0.05	0.00	0.10
	2050	0.83	0.00	0.19
	2100	4.52	4.79	2.75
Level 2	2030	0.12	0.00	0.13
	2050	1.96	0.44	0.38
	2100	4.97	6.63	5.56
Level 1	2030	0.37	0.66	0.82
	2050	2.76	2.24	2.95
	2100	4.44	7.17	6.23

Table 4.4. Global Cumulative CO_2Captureand Storage in 2050 and2100 for Four Stabilization Levels

		Cumulative Global Carbon Capture and Storage (GtC)		
Stabilization Level	**Year**	**IGSM**	**MERGE**	**MiniCAM**
Level 4	2030	0.0	0.0	1.1
	2050	3.6	0.0	3.4
	2100	91.7	21.1	20.7
Level 3	2030	0.2	0.0	1.2
	2050	8.5	0.0	4.0
	2100	152.8	64.2	51.8
Level 2	2030	0.5	0.0	1.5
	2050	19.5	3.2	6.4
	2100	208.0	187.7	144.2
Level 1	2030	1.8	7.4	6.9
	2050	36.7	32.4	43.0
	2100	230.6	272.5	278.0

There are substantial differences in the scale of global electricity production across the three reference scenarios, as shown in Chapter 3 and repeated at the top of Figure 4.12. Electricity production increases from about 50 EJ/yr in the year 2000 to between 230 EJ/yr (IGSM) to 310 EJ/yr (MiniCAM) by 2100. In all three reference scenarios, electricity becomes an increasingly important component of the global energy system, fueled by growing quantities of fossil fuels. Despite differences in the relative contribution of different fuel sources across the three reference scenarios, total production of electricity from fossil fuel rises from about 30 EJ/yr in 2000 to between 150 EJ/yr and 190 EJ/yr in 2100. Thus, the difference in total reference scenario electricity production among the models largely reflects differences in the deployment of non-fossil energy forms: bio-fuels, nuclear power, fuel cells, and other renewables such as wind, geothermal, and solar power.

The imposition of radiative forcing limits dramatically changes the electricity sector. Common characteristics of the stabilization scenarios across models are that CCS (with coal, gas, and, where present, oil-generated power) is deployed at a large scale by the end of the century and that use of coal without CCS declines and eventually is not viable. The IGSM scenarios, as has been noted, assume restrictions on the expansion of nuclear power, and other renewables are either resource limited (hydro power and electricity from bio-fuels) or become more costly to integrate into the grid as their share of electricity production rises because they are intermittent (wind and/or solar). Partly as a result, natural gas use in electricity production increases in the IGSM stabilization scenarios, especially in the nearer term before CCS becomes economically viable. In the MERGE stabilization scenarios, carbon-free technologies, including non-biomass renewables and nuclear, are viable and, thus, are favored over natural gas, the use of which falls relative to the reference scenario. In the MiniCAM stabilization scenarios, nuclear and non-biomass renewable energy technologies capture a larger share of the market. At the less stringent levels of stabilization, Level 3 and Level 4, additional biofuels are deployed in electricity production, and total electricity production declines. Under the most stringent stabilization level, commercial bio-fuels used in electricity production in the MiniCAM stabilization scenarios are diverted to the transportation sector, and use in electricity production actually declines relative to the reference toward the end of the century. In all of the IGSM scenarios, bio-fuels are used preferentially for transportation rather than for electricity generation. The difference between MiniCAM and IGSM scenarios in this regard is in part a reflection of the higher fuel prices in the IGSM scenarios discussed in Section 4.6.3.

All modeling groups assumed that CO_2 could be captured and stored in secure repositories, and as noted, in all scenarios CCS becomes a large-scale activity. Annual capture quantities are shown in Table 4.3. CCS is always one of the largest single changes in the electricity production system in response to stabilization, as can be seen in Figure 4.13. As with mitigation in general, CCS starts relatively modestly in all the scenarios, but grows to large levels. The total storage over the century is recorded in Table 4.4, spanning a range from 20 GtC to 90 GtC for Level 4 and 230 GtC to 280 GtC for Level 1. The modeling groups did not report either location of storage sites for CO_2 or the nature of the storage reservoirs, but these scenarios are within the range of the estimates of global geologic reservoir capacity (Edmonds et al. 2001, Dooley et al. 2004).

Deployment rates for CCS depend on a variety of circumstances, including capture cost, new plant construction versus retrofitting for existing plants, the scale of power generation, the price of fuel inputs, the cost of competing technologies, and the level of the CO_2 price. It is clear that the constraints on radiative forcing considered in these scenarios are sufficiently stringent that, if CCS is available at a cost and performance similar to that considered in these scenarios, and that it successfully navigates other potential obstacles to widespread deployment, it could be a crucial component of future power generation.

Yet CCS is hardly ordinary today. Geologic storage is largely confined to experimental sites or enhanced oil and gas recovery. There are as yet no clearly defined institutions or accounting systems to reward such technology in emissions control agreements, and long-term liability for stored CO_2 has not been determined. All of these issues and more must be resolved before CCS could deploy on the scale envisioned in these stabilization scenarios. If CCS were unavailable, the effect would be to increase the cost of achieving stabilization in all of the scenarios. These scenarios tend to favor CCS, but that tendency could easily change with different assumptions about technologies such as nuclear power that are well within the range of uncertainty about future costs and the policy environment. Nuclear power carries with it issues of safety, waste, and proliferation. Thus, the viability of both CCS and nuclear power depends on regulatory and public acceptance issues. For example, global nuclear power in the reference scenarios ranges from about 1½times current levels (if non-climate concerns such as safety, waste, and proliferation constrain its growth as is the case in one reference scenario), to an expansion of almost an order of magnitude assuming relative economics as the only constraint.

Absent CCS and nuclear power, these models would need to deploy other emissions reduction options that could potentially be more costly, or would need to assume large breakthroughs in cost, performance, and reliability. This study has not attempted to quantify the increase in costs or the reorganization of the energy system that would be required to achieve stabilization without CCS. This sensitivity is an important item in the agenda of future research.

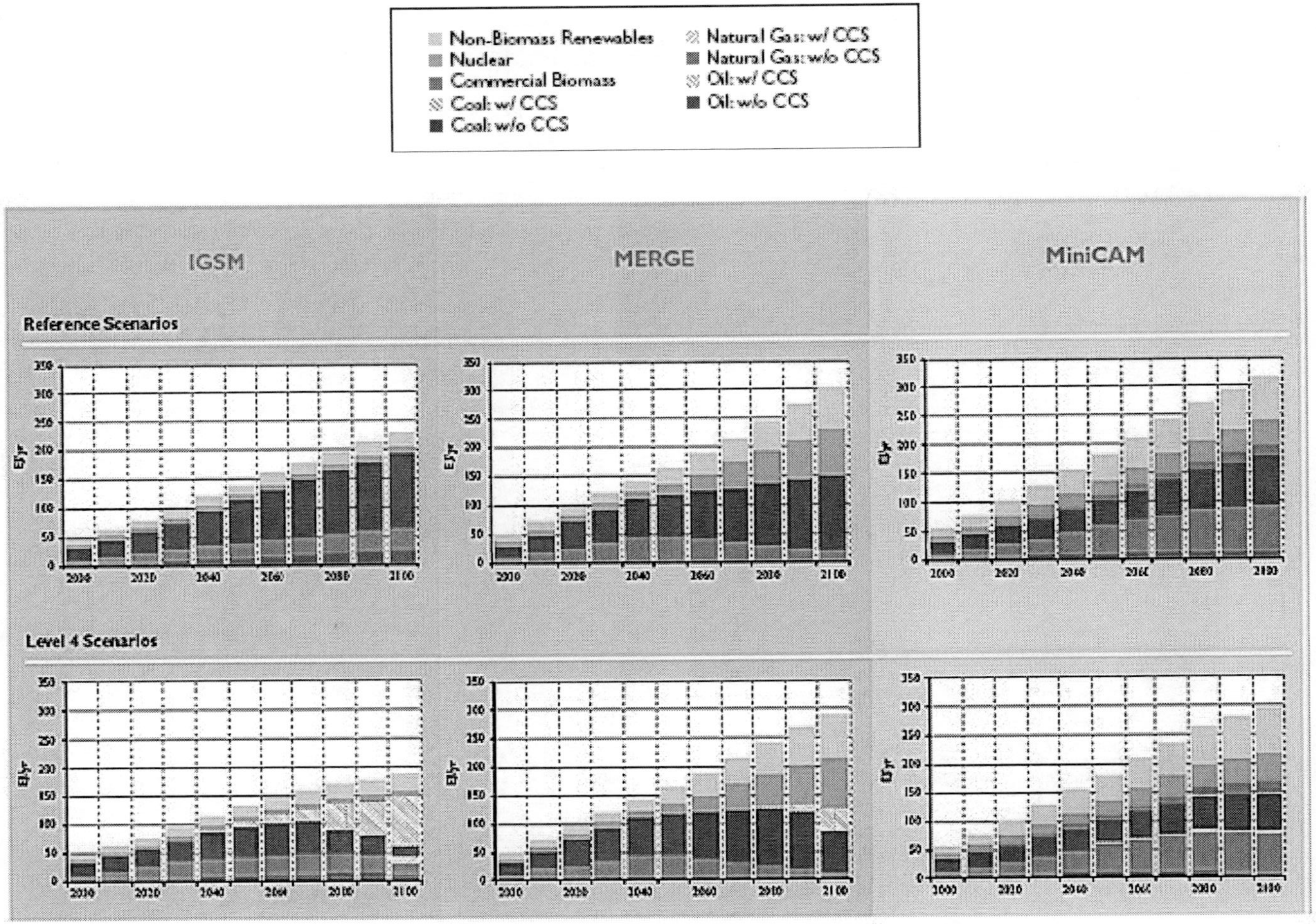

Figure 4.12 continued

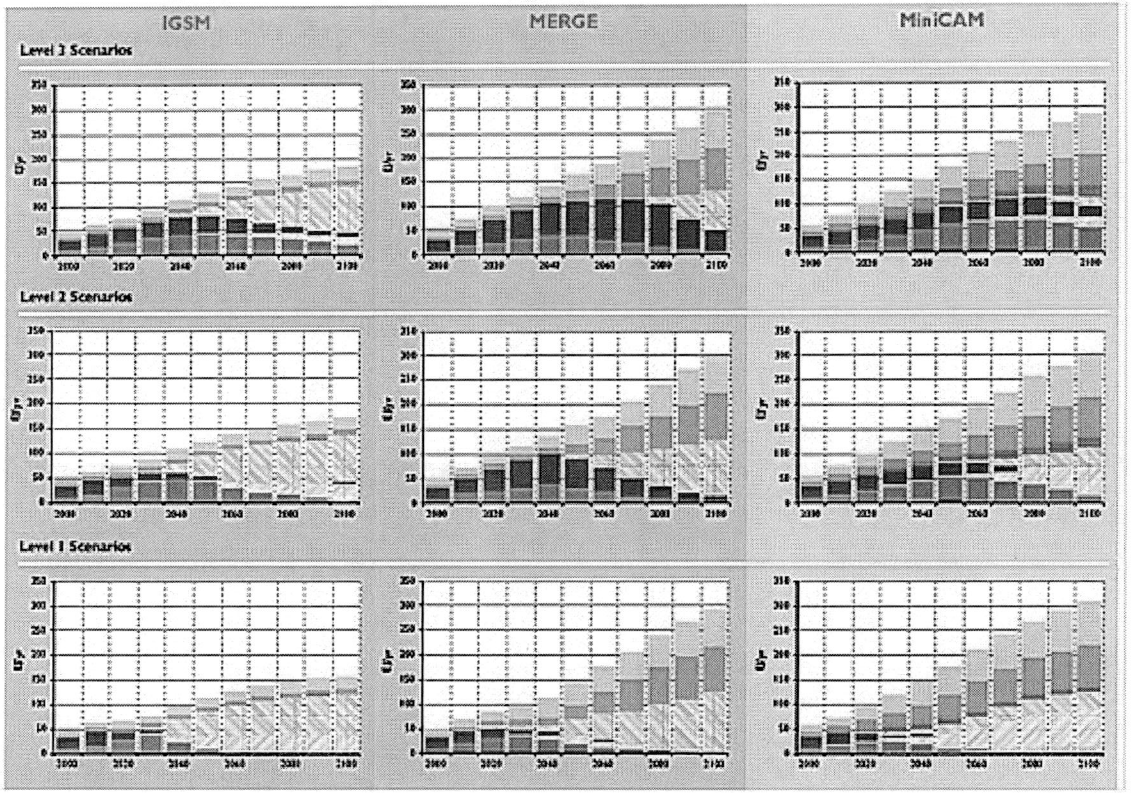

Global electricity production would need to be transformed to meet the four stabilization levels. CCS is important in the scenarios from all three modeling groups; thus, while coal use is reduced in all the stabilization scenarios relative to the reference scenarios, it remains an important fuel for electricity production. Use of CCS is the main supply response in the IGSM stabilization scenarios, in part because nuclear power is limited by assumption to reflect non-climate policy concerns. Nuclear power and renewable electricity sources play a larger role in the MERGE and MiniCAM stabilization scenarios.

Figure 4.12. Global Electricity Production by Fuel Across Scenarios (EJ/yr).

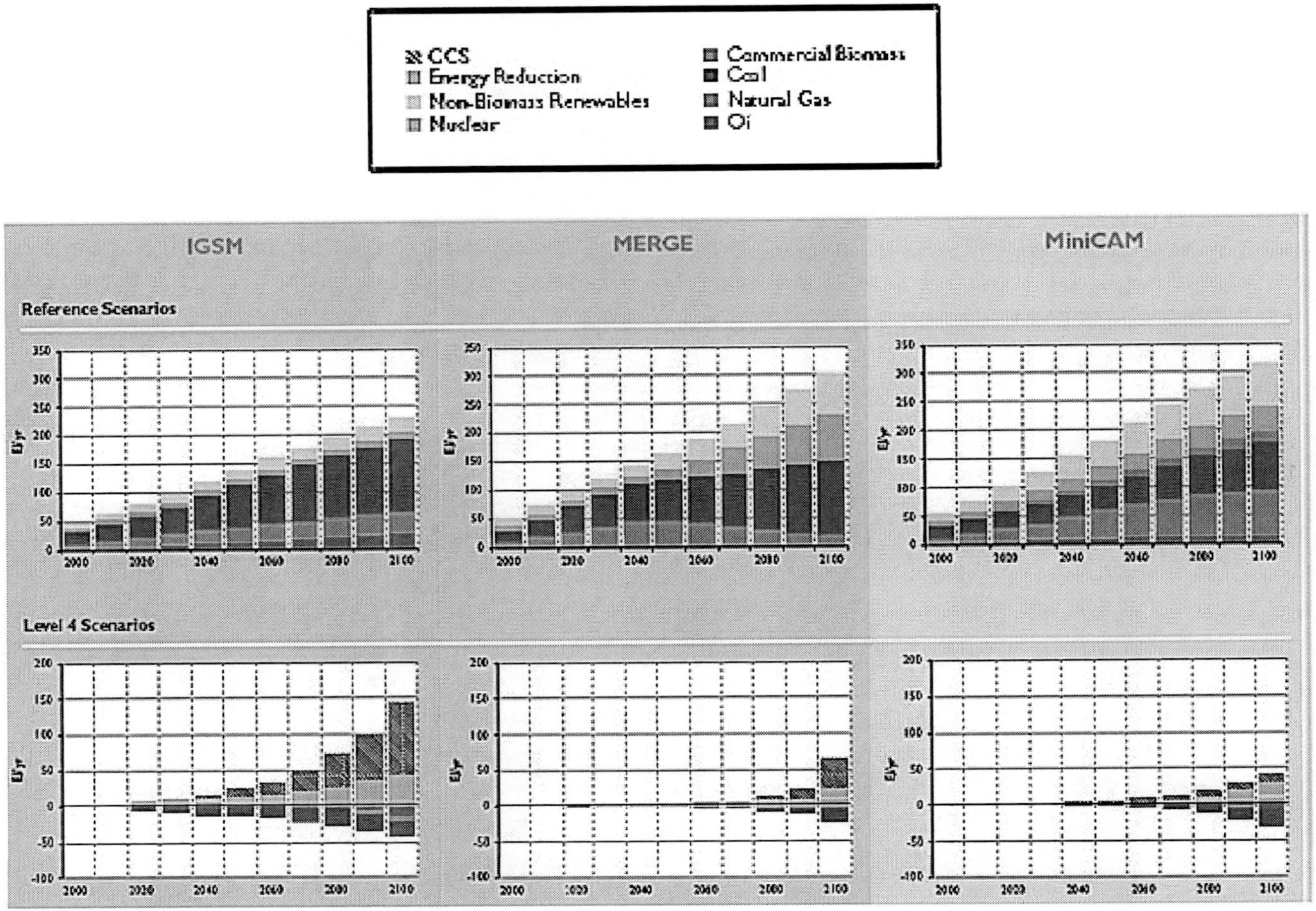

Figure 4.13 continued

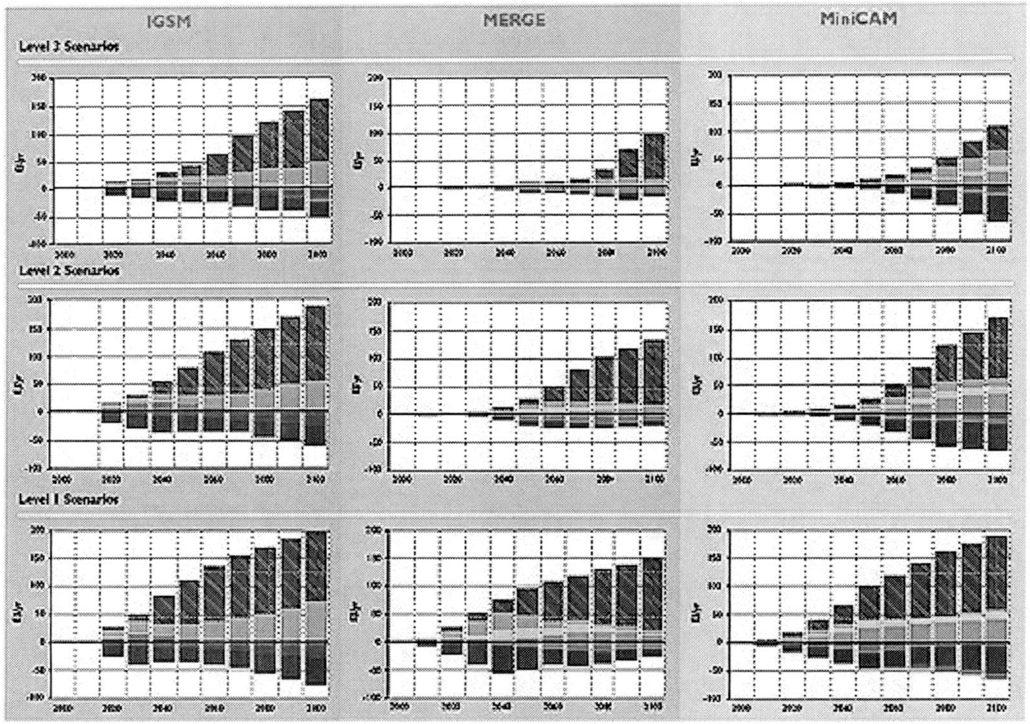

There are multiple electricity technology options that could be competitive in the future, and different assumptions about their relative economic viability, reliability, and resource availability lead to different scenarios for the global electricity sector in reference and stabilization scenarios across the models. In the IGSM reference scenario, there is relatively little change in the fuel mix in the electricity sector, with continued reliance on coal. In the MERGE and MiniCAM reference scenarios, there are large transformations from the present. In the stabilization scenarios from all three modeling groups, large changes relative to the reference scenario are required to meet the stabilization levels. Under less stringent stabilization levels, many of these changes would be pushed into the next century. In most cases, the relative proportion of electricity in energy consumption increases in the stabilization scenarios, so the relative reductions in electricity production are generally smaller than for primary energy.

Figure 4.13. Changes in Global Electricity Production by Fuel Across Stabilization Scenarios, Relative to Reference Scenarios (EJ/yr).

Changes in Energy Patterns in the United States

Changes in U.S. energy patterns are similar to those observed for the world in general. This reflects the facts that the mitigation policy is implemented globally, there are international markets in fuels, each model makes most technologies globally available over time, and the U.S. primary energy consumption in 2000 represented roughly a quarter of the world total.

Changes in the U.S. energy system are modest for stabilization Level 4, but even with this loose constraint, significant changes begin upon implementation of the stabilization policy (the first period shown is 2020) in the IGSM Level 4 scenario (Figure 4.14 and Figure 4.15). Near- term changes are more modest in the MERGE and MiniCAM Level 4 scenarios. At more stringent stabilization levels, the changes are more substantial in the scenarios from all three modeling groups. In the Level 1 scenarios, the reduction is in U.S. primary energy consumption ranges from 8 EJ/yr to over 25 EJ/yr in 2020.

Near-term changes in the U.S. energy system vary more among models than the long-term adjustments. While oil consumption declines at higher carbon prices for all the models and all stabilization levels, near-term changes in oil consumption do not follow a consistent pattern. However, there is no ambiguity regarding the effect on coal consumption, which declines relative to the reference scenario in all stabilization scenarios for all models in all time periods. Similarly, total primary energy consumption declines in all the stabilization scenarios. Nuclear power, commercial biomass, and other renewable energy forms are advantaged with at least one of them always deployed to a greater extent in stabilization scenarios than in the reference scenario. The particular form and timing of expanded development varies across models.

The stabilization scenarios from the three modeling groups exhibit different energy sector responses reflecting differences in underlying reference scenarios and technology assumptions. The largest change in the U.S. energy system in the IGSM stabilization scenarios is the reduction in total primary energy consumption augmented by an expansion in the use of commercial biomass fuels and deployment of CCS. Similarly, the largest change in the MERGE scenarios is the reduction in total primary energy consumption augmented by deployment of CCS and bioenergy. The MiniCAM stabilization scenarios also exhibit reductions in primary energy consumption and increases in nuclear power, along with smaller additions of commercial biomass and other renewable energy forms. The adjustment of the U.S. electric sector to the various stabilization levels is similar to that for the world electricity sector. (Figure 4.16 and Figure 4.17).

Implications for Agriculture, Land-Use, and Terrestrial Carbon

In the stabilization scenarios, increased use is made of biomass energy crops, the contribution of which is ultimately limited by competition with agriculture and forestry. Two of the modeling groups employed explicit agriculture-land-use models to represent this competition and represent land constraints on the use of bio-energy. In the scenarios from one modeling group, increased use of bio-energy at more stringent stabilization levels leads to substantial land use change emissions as previously unmanaged lands are shifted to biomass production.

The three modeling groups employed different approaches to the treatment of the terrestrial carbon cycle, ranging from a simple neutral biosphere model to a state-of-the art terrestrial carbon-cycle model. In two of the models, a CO_2 fertilization effect plays a significant role. As stabilization levels become more stringent, CO_2concentrations decline and terrestrial carbon uptake declines, with implications for emissions mitigation in the energy sector. Despite the differences acrossthe modeling groups' treatments of the terrestrial carbon cycle, the aggregate behavior of the carbon cycles across models is similar.

In the stabilization scenarios, the cost of using fossil fuels and emitting CO_2 rises, providing an increasing motivation for the production and transformation of bioenergy, as shown in Figure 4.18. In all of the stabilization scenarios, production begins earlier and produces a larger share of global energy as the stabilization level becomes more stringent. Under less stringent stabilization levels, production of bio-crops is lower in the second half of the century in the MERGE and MiniCAM scenarios than in the IGSM scenarios. Differences between the models with respect to biomass deployment are not simply due to different treatments of agriculture and land use but also result from the full suite of competing technologies and behavior assumptions.

Although total land areas allocated to bioenergy crops are not reported in these scenarios, the extent of land areas engaged in the production of energy becomes substantial. This is possible only if appropriate land is available, which hinges on future productivity increases for other crops and the potential of bioenergy crops to be grown on lands that are less suited for food, pasture, and forests. In both the MiniCAM and IGSM scenarios – MiniCAM and IGSM are the two models with agriculture and land-use sub- models – demands on land for bio-fuels cause land prices to increase substantially as compared with the reference scenarios because of competition with other agricultural demands.

Stabilization scenarios limit the rise in CO_2 concentrations and reduce the CO_2 fertilization effect below that in the reference scenarios, which in turn leads to smaller CO_2 uptake by the terrestrial biosphere in the IGSM and MiniCAM stabilization scenarios (Figure 4.19). The effect is larger and begins earlier the more stringent the stabilization level. For example, in the IGSM Level 4 scenario, the effect becomes substantial after 2070 and amounts to about 0.8 GtC/yr in 2100.

The IGSM Level 1 scenario begins to depart markedly from the reference before 2050, and the departure from reference grows to approximately 2.0 GtC/yr by 2100. The effect of the diminished CO_2 fertilization effect is to require emissions mitigation in the energy-economy system to be larger by the amount of the difference between the reference aggregate net terrestrial CO_2 uptake and the uptake in the stabilization scenario. The MiniCAM stabilization scenarios exhibit similar carbon cycle behavior. The MERGE stabilization scenarios maintain the assumption of a neutral terrestrial biosphere as in the MERGE reference scenario.

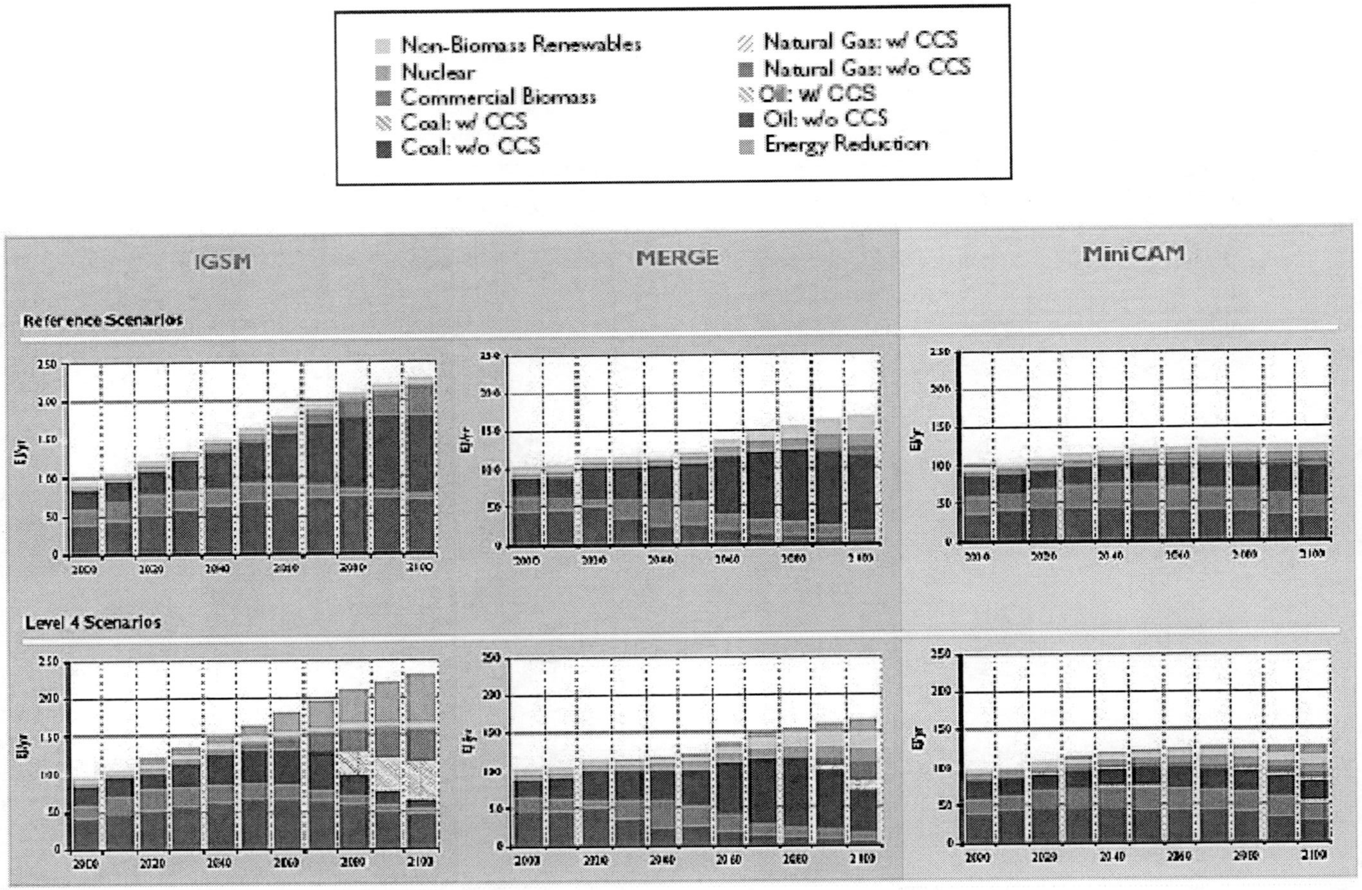

Figure 4.14 continued

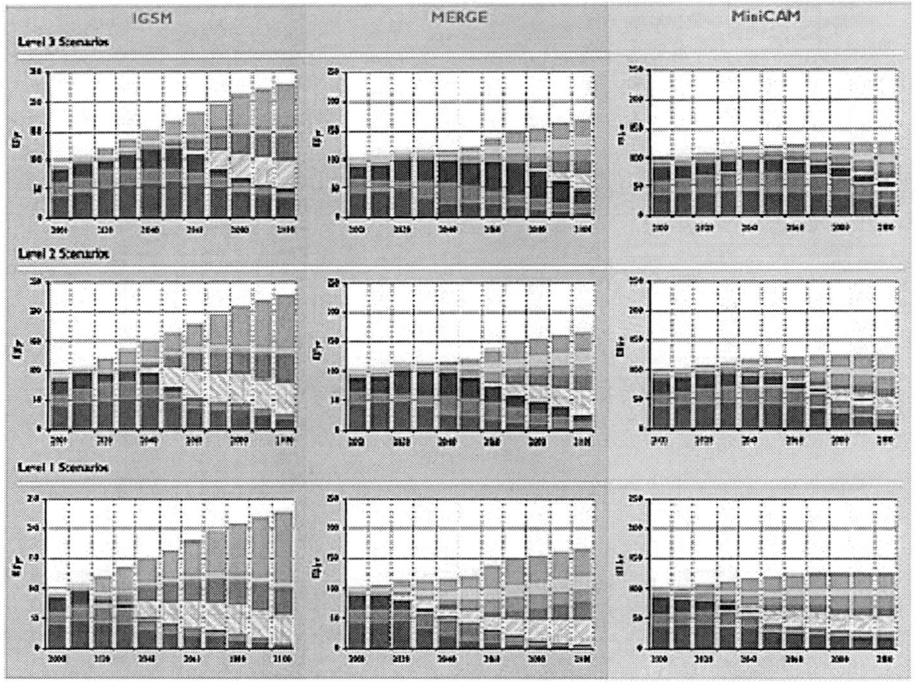

U.S. primary energy consumption under the four stabilization levels differs considerably among the three models. All the scenarios exhibit a diverse energy mix throughout the century, although the IGSM scenarios include relatively less nuclear power and non-biomass renewables than the other models. The relative contributions of different technologies over the course of the century depend on the specific cost and performance characteristics of the competing technologies represented in the scenarios.

[Notes. i. Oil consumption includes that derived from tar sands and oil shales, and coal consumption includes that used to produce synthetic liquid and gaseous fuels. ii. Primary energy consumption from nuclear power and non-biomass renewable electricity are accounted for at the average efficiency of fossil-fired electric facilities, which vary over time and across scenarios. This long-standing convention means that, all other things being equal, increasing efficiency of fossil-electric energy lowers the contribution to primary energy from these sources.]

Figure 4.14. U.S. Primary Energy Consumption by Fuel Across Scenarios (EJ/yr).

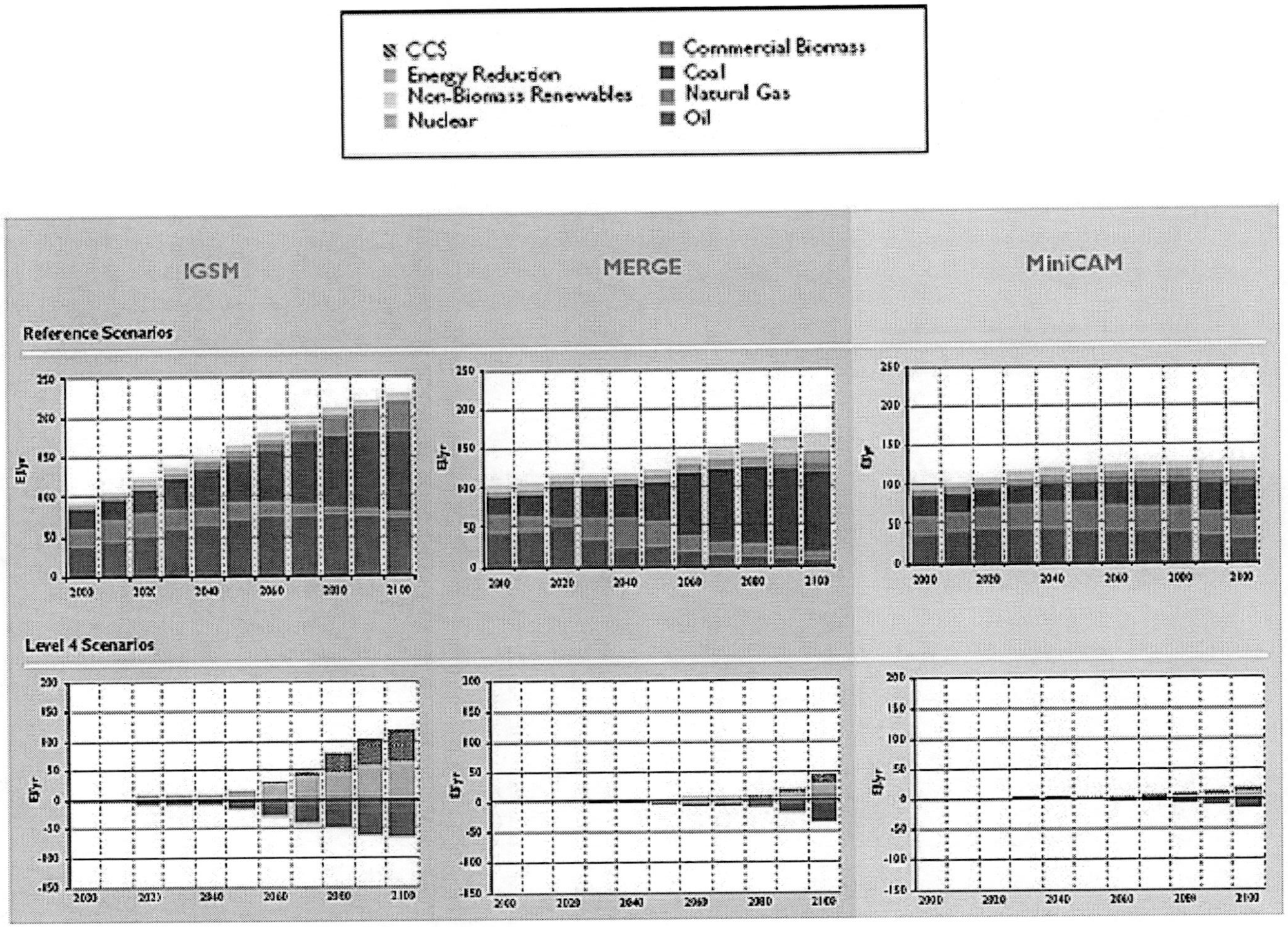

Figure 4.15 continued

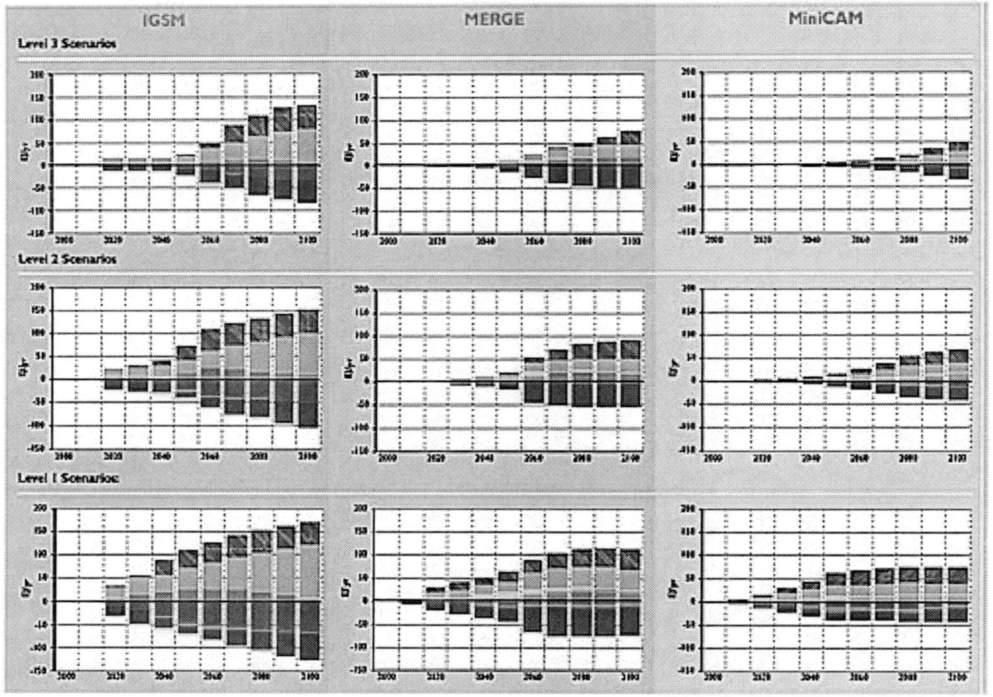

The transformations in the U.S. energy system in the stabilization scenarios are similar to those for the global energy system. Although it is not illustrated in this figure, one difference is the transformation from conventional oil and gas to synthetic fuel production derived from shale oil or coal. The IGSM reference scenario includes heavy use of shale oil with some coal gasification, whereas the MERGE reference scenario is based more heavily on synthetic liquid and gaseous fuels derived from coal. The MiniCAM reference scenario includes moderate levels of both.

[Notes. i. Oil consumption includes that derived from tar sands and oil shales, and coal consumption includes that used to produce synthetic liquid and gaseous fuels. ii. Primary energy consumption from nuclear power and non-biomass renewable electricity are accounted for at the average efficiency of fossil-fired electric facilities, which vary over time and across scenarios. This long-standing convention means that, all other things being equal, increasing efficiency of fossil-electric energy lowers the contribution to primary energy from these sources.]

Figure 4.15. Changes in U.S. Primary Energy Consumption by Fuel Across Stabilization Scenarios, Relative to Reference Scenarios (EJ/yr).

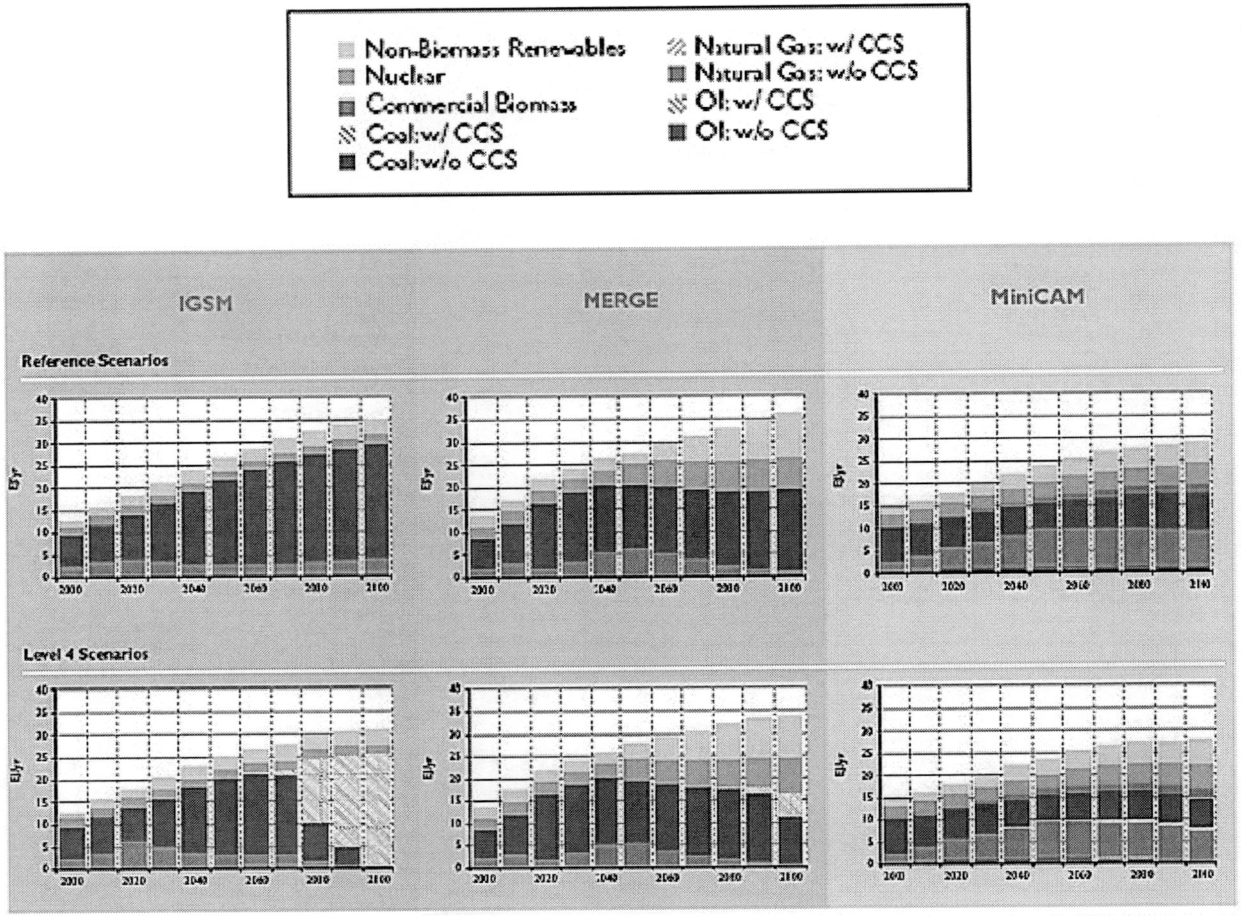

Figure 4.16 continued

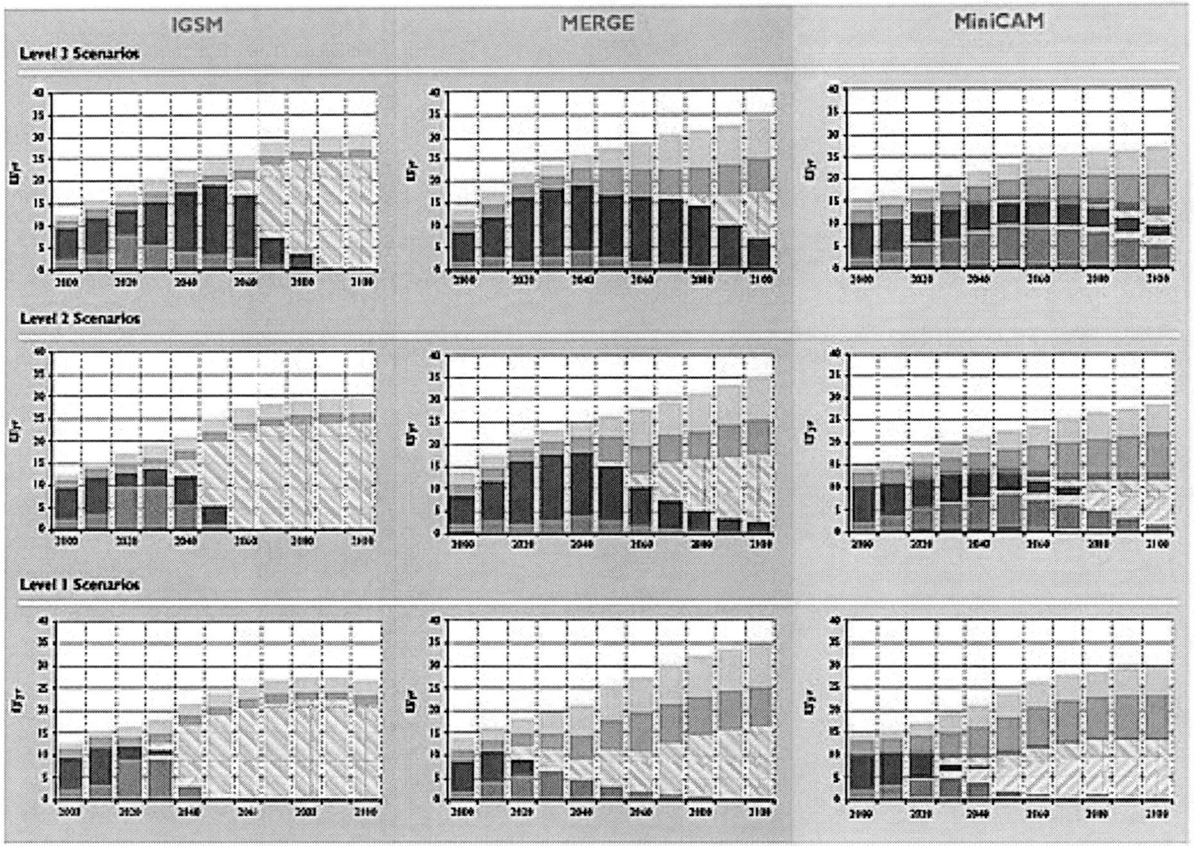

In these scenarios, U.S. electricity production sources and technologies are substantially transformed to meet the four stabilization levels. CCS figures in all the stabilization scenarios, but the contribution of other sources and technologies as well as the total production electricity differs substantially among the three models.

Figure 4.16. U.S. Electricity Production by Fuel Across Scenarios (EJ/yr).

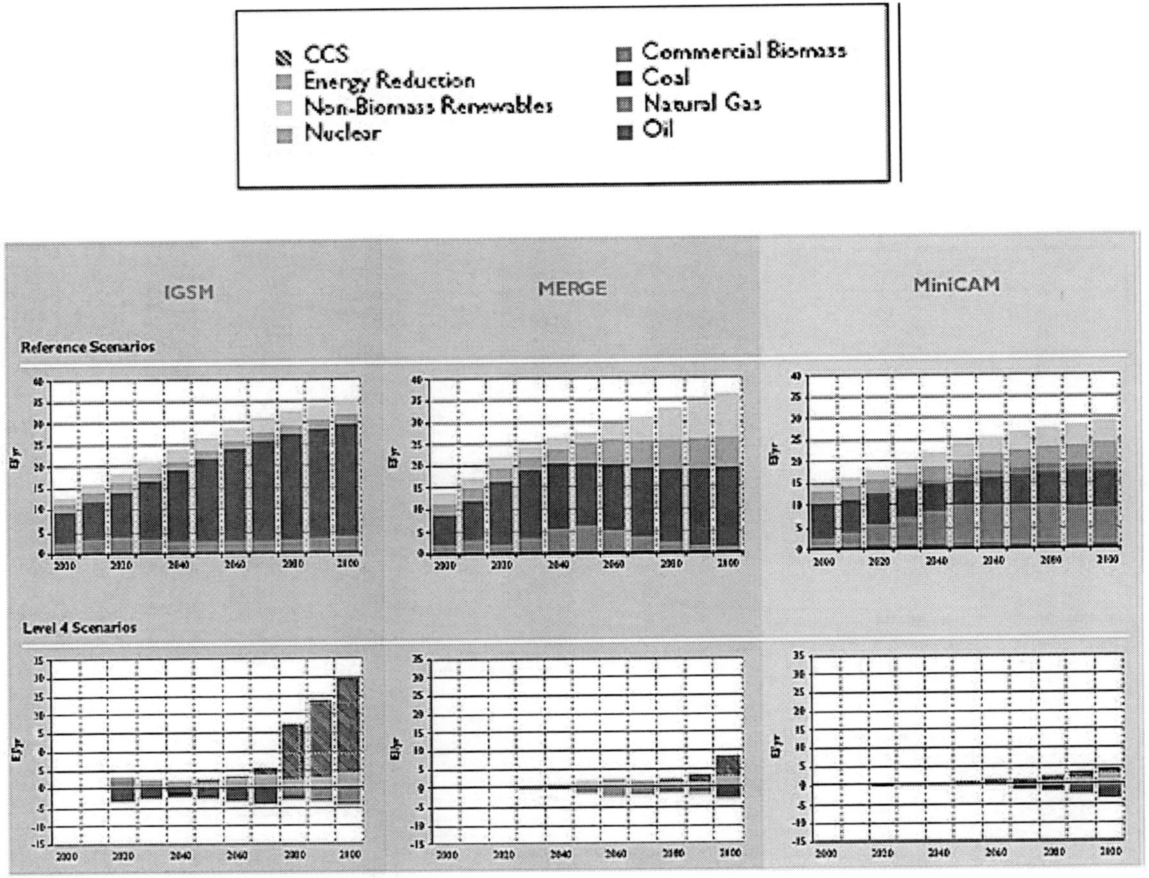

Figure 4.17 continued

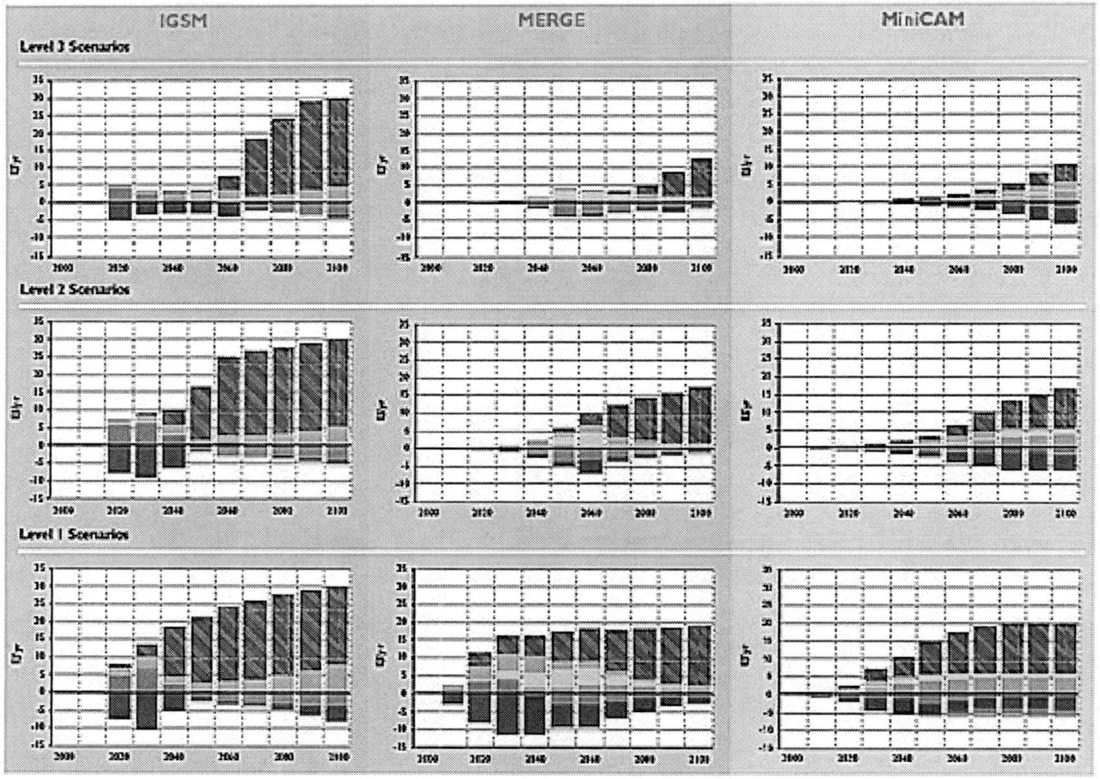

Transformation of the U.S. electricity sector in these scenarios implies increasing use of low- or zero-carbon technologies, such as renewable electricity sources, nuclear power, and fossil generation with CCS, and decreasing use of fossil fuel technologies that freely emit CO_2 to the atmosphere. Natural gas use increases in the early part of the century in several stabilization scenarios as a lower carbon substitute for coal-fired electricity. In most cases, the relative proportion of electricity in energy consumption increases in the stabilization scenarios, so the relative reductions in electricity production are generally smaller than for primary energy.

Figure 4.17. Change in U.S. Electricity Production by Fuel Across Stabilization Scenarios, Relative to Reference Scenarios (EJ/yr).

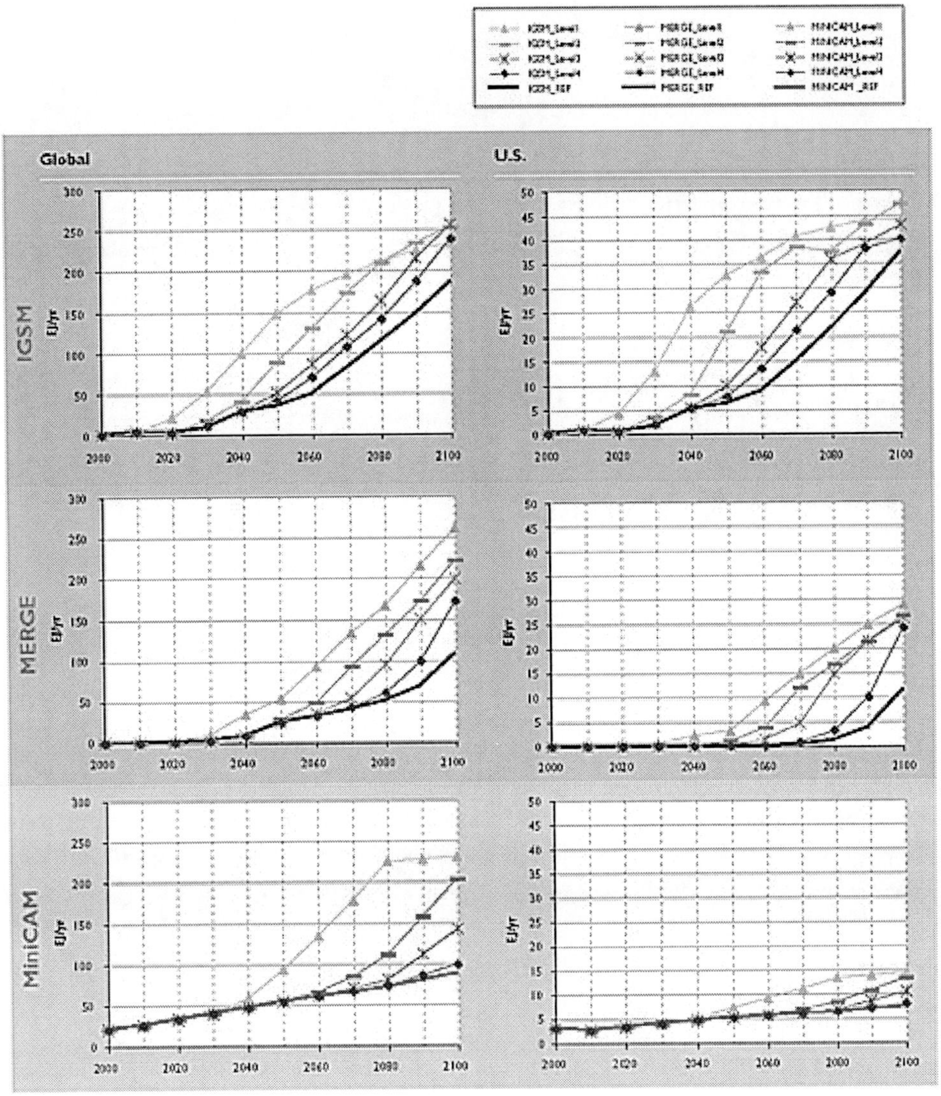

Scenarios of the potential for commercial biomass production for the world and the U.S. are similar in magnitude and behavior among the models. Commercial biomass production increases over time in the reference scenarios due in large part to technological improvements in bioenergy crop production and increasing demand for liquid fuels. Stabilization increases the demand for bioenergy crops, causing production to increase more rapidly and to reach higher levels than in the reference scenarios. Dramatic growth in bioenergy crop production raises important issues about the attendant increases in the land that is devoted to these crops, including competition with other agricultural crops, encroachment into unmanaged lands, and water and other resource and environmental impacts.

Figure 4.18. Global and U.S. Commercial Biomass Production Across Scenarios.

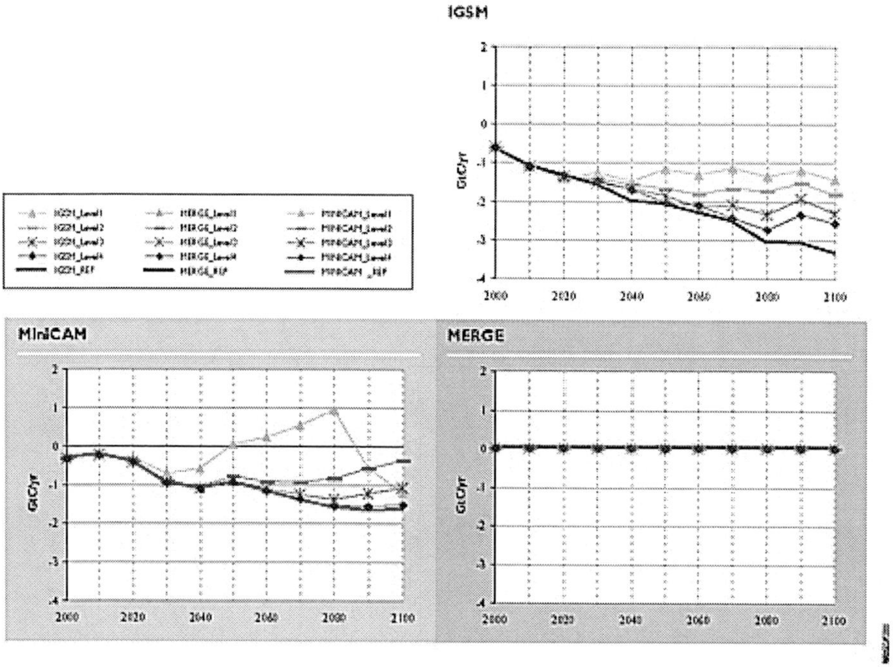

Net terrestrial carbon emissions to the atmosphere, under reference and stabilization levels, reflect differences in the model structures for processes that remain highly uncertain. The MERGE scenarios are based on the assumption of a neutral biosphere. The IGSM and MiniCAM scenarios generally represent the land as a growing carbon sink, with the exception of the Level 1 MiniCAM stabilization scenario, in which increased demand for land for biomass production leads to conversion and carbon loss. This effect is particularly strong prior to 2080 in the Level 1 MiniCAM stabilization scenario.

Figure 4.19. Net Terrestrial Carbon Emissions Across Scenarios (GtC/yr).

The MiniCAM scenarios also include a second effect that results from the interaction between the energy system and emissions from changes in land use, such as converting previously unmanaged lands to bioenergy crop production. As in the IGSM scenarios, economic competition among alternative human activities, crops, pasture, managed forests, bioenergy crops, and unmanaged ecosystems determine land use. In the MiniCAM scenarios, this competition also determines land-use change emissions. One implication is increasing pressure to deforest under stabilization in order to clear space for biomass crops (Sands and Leimbach 2003). This effect is best exhibited in the Level 1 scenarios, in which the terrestrial biosphere becomes a net source of carbon rather than a sink from 2050 to past 2080. The effect subsides after 2080 because commercial biomass production ceases to expand beyond 2080, reducing any further pressure to deforest for biomass crops. Thus, terrestrial uptake in the MiniCAM scenarios is reduced because of the lower CO_2 fertilization effects as in the IGSM scenarios, and it is also reduced by any land use change emissions that derive from the increasing demand for bioenergy crops.

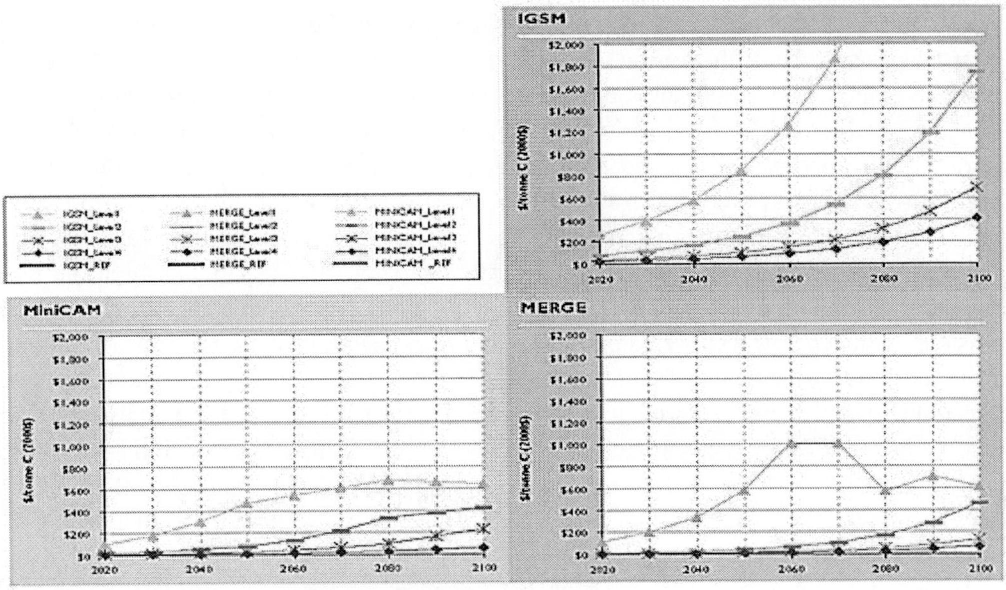

In all the stabilization scenarios, the carbon price rises, by design, over time until stabilization is achieved (or the end-year 2100 is reached), and the prices are higher the more stringent is the stabilization level. There are substantial differences in carbon prices between MERGE and MiniCAM stabilization scenarios, on the one hand, and the IGSM stabilization scenarios on the other. Differences between the models reflect differences in the emissions reductions necessary for stabilization and differences in the technologies that might facilitate carbon emissions reductions, particularly in the second half of the century

Figure 4.20. Carbon Prices Across Stabilization Scenarios ($/tonne C, 2000$).

Table 4.5. Carbon Prices in 2020, 2030,2050, and 2100 for Each Stabilization Scenario and Model

	2020 ($/tonne C)			2030 ($/tonne C)		
Stabilization Level	**IGSM**	**MERGE**	**MiniCAM**	**IGSM**	**MERGE**	**MiniCAM**
Level 4	$18	$1	$1	$26	$2	$2
Level 3	$30	$2	$4	$44	$4	$7
Level 2	$75	$8	$15	$112	$13	$26
Level 1	$259	$110	$93	$384	$191	$170
Level 4	$58	$6	$5	$415	$67	$54
Level 3	$97	$11	$19	$686	$127	$221
Level 2	$245	$36	$69	$1,743	$466	$420
Level 1	$842	$574	$466	$6,053	$609	$635

The terrestrial emissions reported in Figure 4.19 for the MiniCAM scenarios assume a policy architecture that places a value on energy and industrial emissions as well as carbon in terrestrial systems. Thus, there is an economic incentive to maintain and/or expand stocks of terrestrial carbon as well as an incentive to bring more land under cultivation to grow bioenergy crops. Pricing terrestrial carbon exerts an important counter-pressure to deforestation and other land-use changes that generate increased emissions. To illustrate this effect, sensitivity cases were run by the MiniCAM modeling group in which no price was applied to terrestrial carbon emissions. These sensitivity analyses showed increased levels of land-use change emissions when terrestrial carbon was not valued, particularly at the more stringent stabilization levels, and the potential for a vicious cycle to emerge. Efforts to reduce emissions in the energy sector create an incentive to expand bioenergy production without a counter incentive to maintain carbon in terrestrial stocks. The resultant deforestation increases terrestrial CO_2 emissions, requiring even greater reductions in fossil fuel CO_2 emissions, even higher prices on fossil fuel carbon, and further increases in the demand for bioenergy, leading, in turn, to additional deforestation. The net terrestrial emissions for the MiniCAM scenarios reported here avoid this vicious cycle because they include a policy architecture that places a value on terrestrial carbon.

Despite the significant differences in the treatment of terrestrial systems in the three models, it is interesting to recall from Figure 3.20 that the overall behavior of the three carbon-cycle models is similar.

Economic Implications of Stabilization

The economic implications of stabilization include increases in the prices of fossil fuels and electricity, along with reductions in economic output. Substantial differences in GHG emissions prices and associated economic costs arise among the modeling groups for each stabilization level. Among the most important factors influencing the variation in economic costs are: (1) differences in assumptions – such as those regarding economic growth over the century, the behavior of the oceans and terrestrial biosphere in taking up CO_2, and opportunities for reduction in non-CO_2 GHG emissions – that determine the amount that CO_2 emissions that must be reduced to meet the radiative forcing stabilization levels; and (2) differences in assumptions about technologies, particularly in the second half of the century, to shift final demand to low-CO_2 sources such as biofuels and low-carbon electricity or hydrogen in transportation, industrial, and buildings end uses. Although differences in technology do not strongly emerge until the second half of the century, they cast a shadow over the full century because of the manner in which all three the modeling groups allocated carbon emissions reductions over time.

In most scenarios, carbon prices depress demand for fossil fuels and therefore their producer prices. Electricity producer prices generally increase because of increasing de-mand for electricity along with substitution to higher cost, lower emitting electricity production technologies. Consumer prices for all fuels (fuel price plus the carbon price for emitted carbon plus any added cost of capturing and storing carbon) are generally higher under the stabilization scenarios due to carbon price. The approaches to Non-CO_2 GHG prices differs among the modeling groups, reflecting differing approaches tothe tradeoffs between reductions in the emissions of these GHGs and reductions inCO_2 emissions.

Stabilization and Carbon Prices

As discussed earlier, all of the modeling groups implemented prices or constraints that provide economic incentives to reduce GHG emissions. The instruments used to reduce CO_2 emissions in the models can be interpreted as the carbon price that would be consistent with either a universal cap-and-trade system or a harmonized carbon tax.

Across models, the more stringent stabilization levels require higher carbon prices because they require larger emissions reductions (Figure 4.20 and Table 4.5). Stabilization becomes increasingly difficult at the more stringent stabilization levels as can be seen in the difference in carbon prices between Level 2 and Level 1 as compared to that between Level 3 and Level 4. (Note that $100/tonne C is equivalent to $27/tonne CO_2. See Box 3.2 for more on converting between units of carbon and units of $CO_ç$.)

Table 4.6. Cumulative Emissions Reductions Across Scenarios(GtC through 2100)

	IGSM	MERGE	MiniCAM
Level 4	472	112	97
Level 3	674	258	267
Level 2	932	520	541
Level 1	1172	899	934

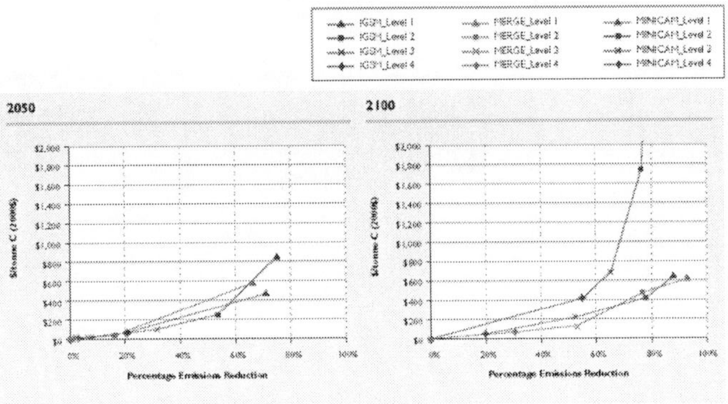

The relationship between carbon price and percentage reductions in carbon emissions is similar among the models in 2050. In 2100, a given percentage emissions reduction is generally more expensive in the IGSM stabilization scenarios than in the MERGE and MiniCAM stabilization scenarios. The difference in 2100 is due, in large part, to different assumptions regarding the technologies available to facilitate emissions reductions in the second half of the century, with IGSM scenarios assuming relatively fewer or more costly options than the scenarios from the other two modeling groups.

[Note. CO_2 emissions vary across the reference scenarios from the three modeling groups, so that similar percentage reductions, as shown in this figure, imply differing levels of total emissions reduction.]

Figure 4.21. Relationship Between Carbon Price and Percentage Emissions Reductions in 2050 and 2100.

Across models, the carbon prices rise exponentially throughout the century (in the IGSM scenarios) or until stabilization is reached (in the MERGE and MiniCAM scenarios). This similarity in the qualitative structure of the carbon price paths reflects the similarity in the approach that the modeling groups took to allocate emissions reductions over time, or *when* flexibility, as discussed in Section 4.2. This approach to *when* flexibility, with a carbon price that rises over time, tends to minimize the present discounted cost of emissions mitigation over the whole century. It also has the effect of linking future carbon prices to near-term carbon prices in a predictable way. Thus, when there are differences in technology assumptions that mostly appear in the second half of the century or in reference emissions that occur mostly in the middle of the century, the assumption imposed on the price path means that the burden of emissions reduction is spread over the entire century. In this way, forces that do not emerge until mid- century or beyond cast a shadow onto the present.

At every stabilization level, there is variation in the carbon prices among the models. For example, the carbon price in 2100 exceeds \$1700/tonne C in the IGSM Level 2 scenario while the carbon prices in the MERGE and MiniCAM Level 2 scenarios are \$420 to \$460/tonne C. The ratio among the models of carbon prices for other stabilization levels follows the same pattern. The range of carbon prices shown in these scenarios is consistent with other studies in the literature (IPCC 2001).

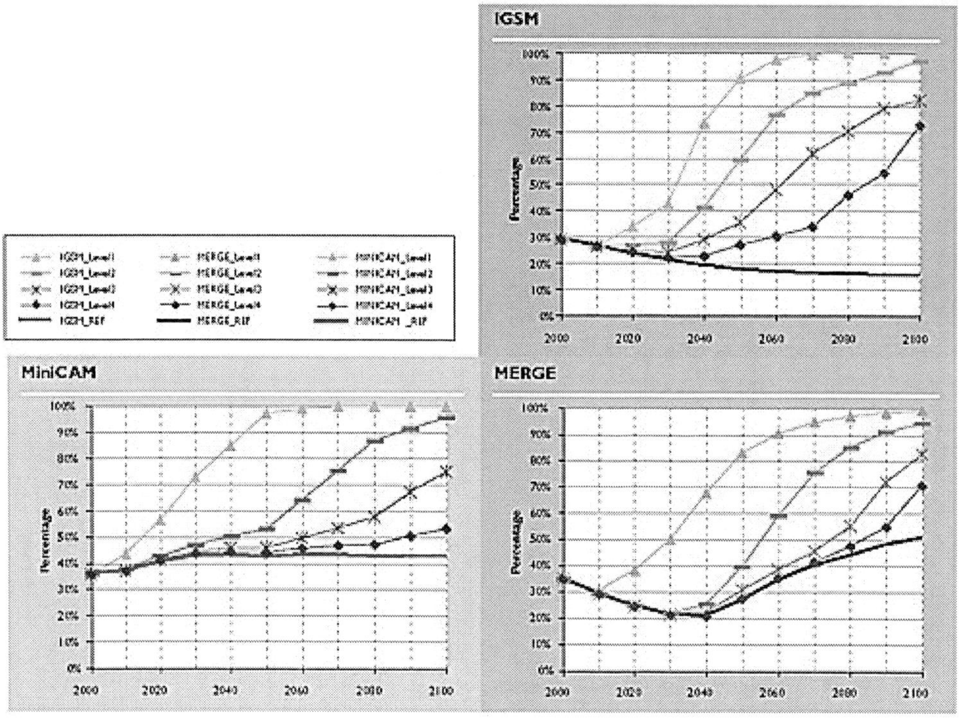

All three modeling groups assumed sufficient technological options to allow for substantial reductions in carbon emissions from electricity production. Options include fossil power plants with CCS, nuclear power, and renewable energy such as hydro electric power, wind power, and solar power. In all of the Level 1scenarios, the electricity sector is almost fully de carbonized byt he end of the century.

Figure 4.22. Percentage of Global Electricity Production from Low-or Zero-Emissions Technologies Across Scenarios (percent).

The carbon prices in the scenarios in this study are the result of a complex interplay of differing structural characteristics of the participating models and variation in key parameter values. Nonetheless major differences among carbon prices can be attributed to two influences: (1) the amount that emissions must be reduced to achieve an emissions path to stabilization, and (2) the technologies that are available to facilitate these changes in the economy.

On the first point, Table 4.6 shows the cumulative CO_2 emissions reductions required over the century across all four stabilization scenarios from each modeling group. Differences in total reductions come principally from three aspects of model behavior and assumptions: differences in forces, such as economic growth, that determine emissions in the reference scenario (Tables 3.2 and 3.3, and Figure 3.2); the behavior of the ocean and terrestrial systems in taking up carbon (Figure 4.6 and Figure 4.19); and the technological options available for constraining the emissions of non-CO_2 GHGs (Figure 4.8 and Figure 4.9). At all stabilization levels, the IGSM stabilization scenarios require greater CO_2 emissions reductions than the MERGE or MiniCAM stabilization scenarios. Indeed, the emissions reductions in the IGSM Level 2 scenario are commensurate with those of the MERGE and MiniCAM Level 1 scenarios. All other things being equal, the greater the required emissions reductions the higher will be the emissions prices required to meet each target.

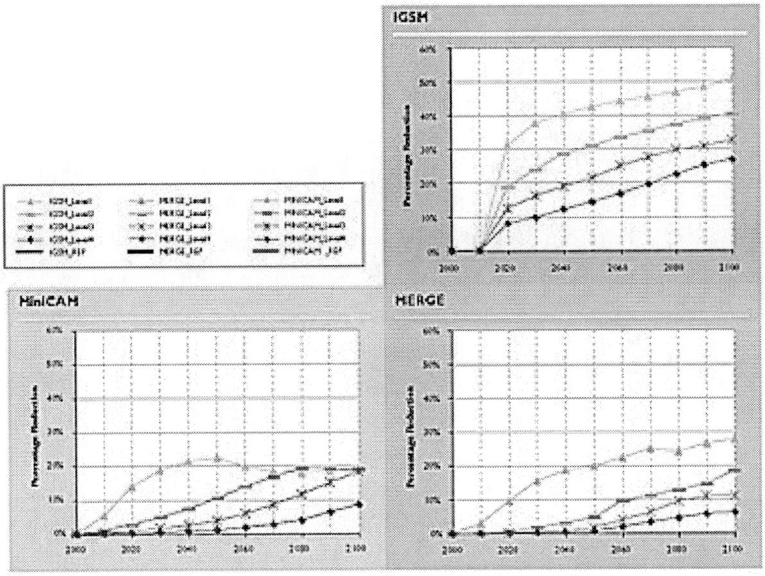

Differences in assumptions about technological opportunities result in different aggregate approaches to emissions reductions in the stabilization scenarios from the three modeling groups. The IGSM stabilization scenarios include greater reductions in primary energy consumption than the MERGE and MiniCAM stabilization scenarios because fewer technological opportunities, on both the demand and supply side, are available for emissions reductions through substitution to low or zero-carbon energy sources.[Note. Primary energy consumption from nuclear power and non-biomass renewable electricity are accounted for at the average efficiency of fossil-fired electric facilities, which vary over time and across scenarios.]

Figure 4.23. Percentage Reduction in World Primary Energy Consumption Across Scenarios (percent).

The second factor, the modeling of technology, also contributes to the differences among costs. The aggregate effect of differing technological assumptions is illustrated in Figure 4.21, which shows the relationship between the carbon price and percentage emissions reductions in 2050 and 2100 across all four stabilization scenarios from each modeling group. Roughly speaking, these figures represent what economists refer to as the marginal abatement cost functions for these periods. They broadly capture the technological opportunities for emissions reductions represented in the models. The similarity between the marginal abatement cost functions in 2050 implies that the technological opportunities represented by the three modeling groups are similar in 2050. The implication is that if the three modeling groups were to determine the carbon price associated with, for example, a 50% reduction in emissions in 2050, the results would be similar.

It is in the second half of the century that substantial differences in the marginal abatement cost functions emerge, particularly when the required abatement pushes towards and beyond 60% below the reference level as is the case in the Level 1 and Level 2 scenarios. There is no small set of technology assumptions used by the modeling groups that determines these differences. Among the modeling groups, assumptions about technology vary along a range of dimensions such as the rate of growth in labor productivity, the cost and performance of particular energy supply technologies, the productivity of agriculture and the associated costs of bioenergy, and the ability to substitute among various fuels and electricity in key demand sectors such as transportation. These assumptions are embodied not just in model parameters, but also, as discussed in Chapter 2, in the underlying mathematical structures of the models. As can be seen in Table 2.1, end-use technologies, are, in general, not represented explicitly. None of the participating models, for example, identify multiple steel production technologies or a wide range of vehicle options each with different energy using characteristics. Instead, energy demand responses are represented in relatively aggregate economic sectors (e.g., energy intensive industry or transportation). Other technologies, particularly in energy supply (e.g., CCS) are more likely to be identified specifically.

Three general characteristics of technology bear note with respect to the variation in carbon prices: (1) the availability of low- or zero-carbon electricity production technologies, (2) the supply of non-electric energy substitutes such as biofuels and hydrogen, and (3) the availability of technologies to facilitate substitution toward the use of electricity.

All three modeling groups assumed a variety of cost-effective technology options would be available to limit CO_2 emissions from electricity production. For example, the electric sector is almost fully de-carbonized by the end of the century in all three Level 1 scenarios (Figure 4.22). Electricity is produced with non-fossil technologies (nuclear or renewables) or fossil- fired power plants with CCS. Thus, although low carbon technologies in the electric sector do influence the carbon prices, it is forces outside of electricity production that drive costs at higher levels of abatement because options available to the electric sector can support its almost complete de-carbonization.

The second technology factor is the set of options available to substitute alternative, non-electric fuels for fossil energy in end-use sectors, most importantly in transportation. All three modeling groups assumed biofuels as a substitute for fossil fuels in non-electric applications. As discussed in Section 2 and Section 3, production of bioenergy crops must compete with other uses of agricultural lands in the IGSM and MiniCAM scenarios, which constrains total production of these substitutes. MERGE uses an aggregate parameterization to represent these same constraints. Even with these differing approaches, bioenergy production is similar across the stabilization scenarios. However, because of higher oil prices (Figure 3.7), the IGSM reference scenario includes substantial biofuels (Figure 4.10) so that expansion of biofuels is more limited in the IGSM stabilization scenarios.

In addition to biofuels, the MiniCAM and MERGE scenarios include other non-electric alternatives, and these become important for more stringent emissions reductions. The MERGE scenarios include a generic alternative fuel generated from renewable sources; which could be, for example, hydrogen from solar or wind power. In the MERGE Level 1 scenario, this alternative fuel provides roughly 80% as much non-electric energy as biofuels by 2100. The MiniCAM scenarios include hydrogen production using electricity, nuclear thermal dissociation, and fossil fuels with and without CCS. Though smaller than biofuels, the contribution of hydrogen rises to a little over 15% of global non-electric energy consumption in the Level 1 MiniCAM scenario. Without these additional options included in the MERGE and MiniCAM scenarios, the marginal cost of emissions reductions is higher in the IGSM scenarios, and more of the abatement is met through reductions in energy use (Figure 4.23).

Another factor influencing carbon prices at higher levels of emissions reduction is the ability to substitute to electricity in end-use sectors, through technologies such as heat pumps, electrically-generated process heat, or electric cars. Were all end uses to easily switch to electricity, then the availability of nearly carbon-free electricity production options in these scenarios would allow complete CO_2 emissions reduction at no more than the cost of these generation options. However, assumptions about technologies for electrification differ substantially among the modeling groups. The MERGE and MiniCAM modeling groups assumed greater opportunities for substitution to electricity than did the IGSM modeling group in the second half of the 21st century. As a result the electricity fraction of primary energy consumption is higher in the MERGE and MiniCAM scenarios in both the reference scenario and the stabilization scenarios, as shown in Figure 4.24. This means that low- or zero-carbon electricity production technologies can serve more effectively as a low-cost option for emissions reduction, reducing costs. In the IGSM scenarios, fuel demand for transportation, where electricity is not an option and for which biofuels supply is insufficient, continues to be a substantial source of emissions.

Although the main technological influences discussed above do not emerge for many decades, they influence carbon prices and economic costs from the outset because of the approach the modeling groups took to *when* flexibility, as discussed above. This dynamic view of the stabilization challenge reinforces the fact that actions taken today both influence and are influenced by the possible ways that the world might evolve in the future.

Finally, there are other structural differences among the approaches taken by the modeling groups that likely play a role in the variation in carbon prices. For example, MERGE is a forward-looking model and that behavior allows it to more fully optimize investments over time than the other two models, including investments in emissions reductions. Another difference is that the MiniCAM scenarios include CCS in cement production, which allows for cement emissions to be reduced to almost zero at more stringent stabilization levels. The IGSM scenarios include cement production within an aggregate sector so that mitigation options that may be specific to this industry are not explicitly modeled. The MERGE scenarios explicitly include emissions from cement production, but do not include options for reducing these emissions. This omission puts more pressure on emissions reductions elsewhere in the IGSM and MERGE stabilization scenarios and would tend to raise carbon prices relative to the Mini- CAM scenarios. Finally, IGSM and MERGE explicitly track savings and investment, whereas MiniCAM does not. In IGSM and MERGE, investments in emissions reductions lower savings and investment in other sectors, affecting the scale of economic output in future periods, and this effect accumulates over time. The most direct effect of this dynamic is felt on economic output, and therefore stabilization costs (addressed later in this chapter), but it may also affect carbon prices through reductions in the scale of economic activity.

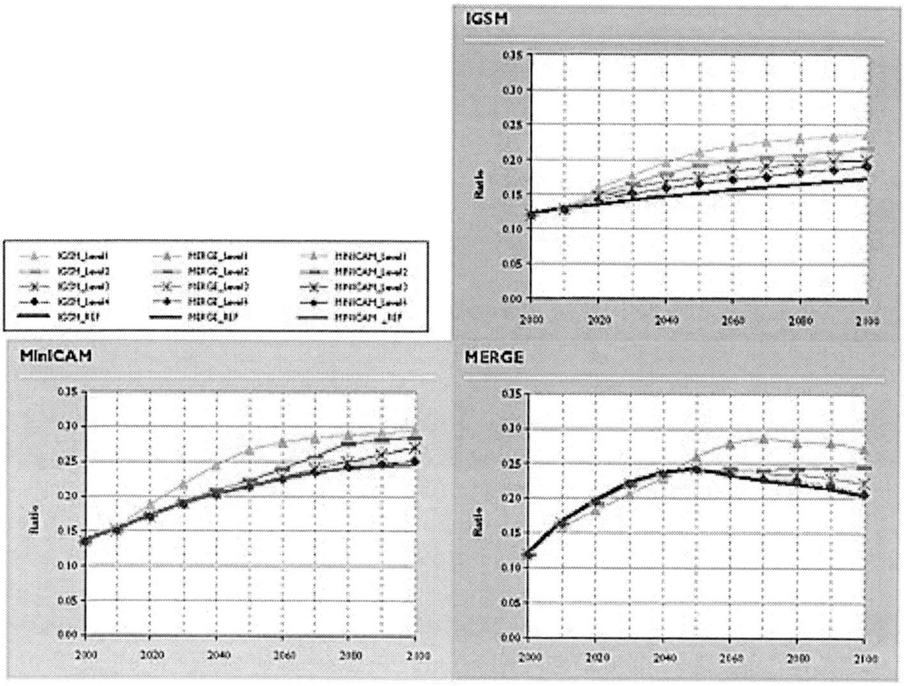

Scenarios. Efforts to constrain CO_2emissions result in increased use of electricity as a fraction of total primary energy in the scenarios from all three modeling groups. This is because all three modeling groups assumed lower cost technology options for reductions in emissions from electricity production than for substitution away from fossil fuels in direct uses such as transportation. The MERGE and MiniCAM scenarios generally include greater electrification than the IGSM scenarios, with MiniCAM having the highest proportion of electricity to primary energy. Greater opportunities to electrify reduce the economic impacts of stabilization.

[Note. Primary energy consumption from nuclear power and non-biomass renewable electricity are accounted for at the average efficiency of fossil-fired electric facilities, which vary over time and across scenarios.]

Figure 4.24. Ratio of Global Electricity Production to Primary Energy Consumption Across

Stabilization and Non-CO_2 Greenhouse Gas Prices

Each of the three modeling groups employed a different approach to reductions in the emissions the non-CO_2 GHGs. After CO_2, CH_4 is the next largest component of radiative forcing in all three reference scenarios. Emissions of CH_4 vary among the reference scenarios. The IGSM reference scenario starts in the year 2000 at about 350 Mt/yr and rises to more than 700 Mt/yr (Figure 4.8), while the MERGE and MiniCAM scenarios begin with 300 Mt/yr in the year 2000. These are anthropogenic CH_4 emissions, and the differences reflect existing uncertainties in how much of total CH_4 emissions are from anthropogenic and natural sources. CH_4 emissions grow to almost 600 Mt/yr in the MERGE reference scenario. The MiniCAM reference scenario is characterized by a peak in CH_4 emission at less than 400 Mt/yr, followed by a decline to about 300 Mt/yr.

Each of the modeling groups took a different approach to setting a stabilization constraint on CH_4. The MiniCAM stabilization scenarios are based on constant GWP coefficients, so the price of CH_4 is simply the price of CO_2 multiplied by the GWP. This means that the price of CH_4 relative to the carbon price (the relative CH_4 price) is constant over time, as shown in Figure 4.25.

In contrast, MERGE determines the price of CH_4 to carbon through inter-temporal optimization. The relative price of CH_4 begins very low, although it is higher the more stringent the stabilization level. The relative price then rises at a roughly constant exponential rate of between 8% and 9% per year until stabilization is reached, at which point, the relative price remains approximately constant at around 10 times the carbon price. These characteristics of the CH_4 price and its relationship to the carbon price are the product of the inter-temporal optimization in which the long-term limit on radiative forcing is the only goal. Manne and Richels (2001) have shown that different patterns are possible if other formulations of the policy goal, such as limiting the rate of change of radiative forcing, are taken into account.

The IGSM stabilization scenarios are based on a third approach. CH_4 emissions are limited to a maximum value in each stabilization scenario: 425 Mt/yr at Level 4, 385 Mt/yr at Level 3, 350 Mt/yr at Level 2, and 305 Mt/yr at Level 1. As a consequence, the relative price of CH_4 initially grows from one-tenth to a maximum of between 3 and 14 between the years 2050 and 2080 and then declines thereafter. As previously discussed, this reflects an implicit assumption that a long-run requirement of stabilization means that eventually each substance must be (approximately) independently stabilized, and absent an explicit evaluation of damages of climate change, any time path of relative GHG prices cannot be determined.

As with CH_4, emissions of N_2O in the reference scenarios vary across the three modeling groups (Figure 4.9). The IGSM reference trajectory roughly doubles from approximately 11 Mt/yr to approximately 25 Mt/yr. In contrast, the MERGE and MiniCAM reference scenarios are roughly constant over time.

MERGE also sets the price of N_2O as part of the inter-temporal optimization process. The relative price trajectory for N_2O begins at roughly the level of the GWP-based relative price used in the MiniCAM stabilization scenarios and then rises, roughly linearly with time (Figure 4.25). The relative N_2O price approximately doubles in the MERGE Level 4 scenario, but is almost constant in the MERGE Level 1 scenario. Thus, in the Level 1 scenarios, the relative N_2O price path is virtually the same in the MERGE and MiniCAM scenarios.

In contrast, in the IGSM stabilization scenarios, stabilization sets a path to a predetermined N_2O concentration for each stabilization level, and the complexity of the price paths in Figure 4.25 shows the difficulty of stabilizing the atmospheric level of this GHG. Natural emissions of N_2O are calculated, which vary with the climate consequences of stabilization. The main anthropogenic source, agriculture, has a complicated relationship with the rest of the economy through the competition for land use.

The approaches employed by the three modeling groups do not necessarily lead to the stabilization of the concentrations of the non-CO_2 GHGs before the end of the twenty-first century, as concentrations are still rising slowly in some scenarios but below a long-term stabilized level (Figure 4.4 and Figure 4.5). How longterm stabilization was approached was independently developed by each modeling group.

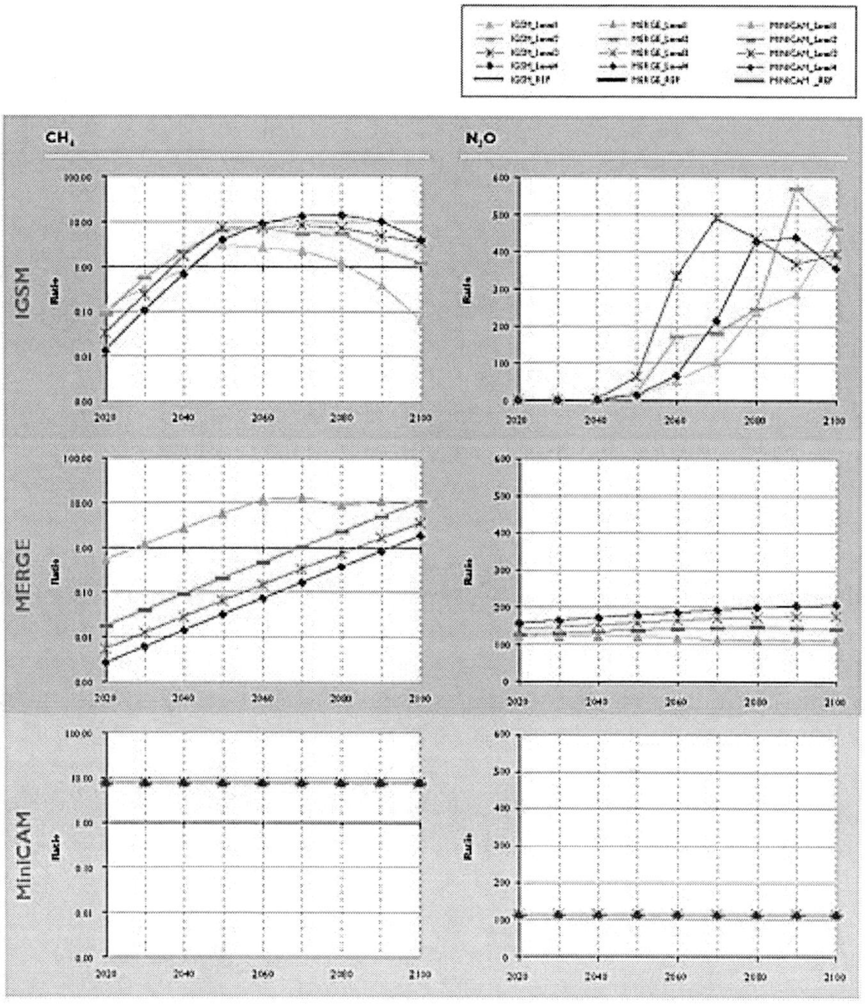

Differences in the prices of CH_4 and N_2O relative to the carbon price reflect different treatments of this tradeoff among the modeling groups, often referred to as what flexibility. In the MiniCAM stabilization scenarios, the tradeoff is based on the GWPs of the non-CO_2 GHGs, which are constants, leading to constant relative prices of the non-CO_2 GHGs. In the MERGE stabilization scenarios, relative prices are optimized with respect to meeting the long-run stabilization levels. In the IGSM stabilization scenarios, stabilization was forced for each GHG independently. Emissions were set so that concentrations of CH_4 would stabilize and allowed the CH_4 price path to be determined by changing opportunities for reducing emissions. Given N2O emissions from agriculture, the relative price of N_2O is higher in the IGSM stabilization scenarios, in part because emissions were higher in the IGSM reference scenario than in the reference scenarios from the other two modeling groups. Lower emissions of N_2O for the MERGE and MiniCAM reference scenarios allowed the corresponding stabilization scenarios to achieve relatively low emissions at lower N_2O prices.

Figure 4.25. Relative Prices of CH_4 and N_2O to Carbon Price Across Scenarios (CH_4 in log scale).

Table 4.7. Relationship Between a$100/tonne Carbon Price and Energy Prices

Fuel	Base Cost ($2005)	Added Cost ($)	Added Cost (%)
Crude Oil ($/bbl)	$60.0	$12.2	20%
Regular Gasoline ($/gal)	$2.39	$0.26	11%
Heating Oil ($/gal)	$2.34	$0.29	12%
Wellhead Natural Gas ($/tcf)	$10.17	$1.49	15%
Residential Natural Gas ($/tcf)	$15.30	$1.50	10%
Utility Coal ($/short ton)	$32.6	$55.3	170%
Electricity (c/kWh)	9.6¢	1 .76¢	18%

(In most cases,stabilization depressesproducer prices and sothe percentage rise in thefuel cost seen byconsumers would be lessthan indicated here. Thechange in producer priceis highly scenario andmodel dependent.)

Source: Bradley et al. (1991), updated with U.S. average prices for the 4th quarter of 2005 as reported by DOE (2006).

Stabilization and Energy Prices

The carbon price drives a wedge between the producer prices of fuels and the costs to consumers. Table 4.7 provides an approximation of that of the relationship. A given carbon price has the largest impact on consumer cost of coal in percentage terms because the fuel price per unit of energy is low, and carbon emissions are relatively high per unit of energy. In comparison, natural gas prices were at historic highs in recent years and CO_2 emissions per unit of energy are lower than oil or coal. This means that the carbon price has a relatively smaller effect in comparison to the fuel price.

Stabilization scenarios tend to result in a lower producer price for oil (Figure 4.26). Stabilization at Level 4 has a relatively modest effect on the oil producer price, particularly prior to 2040; the effect is stronger the more stringent the stabilization level. Oil producer price reductions vary across the three models, ranging from the IGSM stabilization scenarios, which show the most pronounced effects, to the MERGE stabilization scenarios, which show a substantial effect only in the Level 1 scenario. The effect on world oil producer prices, in turn, depends on many factors, including how the supply of oil is characterized; the carbon price; and the availability of substitute technologies for providing transportation liquids, such as bio-fuels or hydrogen.

Coal producer prices are similarly depressed in the IGSM and MiniCAM stabilization scenarios (Figure 4.27). The effect is mitigated by two features: (1) the assumed availability of CCS technology, which allows the continued large- scale use of coal in electricity production in the presence of a positive carbon price and (2) a coal supply schedule that is highly elastic. That is, demand for coal can exhibit large increases or decreases without much change in price. The high elasticity of supply in the MERGE scenarios leaves coal producer prices unchanged across the stabilization scenarios, whereas the MiniCAM and IGSM scenarios have lower supply price elasticities and, hence, greater producer price responses.

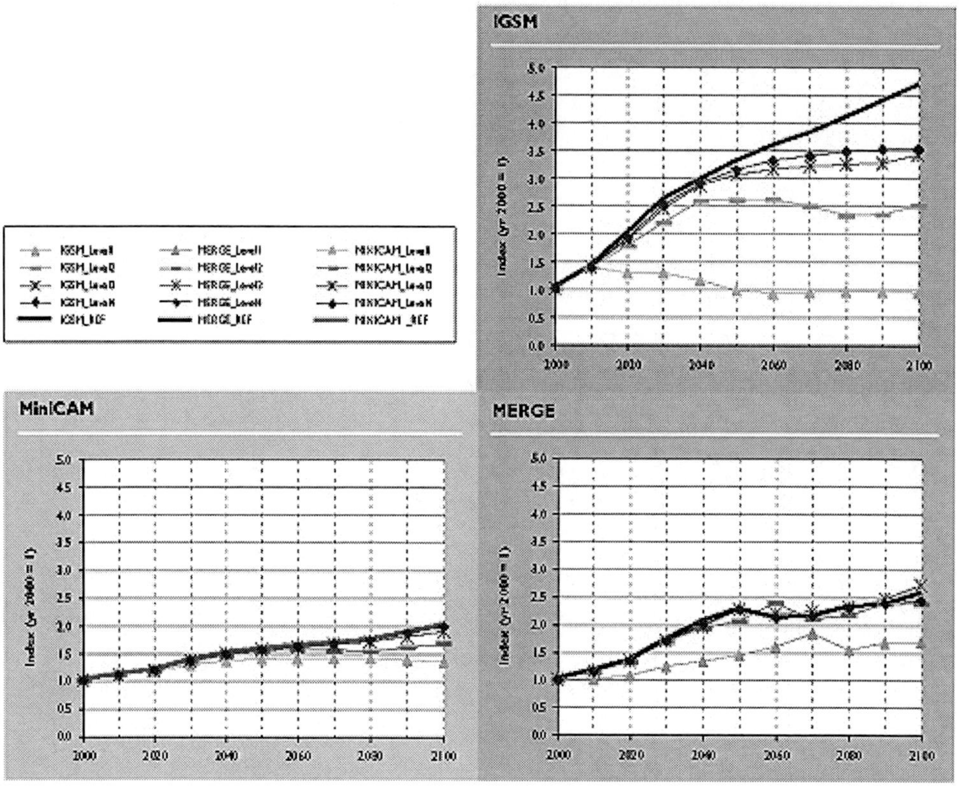

World oil prices (producer price) vary considerably across the reference scenarios. In all three models, stabilization tends to depress the producer prices of oil relative tothe reference scenarios. [Note. Producer prices as defined here do not include additional costs associated with carbon emissions to the atmosphere through the combustion of fossil fuels, as shown in Table 4.7.]

Figure 4.26. World Oil Price Across Scenarios (Index, yr 2000 = 1).

The impact on the natural gas producer price is more complex (Figure 4.28). Natural gas has roughly one-half the carbon-to-energy ratio of coal. Thus, emissions can be reduced without loss of available energy simply by substituting natural gas for coal or oil. As a consequence, two effects on the natural gas producer price work in opposite directions. With a postive carbon price, natural gas tends to substitute for other fossil fuels, increasing its demand. But a positive carbon price also means that a low- or zero-carbon substitutes, such as electricity, bioenergy, or energy-efficiency technologies, will tend to displace natural gas from markets, as happens for the more carbon-intensive fuels.

Thus, depending on the strength of these two effects, the producer price of natural gas can either rise or fall.

The natural gas producer price is most affected in the IGSM stabilization scenarios, reflecting the greater substitution of natural gas for coal in IGSM Level 2, 3, and 4 stabilization scenarios. In the IGSM Level 1 stabilization scenario, natural gas consumption is reduced over the entire period. On balance, the natural gas producer price is less affected by stabilization in the MERGE and MiniCAM scenarios in which the substitution and conservation effects are roughly offsetting.

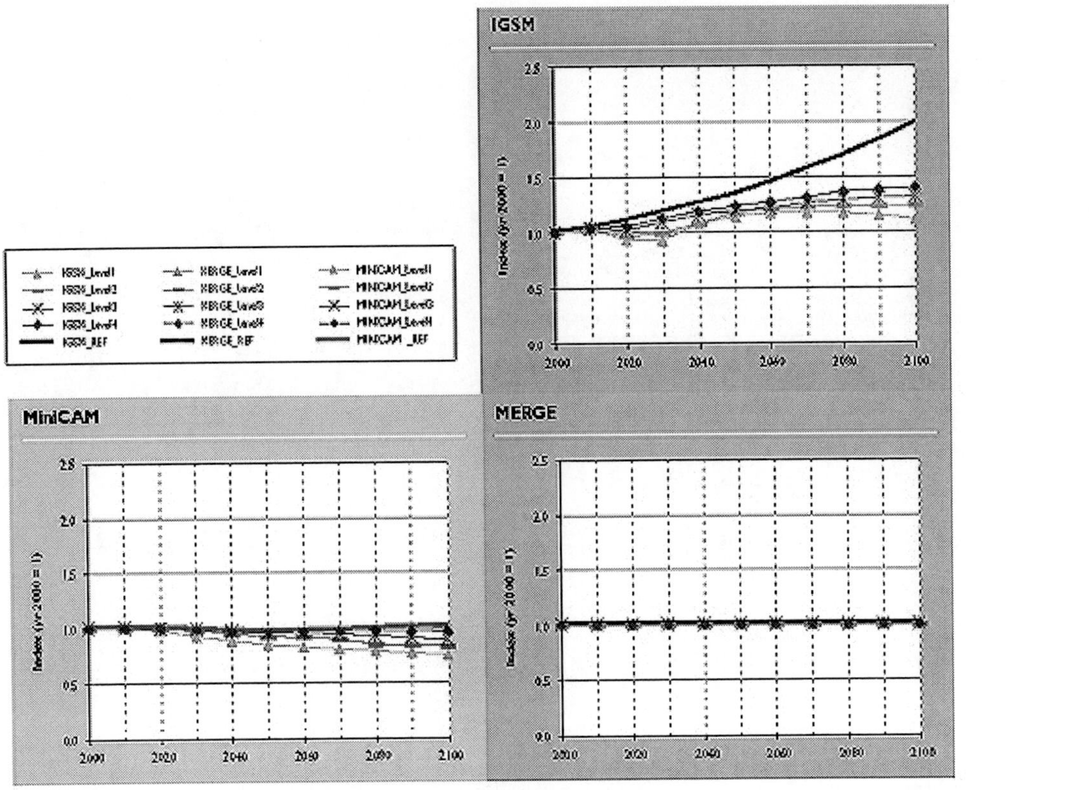

U.S. mine-mouth coal price varies acrosss the reference scenarios. In the IGSM and MiniCAM stabilization scenarios, stabilization depresses coal prices, where as stabilization has no impact on coal prices in the MERGE stabilization scenarios, reflecting characterization of coal supply as an inexhaustible single grade such that there is no rent associated with the resource. Prices in the MERGE scenarios thus reflect the cost capital, labor, and other inputs that are little affected by the stabilization policy.

[Note. Producer prices as defined here do not include additional costs associated with carbon emissions to the atmosphere through the combustion of fossil fuels, as shown in Table 4.7.]

Figure 4.27. U.S. Mine-Mouth Coal Price Across Scenarios (Index, yr 2000 = 1).

Although the price that oil and coal producers receive tends to be either stable or depressed, that is not the full cost of using the fuel. Users, such as households or industrial fuel users, pay the market price plus the value of the carbon emissions associated with the fuel, which is the carbon price times the fuel's carbon-to-energy ratio. If they employ CCS, the carbon emissions are lower, but they face the added cost of CCS. Any additional carbon cost will be reflected in the users' fuel price if the carbon taxes, or required permits in a cap-and-trade system, are placed upstream with fuel producers. On the other hand, the actual fuel price impact they see may be similar to the producer price impact if carbon is regulated downstream where the fuel is consumed. In this case, users would be able to buy fuel relatively inexpensively, but would pay a separate large price for necessary carbon charges associated with emissions.

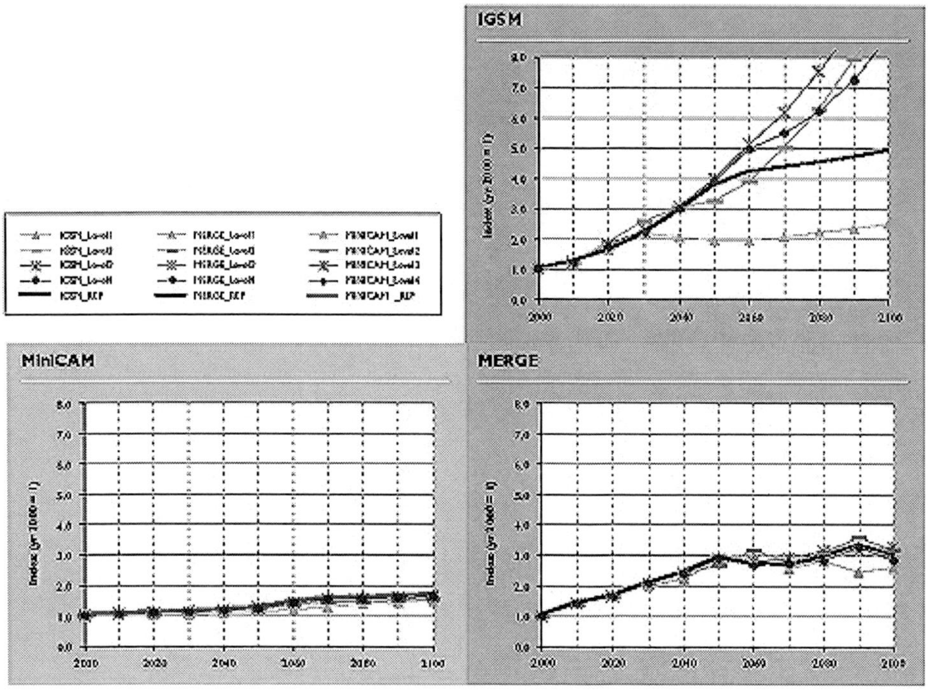

U.S. natural gas producers' prices vary among the reference scenarios. In the MiniCAM and MERGE stabilization scenarios, stabilization has little effect on the natural gas price. Stabilization at Levels 2, 3, and 4 increases theprice of natural gas in the IGSM stabilization scenarios because of substitution toward natural gas and away from coal and oil. Natural gas prices fall relative to reference scenario in the IGSM Level 1 stabilization scenario because natural gas demand isdepressed from the tight carbon constraint.

[Note. Producer prices as defined here do not include additional costs associated with carbon emissions to the atmosphere through the combustion of fossil fuels, as shown in Table 4.7.]

Figure 4.28. U.S. Natural Gas Producers' Price Across Scenarios (Index, yr 2000 = 1).

The effect on the price of electricity is another unambiguous result (Figure 4.29). Because electricity producers are fossil fuel consumers, the price of electricity contains the implicit carbon price in the fuels used for generation. All of the scenarios exhibit upward pressure on electricity prices, and the more stringent the stabilization level, the greater the upward pressure. The pressure is limited by the fact that there are many options available to electricity producers to lower emissions. These options include, for example, the substitution of natural gas for coal; the use of CCS; the expanded use of nuclear power; the use of bioenergy; and the expanded use of wind, hydro, and other renewable energy sources.

The Total Cost of Stabilization

Assessing the macroeconomic cost of stabilization is not a simple task either conceptually or computationally. From an economic perspective, cost is the value of the loss in welfare associated with pursuing stabilization or equivalently, the value of activities that society will not be able to undertake as a consequence of pursuing stabilization. Although the concept is easy enough to articulate, defining an unambiguous measure is problematic. Any measure of cost is a more or less satisfactory compromise.

The task is further complicated by the need to aggregate the welfare of individuals who have not yet been born and who may or may not share present preferences. Even if these problems were not difficult enough, economies can hardly be thought to currently be at a maximum of potential welfare. Preexisting market distortions impose costs on the economy, and mitigation actions may interact with them so as to reduce or exacerbate their effects. Any measure of global cost also runs into the problem of international purchasing power comparisons discussed in Chapter 3. Finally, climate change is only one of many public goods, and measures to address other public goods (like urban air quality) can either increase or decrease cost. To create a metric that is consistent and comparable across the three modeling platforms used in this study, all of these issues would have to be addressed in some way.

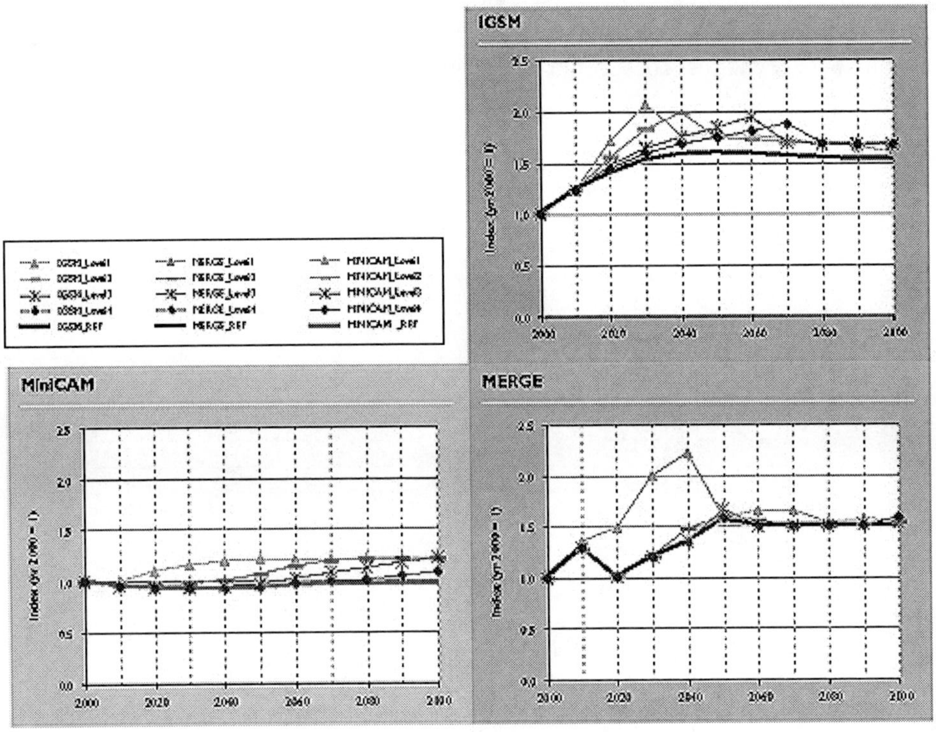

U.S. electricity prices in the reference scenarios range from little change over the century in the MiniCAM reference scenario to about a 50% increase from present levels in the IGSM reference scenario. Under stabilization, producer prices are affected by increasing use of more expensive low- or zero-emissions electricity technologies, including fossil electricity with CCS, nuclear power, and non-biomass renewables such as solar and wind power. Across the scenarios, rising fossil fuel prices are partially offset by increasing efficiency of fossil electric facilities.

[Note. Producer prices as defined here do not include additional costs associated with carbon emissions to the atmosphere through the combustion of fossil fuels, as shown in Table 4.7.]

Figure 4.29. U.S. Electricity Producer Price Across Scenarios (Index, yr 2000 = 1).

Beyond conceptual measurement issues, any metric including gross domestic product, depends on features of the scenario such as the assumed participation by countries of the world, the terms of the emissions limitation regime, assumed efficiencies of markets, and technology availability – the latter including energy technologies, non-CO_2 GHG technologies, and related activities in non-energy sectors (e.g., crop productivity that strongly influences the availability and cost of producing commercial biomass energy). In almost every instance, scenarios of the type explored in this research employ more or less idealized representations of economic structure, political decision, and policy implementation (i.e., conditions that likely do not accurately reflect the real world, and these simplifications tend to lead to lower mitigation costs).

Finally, assessing welfare effects would require explicit consideration of how the burden of emissions reduction is shared among countries and the welfare consequences of income effects on poorer versus wealthier societies. Of course, if the world were to discover and deploy lower cost technology options than those assumed here, these costs could be lower. On the other hand, if society does not deliver the cost and performance for the technologies assumed in these scenarios, costs could be higher.

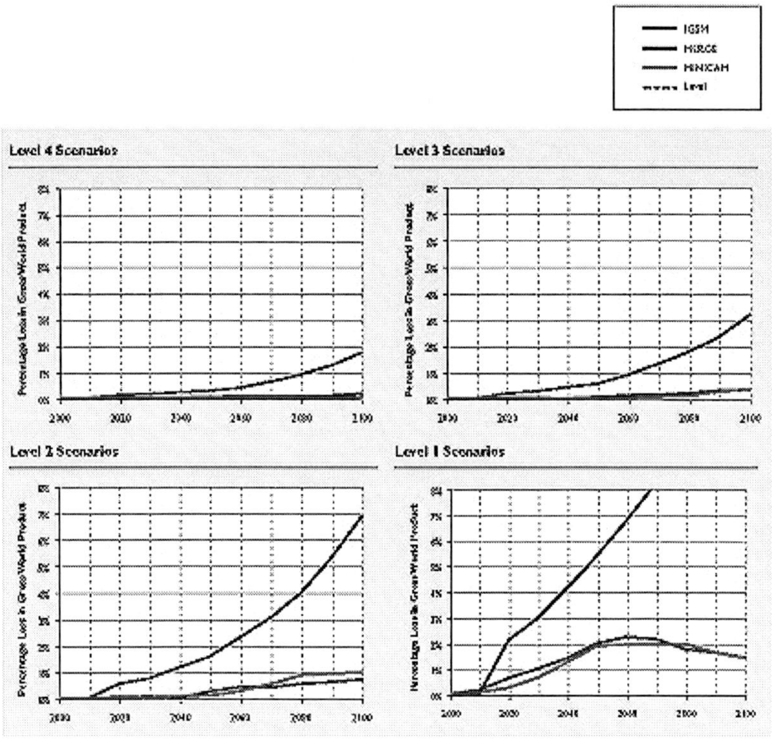

Stabilization imposes costs on the economy, and stated in terms of gross world product, the costs rise over time as ever more stringent emissions restrictions are required. The more stringent the stabilization level, the higher the cost. The variation in costs among the models reflects differences in the emissions reductions necessary for stabilization and differences in the technologies that might facilitate carbon emissions reductions, particularly in the second half of the century.

Figure 4.30. Percentage Reduction in Gross World Product in the Stabilization Scenarios (percentage).

While all of the above considerations have not been extensively investigated in the literature, the implications of less-than-ideal implementation have been investigated, and these analyses show that it could increase the costs substantially. Richels et al. (1996) showed that for a simple policy regime, eliminating international *where* and *when* flexibility, while assuming perfect *where* flexibility within countries, could potentially raise costs by an order of magnitude compared to a policy that employed *where* and *when* flexibility in all mitigation activities. Richels and Edmonds (1995) showed that stabilizing CO_2 emissions could be twice as expensive as stabilizing CO_2 concentrations and leave society with higher CO_2 concentrations. Babiker et al. (2000) similarly showed that limits on *where* flexibility within countries can substantially increase costs – although employing *where* flexibility also can increase costs in the context of tax distortions (Babiker et al. 2003a, Babiker et al. 2003b, Babiker et al. 2004, Paltsev et al. 2005).

Figure 4.30 reports the change of gross world product in the stabilization scenarios during the twenty-first century in the year in which it occurs aggregated using market exchange rates. This information is also displayed in Table 4.8. The use of market exchange rates is a convenient choice given the formulations of the models employed here, but as discussed above and in Chapter 3 the approach has limits (see the Box 3.1 in Chapter 3). Though change in gross world product is not the most intellectually satisfying measure, it serves as a common reference point.

Table 4.8.Percentage Reduction in Gross World Product in the Stabilization Scenarios

Level 1					
	2020	**2040**	**2060**	**2080**	**2100**
IGSM	2.1%	4.1%	6.7%	10.1 %	16.1%
MERGE	0.6%	1.4%	2.2%	1.8%	1.4%
MiniCAM	0.2%	1.2%	1.9%	1.9%	1.4%
Level 2					
IGSM	0.5%	1.2%	2.3%	3.9%	6.8%
MERGE	0.0%	0.0%	0.4%	0.5%	0.7%
MiniCAM	0.0%	0.1%	0.2%	0.8%	1.0%
Level 3					
IGSM	0.2%	0.4% 0	0.9%	1.8%	3.1%
MERGE	0.0%	0.0%	0.1%	0.2%	0.3%
MiniCAM	0.0%	0.0%	0.0%	0.1%	0.4%
Level 4					
IGSM	0.1%	0.2%	0.4%	0.9%	1.7%
MERGE	0.0%	0.0%	0.1%	0.1%	0.2%
MiniCAM	0.0%	0.0%	0.0%	0.0%	0.0%

The effects on gross world product are tightly linked to the carbon prices. Therefore effects on gross world product in the stabilization scenarios follow the same patterns and logic as the carbon prices, which are discussed in substantially greater detail in Section 4.6.1. As with the carbon price, costs rise with increasing stringency of the stabilization level. And, as with the carbon price, there is variation in costs of stabilization among the modeling groups. For example, gross world product in 2100 is reduced by 6.8% in the IGSM Level 2 scenario, while the reduction is less than 1% in the MERGE and MiniCAM Level 2 scenarios. The ratio of stabilization costs among the models at other radiative forcing stabilization levels follows the same pattern.

The differences in stabilization costs among the models can largely be attributed the same influences discussed in Section 4.6.1: (1) the amount that emissions must be reduced to achieve an emissions path to stabilization, and (2) the technologies that are available to facilitate these changes in the economy. A number of additional, structural differences, such as treatment of capital investment, intertemporal model structure, and emissions reductions opportunities in cement production also lead to differences in prices and costs. As with emissions prices, although technology differences emerge primarily in the second half of the century, their influence felt throughout the century because of the common implementation of *when* flexibility in the policy design.

Expressed throughout the report is the view that the development of independent sets of scenarios using three different models helps to inform common understanding of the forces that shape opportunities to stabilize greenhouse gas concentrations. The differences discussed here demonstrate the fundamental importance of technology in facilitating stabilization – particularly the importance of future technology, even developments more than half a century in the future. The scenarios also suggest the particular importance of options that facilitate the production of alternative non-electric fuels and demand-side technologies that will allow the substitution of electricity for current applications of fossil fuels.

5. Findings, Uses, and Future Directions

Introduction

Scenarios based on formal, computer-based models, such as the scenarios developed in this research, can help illustrate how key drivers such as economic and population growth or policy options lead to particular levels of GHG emissions. An important benefit of models such as those used in this research is that they ensure basic accounting identities and consistent application of behavioral assumptions. However, model-based scenarios are only one approach to scenario development, and models designed for one set of purposes may not be the most appropriate for other applications. Thus, the scenarios developed here should be viewed as complementary to other ways of thinking about the future, such as formal uncertainty analyses, story lines, baselines for further model-based scenarios, and analyses using other types of models.

The users of emissions scenarios are many and diverse and include climate modelers and the science community; those involved in national public policy formulation; managers of Federal research programs; state and local government officials who face decisions that might be affected by climate change and mitigation measures; and individual firms, non-governmental organizations, and members of the public. Such a varied clientele implies an equally diverse set of possible needs, and no single scenario exercise can hope to satisfy all of them. Scenario analysis is most effective when its developers can work directly with users, and initial scenarios lead to further *what if* questions that can be answered with additional scenarios or by probing more deeply into particular issues. The Prospectus for this research did not, however, prescribe such an interactive approach with a focused set of users. Instead, it called for a set of scenarios that provide broad insights into the energy, economic, and emissions implications of stabilizing radiative forcing. For the issue of stabilization, these scenarios are an initial offering to potential user communities that, if successful, will generate further questions and more detailed analysis.

This research focuses on three sets of scenarios, each including a reference scenario and four scenarios in which the radiative forcing from a common suite of GHGs is stabilized at four alternative levels. The stabilization scenarios describe a range of possible long-term goals for global climate policy. The stabilization levels imply a range of policy efforts and levels of urgency, from relatively little deviation from reference scenarios over the course of the century to major deviations starting very soon. Although the Prospectus did not mandate a formal treatment of likelihood or uncertainty, such analysis could be a useful follow-on activity. Here, however, the range of outcomes from the different modeling groups helps to illustrate, if incompletely, the range of possibilities.

For this research, a scenario is an illustration of future developments based on a model of the economy and the Earth system, applying a plausible set of model parameters and providing a basis for future work. None of the reference scenarios is a prediction or best-judgment forecast of the future, and none can be said to have the highest probability of being right. Nor does any single stabilization scenario provide the most correct picture of the changes to energy and other systems that would be required for stabilization. Instead, each scenario in this chapter is a thought experiment that helps illuminate the implications of different long-term policy goals.

Overview of the Scenarios

The scenarios are presented in text and figures in Chapters 3 and 4, and here a summary is provided of some of their key characteristics, some of the magnitudes involved, and the assumptions that lie behind them.

Reference Scenarios

The difficulty in achieving any specified radiative forcing stabilization level depends heavily on the emissions that would occur absent actions to address GHG emissions. In other words, the reference scenario strongly influences the stabilization scenarios. If the reference scenario has inexpensive fossil fuels and high economic growth, then larger changes to the energy sector and other parts of the economy may be required to stabilize the atmosphere. On the other hand, if the reference scenario shows lower growth and emissions, and perhaps increased exploitation of non-fossil sources even in the absence of climate policy, then the effort required to stabilize radiative forcing will not be as great.

Energy production, transformation, and consumption are central features in all of these scenarios, although non-CO_2 gases and changes in land use also make a significant contribution to aggregate GHG emissions. Demand for energy over the coming century will be driven by economic growth and will also be strongly influenced by the way that energy systems respond to depletion of resources, changes in prices, and improvements in technology. Demand for energy in developed countries remains strong in all the scenarios and is even stronger in developing countries, where millions of people seek greater access to commercial energy. These developments strongly influence the emissions of GHGs, their disposition, and the resulting change in radiative forcing in the reference scenarios.

The three reference scenarios show the implications of this increasing demand and the improved access to energy. The variation between the reference scenarios reflects the differing assumptions used by the modeling groups.

- Global primary energy consumption rises substantially in all three reference scenarios, from about 400 EJ/yr in 2000 to between roughly 1275 EJ/yr and 1500 EJ/yr in 2100 (Figure ES.1). U.S. primary energy consumption also grows substantially, about 11/4 to 21/2 times present levels by 2100. Primary energy consumption growth occurs despite continued improvements in the efficiency of energy use and energy production technologies. For example, the U.S. energy intensity – the ratio of primary energy consumption to economic output – declines 60% to 75% between 2000 and 2100 across the three reference scenarios.
- All three reference scenarios include a gradual reduction in the consumption of conventional oil resources. However, in all three, a range of alternative fossil-based resources, such as synthetic fuels from coal and unconventional oil resources (e.g., tar sands and oil shales), are available and become economically viable. Fossil fuels provided almost 90% of the global primary energy in 2000, and they remain the dominant energy source in the three reference scenarios throughout the twenty-first century, supplying 70% to 80% of total primary energy in 2100.
- Non-fossil fuel energy use also grows over the century in all three reference scenarios. Contributions to primary energy consumption in 2100 range from 250 EJ to 450 EJ – a range that at the hight end exceeds global primary energy consumption today. Despite this growth, these sources never supplant fossil fuels, although they provide an increasing share of the total, particularly in the second half of the century.
- Consistent with the characteristics of primary energy consumption, global and U.S. electricity production continues to rely on coal, although this contribution varies among the reference scenarios. The contribution of renewable and nuclear energy varies considerably in the different reference scenarios, depending on resource availability, technology, and non-climate policy considerations. For example, global nuclear power in the reference scenarios ranges from about 1½ times current levels (if non-climate concerns such as safety, waste, and proliferation constrain its growth as is the case in one reference scenario), to an expansion of almost an order of magnitude assuming relative economics as the only constraint.
- Oil and natural gas producer prices rise through the century relative to year 2000 levels, whereas coal and electricity prices remain relatively stable. It should be emphasized that the models used in this research were not designed to simulate short-

term fuel-price spikes, such as those that occurred in the 1970s, early 1980s, and more recently in 2005. Thus, price trends in the scenarios should be interpreted as multi-year averages.

- As a combined result of all these influences, CO_2 emissions from fossil fuel combustion and industrial processes in the reference scenarios increase from approximately 7 GtC/yr in 2000 to between 22.5 GtC/yr and 24.0 GtC/yr in 2100; that is, to roughly 3 to 3½ times current levels.

The non-CO_2 GHGs, CH_4, N_2O, SF_6, PFCs, and HFCs, are emitted from various sources including agriculture, waste management, biomass burning, fossil fuel production and consumption, and a number of industrial activities.

- Future global anthropogenic emissions of CH_4 and N_2O vary widely among the reference scenarios, ranging from flat or declining emissions to increases of 2 to 2½ times present levels. These differences reflect alternative assumptions about technological opportunities and about whether current emissions rates will be reduced significantly for non-climate reasons, such as air pollution control and/or higher natural gas prices that would further stimulate the capture of CH_4 emissions for its fuel value.

Increases in emissions from the global energy system and other human activities lead to higher atmospheric GHG concentrations and radiative forcing. This increase is moderated by natural biogeochemical removal processes.

- The oceans are a major sink for CO_2, and the rate at which they take up CO_2 generally increases in the reference scenarios as concentrations rise early in the century. However, processes in the ocean can slow this rate of increase at high concentrations late in the century. Ocean uptake in the three reference scenarios is in the range of 2 GtC/yr in 2000, rising to about 5 GtC/yr to 11 GtC/yr by 2100. The three ocean models behave more similarly in the stabilization scenarios; for example, the difference in ocean uptake among the models at the most stringent stabilization levels is less than 1 GtC/yr in 2100.
- Two of the three participating models include sub-models of the exchange of CO2 with the terrestrial biosphere, including the net uptake by plants and soils and the emissions from deforestation. In the reference scenarios from these modeling groups, the terrestrial biosphere acts as a small annual net sink (less than 1 GtC/yr of carbon) in 2000, increasing to an annual net sink of roughly 2 GtC/yr to 3 GtC/yr by the end of the century. The third modeling group assumed a zero net exchange. Changes in emissions from terrestrial systems over time in the reference scenarios reflect assumptions about human activity (including a decline in deforestation) as well as increased CO_2 uptake by vegetation as a result of the positive effect of CO_2 on plant growth. There remains substantial uncertainty about this carbon fertilization effect and its evolution under a changing climate.
- As a result of the various influences, GHG concentrations rise substantially over the century in the reference scenarios. By 2100, CO_2 concentrations range from about 700 ppmv to 900 ppmv, up from 365 ppmv in 1998. CH_4 concentrations in 2100

range from 2000 ppbv to 4000 ppbv, up from 1745 ppbv in 1998, and N_2O concentrations in 2100 range from about 375 ppbv to 500 ppbv, up from 314 ppbv in 1998.

- As a result, radiative forcing in 2100 ranges from 6.4 W/m^2 to 8.6 W/m^2 from preindustrial, up from a little over 2 W/m^2 today. The non-CO_2 GHGs account for about 20% to 25% of radiative forcing at the end of the century.

Stabilization Scenarios

Important assumptions underlying the stabilization scenarios include the flexibility that exists in a policy design, as represented by the modeling groups, to seek out least cost options for emissions control regardless of where they occur, what substances are controlled, or when they occur. This set of conditions is referred to as *where*, *what*, and *when* flexibility. Equal marginal costs of abatement among regions across time (taking into account discount rates and the lifetimes of substances), and among substances (taking into account their relative warming potential and different lifetimes) will, under specified conditions, lead to least cost abatement. Each modeling group applied an economic instrument that priced GHGs in a manner consistent with the group's interpretation of *where*, *what*, and *when* flexibility. The economic characteristics of the scenarios thus assume a policy designed with the intent of achieving the required reductions in GHG emissions in a least- cost way. Key implications of these assumptions are that: (1) all nations proceed together in restricting GHG emissions from 2012 and continue together throughout the century, and that the same marginal cost is applied across sectors (*where* flexibility); (2) the marginal cost of abatement rises over time in these three sets of scenarios based on each modeling group's interpretation of *when* flexibility, with the effect of linking emissions mitigation efforts over the time horizon of the scenarios; and (3) stabilization of radiative forcing is achieved by combining control of all GHGs, with differences in how modeling groups compared them and assessed the implications of this *what* flexibility.

Although these assumptions are convenient for analytical purposes, to gain an impression of the implications of stabilization, they are idealized versions of possible outcomes. For the abatement costs in these scenarios to be representative of actual abatement costs would require, among other things, that a negotiated international agreement include these flexibility mechanisms. Failure in that regard could have a substantial effect on the difficulty of achieving any of the stabilization levels considered in this research. For example, a delay of many years in the participation of some large countries would require greater effort by the others, and policies that impose differential burdens on different sectors without mechanisms to allow for equalizing marginal costs across sectors can result in a many-fold increase in the cost of any environmental gain. Therefore, *it is important to view these scenarios as representing possible futures under specified conditions, not as forecasts of the most likely outcome within the national and international political system.* Further, none of the scenarios considered the extent to which variation from these least-cost rules might be improved upon given interactions with existing taxes, technology spillovers, or other non-market externalities.

If the developments in the three reference scenarios were to occur, concerted efforts to reduce GHG emissions would be required to stabilize radiative forcing at the levels considered in this research. Such limits would shape technology deployment throughout the century and have important economic consequences. The stabilization scenarios demonstrate that there is no single technology pathway consistent with a given level of radiative forcing. Furthermore, there are other possible pathways than those considered in this research.

- Stabilization efforts are made more challenging by the fact that ocean uptake of CO_2 declines as the stringency of the stabilization level increases, and, in the scenarios from two of the models, because CO_2 uptake in terrestrial systems also declines with the stringency of the stabilization level.
- Stabilization of radiative forcing at the levels examined in this research would require a substantially different energy system globally, and in the U.S., than what emerges in the reference scenarios. The degree and timing of change in the global energy system depends on the level at which radiative forcing is stabilized. The lower the radiative forcing stabilization level, the larger the scale of change in the global energy system relative to the reference scenario and the sooner those changes would need to occur.
- Across the stabilization scenarios, the energy system relies more heavily on non-fossil energy sources, such as nuclear, solar, wind, biomass, and other renewable energy forms, than in the associated reference scenarios. The stabilization scenarios differ in the degree to which these technologies are deployed, depending on assumptions about: technological improvements; the ability to overcome obstacles, such as intermittency in the case of solar and wind power, or safety, waste, and proliferation issues in the case of nuclear power; and the policy environment surrounding these technologies. Energy consumption, while still higher than today's levels, is lower in the stabilization scenarios than in the reference scenarios.
- CCS is widely deployed in the stabilization scenarios because each modeling group assumed that the technology can be successfully developed and that concerns about storing large amounts of carbon do not impede its expansion. Removal of this assumption would make the stabilization levels more difficult to achieve and would lead to greater demand for low-carbon sources such as renewable energy and nuclear power, to the extent that growth of these other sources is not otherwise constrained.
- Significant fossil fuel use continues across the stabilization scenarios, both because stabilization allows for some level of carbon emissions through 2100, depending on the stabilization level, and because of the presence of CCS technology in all the stabilization scenarios.
- Emissions of non-CO_2 GHGs, such as CH_4, N_2O, HFCs, PFCs, and SF_6, are all reduced in the stabilization scenarios.
- Increased use is made of biomass energy crops in all the stabilization scenarios, but their contribution is ultimately limited by competition with agriculture and forestry, and, in one participating model, by the associated impacts of biomass expansion on carbon emissions from changes in land use.
- The lower the radiative forcing stabilization level, the larger the scale of change in the global energy system relative to the reference scenario required over the coming century and the sooner those changes would need to occur.
- Across the stabilization scenarios, the scale of the emissions reductions required relative to the reference scenario increases over time, with the bulk of emissions

reductions taking place in the second half of the century. But emissions reductions occur in the first half of the century in every stabilization scenario.

- The 2100 time horizon of this research limited examination of the ultimate stabilization requirements. Further reductions in CO_2 emissions after 2100 would be required in all of the stabilization scenarios, because stabilization of radiative forcing at any of the levels considered in this research requires human emissions of CO_2 in the long term to be essentially halted. Despite the fact that much of the carbon emissions will eventually make its way into oceans and terrestrial sinks, some will remain in the atmosphere for thousands of years. Only CCS can allow continued burning of fossil fuels. Higher radiative forcing limits can delay the point in time at which emissions must be reduced toward zero, but this requirement must ultimately be met.

Fuel sources and electricity generation technologies change substantially, both globally and in the U.S., in the stabilization scenarios compared to the reference scenarios. There are a variety of technological options in the electricity sector that reduce carbon emissions in these scenarios.

- Nuclear power, renewable energy, and CCS all play important roles in stabilization scenarios. The contribution of each varies, depending on assumptions about technological improvements, the ability to overcome obstacles such as intermittency of supply, and the policy environment surrounding them.
- By the end of the century, electricity produced by conventional fossil technology that freely emits CO_2 is reduced in the stabilization scenarios relative to reference scenarios. Electricity production from technologies that emit CO_2varies substantially with the stabilization level; in the lowest stabilization level, electricity production from these technologies is reduced toward zero.

The economic effects of stabilization are substantial in many of the stabilization scenarios, although much of this cost is borne later in the century. As noted earlier, each of the modeling groups assumed that a global policy was implemented after 2012, with universal participation by the world's nations, and that the time path of reductions approximated a least-cost solution. These assumptions of *where*, *when*, and *what* flexibility lower the economic consequences of stabilization relative to what they might be with other implementation approaches.

- The stabilization scenarios follow a pattern where, in most scenarios, the carbon price rises steadily over time, providing an opportunity for the energy system to adjust gradually.
- Although the general shape of the carbon price trajectory over time is similar across the models, the carbon prices vary substantially across the models. For example, two of the scenarios have prices of \$10 or below per tonne of carbon in 2020 for the less stringent scenarios, with their prices rising to roughly \$100 per tonne in 2020 for the most stringent stabilization level. A third scenario shows higher initial carbon prices in 2020, ranging from around \$20 for the least stringent stabilization level to over \$250 for the most stringent stabilization level.

- Factors contributing to differences in carbon prices include (1) differences in assumptions – such as those regarding economic growth over the century, the behavior of the oceans and terrestrial biosphere in taking up CO_2, and opportunities for reduction in non-CO_2 GHG emissions – that determine the amount that CO_2 emissions must be reduced to meet the radiative forcing stabilization levels; and (2) differences in assumptions about technologies, particularly in the second half of the century, to shift final demand to low-carbon sources such as biofuels and low-carbon electricity or hydrogen, in transportation, industrial, and buildings end uses. Differences among the scenarios reflect the uncertainty that attends the far future.
- Differences in non-CO_2 gases also contribute to differences in abatement costs. Scenarios that assume relatively better performance of non-CO_2 emissions mitigation require less CO_2 abatement and therefore less stringent changes in the energy system, to meet the same overall radiative forcing goal.
- These differences in carbon prices, along with other model features, lead to similar variation in the costs of stabilization. At the most stringent radiative forcing stabilization level, for example, gross world product (aggregating country figures using market exchange rates) is reduced in 2050 by around 1% in the scenarios from two of the modeling groups and approximately 5% in the scenario from the third, and in 2100 it is reduced by less than 2% in two of the scenarios and over 16% in the third.
- The assumption of *when* flexibility links elements of the stabilization scenarios through time. This in turn means that, in addition to near-term technology availability, differences in assumptions about technology in the post- 2050 period are also reflected in near-term emissions reductions and GHG prices.
- In all of the stabilization scenarios, emissions reductions in electric power sector come at relatively lower prices than in other sectors (e.g., buildings, industry, and transport) so that the electricity sector is essentially decarbonized in the most stringent scenarios. At somewhat higher cost other sectors can respond to rising carbon prices by reducing demands for fossil fuels, applying CCS technologies where possible, and substituting low-carbon energy sources such as bioenergy and low-carbon electricity or hydrogen. The amount of electricity used per unit of total primary energy increases in all of the stabilization scenarios, but those scenarios with the highest relative use of electricity tend to exhibit lower stabilization costs in part because of the larger role of decarbonized power generation. Assumptions regarding costs and performance of technologies to facilitate these adjustments, particularly in the post-2050 period, play an important role in determining stabilization costs
- As noted earlier, the overall cost levels are strongly influenced by the idealized policy scenario that has all countries participating from the start, the assumption of *where* flexibility, an efficient pattern of emissions reductions over time, and integrated reductions in emissions of the different GHGs. Assumptions in which policies are implemented in a less efficient manner would lead to higher cost. Thus, these scenarios should not be interpreted as applying beyond the particular conditions assumed.
- GHG mitigation would also affect fuel prices. Generally, producer prices for fossil fuels fall as demand for them is depressed by the stabilization measures. Consumers

of fossil fuels, on the other hand, pay for fuel plus a carbon price if the CO_2 emissions are freely released to the atmosphere. Therefore, consumer costs of energy rise with more stringent stabilization levels in these scenarios.

Achieving stabilization of atmospheric GHGs poses a substantial technological and policy challenge. It would require important transformations of the global energy system. The cost and feasibility of such a goal depends on the evolution of technology and its ability to overcome existing limits and barriers to adoption, and it depends on the efficiency and effectiveness of the policy instruments employed to achieve stabilization.

Application of the Scenarios in Further Analysis

These scenarios, supported by the accompanying database described in the Appendix, can be used as the basis of further analysis. There are a variety of possible applications for these scenarios. For example, the scenarios could be used as the basis for analysis of the climate implications, and then follow-on studies of potential climate impacts. Such studies might begin with the radiative forcing levels of each scenario, with the individual GHG concentrations (applying separate radiation codes) or with the emissions (applying separate models of the carbon cycle and of the atmospheric chemistry of the non-CO_2 GHGs). Such applications could be made directly in climate models that do not incorporate a three-dimensional atmosphere and detailed biosphere model. For the larger models, some approximation would need to be imposed to allocate the short-lived gases by latitude or grid cell. Such an effort would need to include scenarios of the emissions (or concentrations) of the reflecting and absorbing aerosols. This could be achieved by the use of sub-models linked to scenario for energy use by fuel.

The scenarios could also be used as a point of departure for partial equilibrium analysis of technology development. Because these models compute energy prices, the scenarios can be used for analysis of the cost performance of new technologies and to serve as a basis for analysis of rates of market penetration. Differences in the scenarios among the three modeling groups give an impression of the types of market challenges that new options will face.

In addition, these studies could form the foundation of analysis of the non-climate environmental implications of implementing potential new energy sources at a large scale. Such analysis was beyond the scope of the present research, but information is provided that could form a basis for such analysis, for example, the potential effects on the U.S. and the globe of implied volumes of CCS and biomass production or of nuclear power expansion in some of the scenarios.

The scenarios could also be used in comparative mode. Just as many lessons were learned by comparing the differences between the three modeling groups' scenarios, still more could be learned by extending the comparison to scenarios that predate these or come after, including scenarios developed using entirely different approaches. For example, some scenario exercises do not apply economic models with detailed analysis of energy markets of the type used here. Such scenarios could be compared against those presented here to gain insight into the role of economic factors.

Finally, these scenarios might be used to explore the economic effects of stabilization at different levels. Such work was beyond the scope of the research specified in the Prospectus. However, the scenarios do contain information that can be used to calculate indicators of consumer impact in the U.S., for example, by using the changes in prices and quantities of fuels in moving from one stabilization level to another. (The reader is reminded, however, that these welfare effects do not include the benefits that alternative stabilization levels might yield in reduced climate change risk or ancillary effects, such as effects on air pollution).

Moving Forward

As noted earlier, this work is neither the first nor is it likely to be the last of its kind. Throughout the report, a number of limitations to the approach and the participating models have been highlighted. Studies such as the one presented here would benefit from further research and model development and this section suggests several productive paths to pursue.

Technology Sensitivity Analysis

The importance of future technology development is clear in this chapter, and sensitivity testing of key assumptions would be of use. For example, what are the implications of various non-climate constraints on nuclear power or on the large-scale expansion of CCS or biofuels production? If particular supply technologies – nuclear, wind, natural gas combined cycle generation, and biomass – were assumed to be more or less expensive, how would that affect market penetration and policy cost? On the demand side, what are the effects of alternative views of the technical developments needed to facilitate substitution of electricity for liquid and gaseous fuels in various sectors, particularly in transport? Since technology deployment will be influenced by the policy environment, how would the consideration of less optimistic policy regimes affect this aspect of the scenarios?

Consideration of Less Optimistic Policy Regimes

The discussion in Chapter 4 emphasizes that the difficulty of the stabilization task emerging from any scenario research is crucially dependent on underlying institutional assumptions, and the insight to be gained from a single representation of control policy such as the one adopted in this research is limited. The scenarios assume a wide array of idealized institutions both in individual nations and in the international community. Both developed and developing economies are assumed to possess markets that efficiently pass price information to decision makers. Rules and regulations ranging from accounting and property rights to legal and enforcement systems are assumed to operate efficiently. While such assumptions provide a well-defined reference scenario and lower- bound information on potential costs, the probability is low that the world will actually implement such an idealized architecture. In that light, a natural direction for future research is to supplement the analysis presented here with analyses of policy regimes that are under discussion by nations and international organizations and that have a greater potential for being implemented. Such research would broaden the understanding of the stabilization challenge in areas ranging from technology development to the economics of global mitigation.

Expansion and/or Improvement of the Land-Use Components of the Models

A significant weakness in this research is the handling of the role of forest and agricultural sinks and sources. The major reason for this gap is that the models employed here were not well suited to analyze some of the complexities of this aspect of the carbon cycle. Yet, as this analysis has shown, agriculture, land-use and terrestrial carbon cycle issues play an important role in shaping the long-term radiative character of the atmosphere. Research that would improve the characterization of land use and land cover as well as improve the linkages among energy and economic systems, land use, land cover, terrestrial carbon processes, and other bio-geochemical cycles has potentially high payoff.

Inclusion of other Radiatively Important Substances

The focus in this research is on the relatively long-lived GHGs, but shorter-lived substances, such as ozone and aerosols, have strong radiative effects as well. More complete analysis would include these short-lived contributors, and their control possibilities, directly within the scenario analysis.

Decision Making under Uncertainty

Finally, the problem of how to respond to the threat of climate change is ultimately a problem of decision making under uncertainty that requires an assessment of the risks of climate change and how policies might reduce the odds of extremely bad outcomes. One would like to compare the expected benefits of policies to reduce GHG emissions against the expected costs of achieving those reductions. By focusing only on emission paths that would lead to stabilization, this research considers only the costs of stabilization without consideration of the benefits. Moreover, given the direction provided in the Prospectus, this research focused on scenarios and not on uncertainty analysis. It is not possible to attach probabilities to scenarios constructed in this way; formal probabilities can only be attached to a range, which requires exploration of the effects of many uncertain model parameters.

APPENDIX A

Accompanying this chapter are two databases of scenario information collected from the participating modeling teams. These databases can be found at http://www.climatescience.gov/.

The database entitled *CCSP 2_1A Scenario Information.xls* includes all the information that was collected from the modeling teams to support the development of this chapter, including the figures and tables presented in the report. The database entitled *CCSP 2_1A Other Species.xls* includes additional emissions information for substances not included in the radiative forcing stabilization levels used in this research, such as ozone precursors and aerosols. The emissions profiles for these other substances are different from what would emerge in scenarios in which these substances are explicitly included in stabilizing radiative forcing.

The two data files include the following fields:
Region
Category
Sub Category Variable
Units
Run Label
Year
Acronyms

ACRONYMS

AEEI	autonomous energy efficiency improvement
AOGCMs	atmosphere-ocean general circulation models
CCS	carbon capture and storage
CCSP	Climate Change Science Program
CCTP	Climate Change Technology Program
CFCs	chlorofluorocarbons
CGE	computable general equilibrium
CPDAC	Climate Change Science Program Synthesis and Assessment Product
DOE	U.S. Department of Energy
EMF	Energy Modeling Forum
EPPA	Emissions Prediction and Policy Analysis
FCCC	U.N. Framework Convention on Climate Change
GCMs	general circulation models
GDP	gross domestic product
GHGs	greenhouse gases
GWP	Global Warming Potential
HFCs	hydrofluorocarbons
HCFC	hydrochlorofluorocarbons
IAMs	Integrated Assessment Models
IGCC	Integrated Gasification Combined Cycle
IGSM	Integrated Global Systems Model
IPCC	Intergovernmental Panel on Climate Change
MAC	marginal abatement cost
MAGICC	Model for the Assessment of Greenhouse-Gas Induced Climate Change
MER	market exchange rate
MERGE	Model for Evaluating the Regional and Global Effects
NCAR	National Center for Atmospheric Research
NGCC	natural gas combined cycle
NMVOCs	non-methane volatile organic compounds
OECD	Organization for Economic Cooperation and Development
PFCs	perfluorocarbons
PPP	purchasing power parity
SRES	Special Report on Emissions Scenarios

TAR	Third Assessment Report
U.N.	United Nations
U.S.	United States

Units

$2000	U.S. 2000 dollars
bbl	barrel
c/kWh	cents per kilowatt hour
EJ	exajoule
gal	gallon
GJ	gigajoule
Gt	gigatonne
GtC	gigatonne carbon
Mt	megatonne
MtC	megatonne carbon
Kt	kilotonne
PgC	petagram carbon
ppbv	parts per billion by volume
ppmv	parts per million by volume
ppt	parts per trillion
Quad	quadrillion btu
tcf	thousand cubic feet
W/m^2	watts per meter squared
yr	year

Chemical Formulas

CH_4	methane
CO	carbon monoxide
CO_2	carbon dioxide
N_2O	nitrous oxide
NO_x	nitrogen oxides
O_3	ozone
SF_6	sulfur hexafluoride

REFERENCES

CCSP, (2003). Strategic Plan for the U.S. Climate Change Science Program. A Report by the U.S. *Climate Change Science Program and the Subcommittee on Global Change Research*. Washington, D.C.

CCSP, (2005). Final Prospectus for Synthesis and Assessment Product 2.1. A Report by the U.S. *Climate Change Science Program and the Subcommittee on Global Change Research.* Washington, D.C.

CCSP, (2007). Global Change Scenarios: Their Development and Use. A Report by the U.S. *Climate Change Science Program and the Subcommittee on Global Change Research* [Parson, E., Burkett, V., Fisher-Vanden, K., Keith, D., Mearns, L., Pitcher, H., Rosenzweig, C. & Webster, M.]. U.S. Department of Energy, Washington, D.C.

U.S. DOE – Department of Energy, Energy Information Administration, 2006, *Short-Term Energy*, and Winter Fuels Outlook October 10th, 2006 Release.

IPCC – Intergovernmental Panel on Climate Change. (2001). Climate Change 2001: The Scientific Basis. eds J. T., Houghton, Y., Ding, D J., Griggs, N., Noguer, P. J., van der Linden, X., Dai, K. Maskell, & C A. Johnson, Cambridge University Press, Cambridge, U.K.

Babiker, M., Bautista, M., Jacoby, H. & Reilly, J. (2000). *Effects of Differentiating Climate Policy by Sector: A U.S. Example*. Report No. 73, MIT Joint Program on the Science and Policy of Global Change.

Babiker, M., Metcalf, G. & Reilly, J. (2003a). Tax distortions and global climate policy. *Journal of Economic and Environmental Management, 46*, 269-287.

Babiker, M., Viguier, L., Reilly, J., Ellerman, A. & Criqui, P. (2003b). The welfare costs of hybrid carbon policies in the european union. *Environmental Modeling and Assessment, 8*,187-1

Babiker, M., Reilly, J. & Viguier, L. (2004). Is emissions tradingalways beneficial. *The Energy Journal, 25(2)*, 33-56.

Brenkert, A., Smith, S., Kim, S. & Pitcher, H. (2003). *Model Documentation for the Mini CAM*. PNNL-14337, Pacific Northwest National Laboratory, Richland, Washington.

CCSP, (2003). *Strategic Plan for the U.S. Climate Change Science Program*. A Report by the U.S. Climate Change Science Program and the Subcommittee on Global Change Research. Washington, D.C.

CCSP, (2005). *Final Prospectus for Synthesis and Assessment Product 2.1*. A Report by the U.S. Climate Change Science Program and the Subcommittee on Global Change Research. Washington, D.C.

CCSP, (2007). *Global Change Scenarios: Their Development and Use*. A Report by the U.S. Climate Change Science Program and the Subcommittee on Global Change Research [Parson, E., Burkett, V., Fisher-Vanden, K., Keith, D., Mearns, L., Pitcher, H., Rosenzweig, C. & Webster, M.]. U.S. Department of Energy, Washington, D.C.

CEA, (2005). *Economic Report of the President*, Council of Economic Advisors .U.S. Government Printing Office, Washington, D.C.

Clarke, L., Edmonds, J., Kim, S., Lurz, J., Pitcher, H., Smith, S. & Wise, M. (2007). *Documentation for the Mini CAM CCSP Scenarios*. Battelle Pacific Northwest Division Technical Report, PNNL-16735.

Claussen M., Mysak, L., Weaver, A., Crucifix, M., Fichefet, T., Loutre, M., Weber, S., Alcamo, J., Alexeev, V., Berger, A., Calov, R., Ganopolski, A., Goosse, H., Lohman, G., Lunkeit, F., Mokhov, I., Petoukhov, V., Stone, P. & Wang, Z. (2002). Earth System Models of Intermediate Complexity: Closing the Gap in the Spectrum of Climate System Models. *Climate Dynamics, 15*, 579-586.

Cubasch, U., Meehl, G., Boer, G., Stouffer, R., Dix, M., Noda, A., Senior, C., Raper, S. & Yap, K. (2001). Projections of Future Climate Change. In IPCC, 2001: *Climate Change 2001: The Scientific Basis*, 525-582. Cambridge, U.K.

F. de la Chesnaye, & J. Weyant, (eds.), (2006). Multi-greenhouse gas mitigation and climate policy. *The Energy Journal*, Special Issue.

Edmonds, J., Freund, P. & Dooley, J. (2001). The Role of Carbon Management Technologies in Addressing Atmospheric Stabilization of Greenhouse Gases. In: *Greenhouse Gas Control Technologies, Proceedings of the Fifth International Conference on Greenhouse Gas Control Technologies*, D. J., Williams, R. A., Durie, P., McMullan, C. A. J. Paulson, & A. Y. Smith, (eds.) CSIRO, Collingswood, VIC, Australia. 46-51.

DOE, (1985). *Atmospheric Carbon Dioxide and the Global Carbon Cycle*, ed JR Trabalka, DOE/ER-0239. Office of Energy Research, United States Department of Energy, Washington, D.C.

DOE, (2006). *Short-Term Energy, and Winter Fuels Outlook*, October 10th, 2006 Release. United States Department of Energy, Washington, D.C.

Dooley, J. & Friedman, S. (2004). *A Regionally disaggregated global accounting of CO_2 storage capacity: data and assumptions*. Battelle Pacific Northwest Division Technical Report Number PNWD-343 1.

Edmonds, J. & Reilly, J. (1985). *Global Energy: Assessing the Future*. Oxford University Press, New York.

Hotelling H. (1931). The economics of exhaustible resources. *Journal of Political Economy*, *39*,137-175.

IPCC, (1991). *Climate Change: The IPCC Response Strategies*. Intergovernmental Panel on Climate Change, Island Press, Washington, D.C.

IPCC, (1992). *Climate Change 1992: The Supplementary Report to the IPCC Scientific Assessment*. eds J. T., Houghton, B. A Callander, & S. K. Varney, Intergovernmental Panel on Climate Change, Cambridge University Press, Cambridge, U.K.

IPCC, (1996). *Climate Change 1995: Economic and Social Dimensions of Climate Change*. eds J. P., Bruce, H. Lee, & E. F. Haites, Intergovernmental Panel on Climate Change, Cambridge University Press, Cambridge, U.K.

IPCC, (2001). *Climate Change 2001: The Scientific Basis. The Contribution of Working Group I to the Third Assessment Report of the Intergovernmental Panel on Climate Change*, 944. eds J. T., Houghton, Y., Ding, D. J., Griggs, M., Noguer, P. van der Linden, & D. Xiaosu, Intergovernmental Panel on Climate Change, Cambridge University Press, Cambridge, U.K.

Leggett, J., Pepper, W., Swart, R., Edmonds, J., Meira Filho, L., Mintzer, I., Wang, M. & Wasson, J. (1992). Emissions Scenarios for the IPCC: An Update. In: *Climate Change 1992: The Supplementary Report to the IPCC Scientific Assessment*, University Press, Cambridge, U.K.

Maddison, A. (2001). *The World Economy: A Millennial Perspective.* Organization for Economic Cooperation and Development, Paris, France.

Maier-Reimer, E. & Hasselmann, K. (1987). Transport and Storage of CO_2 in the Ocean —An Inorganic Ocean-Circulation Carbon Cycle Model. *Climate Dynamics*, *2*, 63-90.

Manne, A. & Richels, R. (2005). MERGE-A Model for Global Climate Change. In: *Energy and Environment*, eds R., Loulou, J. Waaub, & Zaccour, Springer, New York.

Manne, A. & Richels, R. (1994). The costs of stabilizing global CO_2 emissions: a probabilistic analysis based on expert judgements. *The Energy Journal, 15(1)*, 31-56.

Manne A. & Richels, R. (2001). An alternative approach to establishing tradeoffs among gases. *Nature, 419*, 675-676.

Morita, T. & Lee, H. C. (1998). IPCC SRES Database, Version 0.1, Emission Scenario Database Prepared for IPCC Special Report on Emission Scenarios. Accessed at http://www cger.nies.go.jp/cger e/db/ipcc.html.

Nakicenovic, N., Alcamo, J., Davis, G., de Vries, B., Fenhann, J., Gan, S., Gregory, K., Grubler, A., Jung, T., Kram, T., Rovere, E., Michaelis, L., Mori, S., Morita, T., Pepper, W., Pitcher, H., Price, L., Riahi, K., Roehrl, A., Rogner, A., Sankovski, M., Schlesinger, P., Shukla, S., Smith, R., Swart, S., van Rooijen, N. Victor, & Dadi, Z. (2000). *Special Report on Emissions Scenarios*. Cambridge University Press, Cambridge, U.K.

Nakicenovic N., Kolp, P., Riahi, K., Kainuma, M. & Hanaoka, T. (2006). Assessment of Emissions Scenarios Revisited. *Environmental Economics and Policy Studies, 7(3)*, 13 7-173

Nakicenovic, N., Victor, N. & Morita, T. (1998). Emissions scenarios database and review of scenarios. *Mitigation and Adaptation Strategies for Global Change 3(2-4)*, 95-120.

NAS, (1983). *Changing Climate: Report of the Carbon Dioxide Assessment Committee*. - National Academy of Sciences, National Academy Press, Washington, D.C.

NOAA - National Oceanic and Atmospheric Administration, (2007). Global Warming: Frequently Asked Questions, at http://lwf.ncdc. noaa.gov/oa/climate

Nordhaus, W. & Yohe, G. (1983). Future Carbon Dioxide Emissions from Fossil Fuels. In: *Changing Climate*, 87-153. National Academy Press, Washington, D.C.

NRC, (2005). *Radiative Forcing of Climate Change: Expanding the Concept and Addressing Uncertainties*. National Research Council, National Academy Press, Washington, D.C.

O'Neill, B. (2005). Population Scenarios Based on Probabilistic Projections: An Application for the Millennium Ecosystem Assessment. *Population and Environmen*t, *26(3)*, 229-254.

Paltsev, S., Jacoby, H., Reilly, J., Viguier, L. & Babiker, M. (2005). Modeling the Transport Sector: The Role of Existing Fuel Taxes. *Energy and Environment*, 211-238 eds R., Loulou, J. P. Waaub, & G. Zaccour, Springer, New York.

Paltsev, S., Reilly, J., Jacoby, H., Eckaus, R., McFarland, J., Sarofim, M., Asadoorian, M. & Babiker, M. (2005). *The MIT Emissions Prediction and Policy Analysis (EPPA) Model: Version 4*. Report 125, MIT Joint Program on the Science and Policy of Global Change, Cambridge, Massachusetts.

Parson, E. & Fisher-Vanden, K. (1997). integrated assessment models of global climate change. *Annual Review of Energy and the Environment, 22*, 589-628.

Peck, S. & Wan, Y. (1996). Analytic Solutions of Simple Greenhouse Gas Emission Models. Chapter 6 in: *Economics of Atmospheric Pollution*, eds E. C. Van Ierland, & K. Gorka, Springer Verlag, New York.

Pepper, W., Leggett, J., Swart, R., Wasson, J., Edmonds, J. & Mintzer, I. (1992). Emissions Scenarios for the IPCC: An Update – Assumptions, Methodology, and Results. *Intergovernmental Panel on Climate Change, Geneva*, Switzerland.

Prinn, R., Reilly, J., Sarofim, M., Wang, C. & Felzer, B. In Press. Effects of air pollution control on climate. In: *Integrated Assessment of Human-induced Climate Change*, ed M Schlesinger, Cambridge University Press, Cambridge, U.K.

Raper, S., Gregory, J. & Osborn, T. (2001). Use of an upwelling-diffusion energy balance climate model to simulate and diagnose a/ogcm results. *Climate Dynamics*, *17*, 601-613.

Reilly, J., Edmonds, J., Gardner, R. & Brenkert, A. (1987). Uncertainty analysis of the aea/orau co_2 emissions model. *The Energy Journal*, *8(3)*, 1-29.

Reilly, J., Jacoby, H. & Prinn, R. (2003). *Multi-Gas Contributors to Global Climate Change: Climate Impacts and Mitigation Costs of Non-CO_2 Gases*. Pew Center on Global Climate Change, Washington, D.C.

Richels, R. & Edmonds, J. (1995). The economics of stabilizing atmospheric co_2 concentrations. *Energy Policy*, *23(4/5)*, 373-78.

Richels, R., Edmonds, J., Gruenspecht, H. & Wigley, T. (1996). The Berlin Mandate: The Design of Cost-Effective Mitigation Strategies. *Climate Change: Integrating Science, Economics and Policy*, CP-96- 1:29-48. N., Nakicenovic, W. D., Nordhaus, R. Richels, & F. L. Toth, (eds.), International Institute for Applied Systems Analysis, Laxenburg, Austria.

Sabine, C., Heiman, M., Artaxo, P., Bakker, D., Chen, C., Field, C., Gruber, N., LeQuéré, C., Prinn, R., Richey, J., Romero-Lankao, P., Sathaye, J. & Valentini, R. (2004). Current Status and Past Trends of the Carbon Cycle. Pages 17-44 in *The Global Carbon Cycle, SCOPE Project*, *62*, eds CB Field and MR Raupach, Island Press, Washington D.C.

Sands, R. & Leimbach, M. (2003). Modeling agriculture and land use in an integrated assessment framework, *Climatic Change*, *56(1)*, 185-210.

Sarofim, M., Forest, C., Reiner, D. & Reilly, J. (2005). Stabilization and Global Climate Policy, *Global and Planetary Change*, *47*, 266-272. Accessed as MIT Joint Program on the Science and Policy of Global Change Reprint No. 2005-5 at http://mit.edu/globalchange/www/.

Scott, M., Sands, R., Edmonds, J., Liebetrau, A. & Engel, D. (2000). Uncertainty in integrated assessment models: modeling with minicam 1.0. *Energy Policy*, *27(14)*, 855-879.

Smith, S. & Wigley, T. (2006). Multi-Gas Forcing Stabilization with Minicam, *The Energy Journal*, Multi-Greenhouse Gas Mitigation and Climate Policy Special Issue.

Smith, S. (2005). Income and pollutant emissions in the objects minicam model. *Journal of Environment and Development*, *14(1)*, 175-1 96.

Sokolov, A., Schlosser, C., Dutkiewicz, S., Paltsev, S., Kicklighter, D., Jacoby, H., Prinn, R., Forest, C., Reilly, J., Wang, C., Felzer, B., Sarofim, M., Scott, J., Stone, P., Melillo, J. & Cohen, J. (2005). *The MIT Integrated Global System Model (IGSM) Version 2*: Model Description and Baseline Evaluation. Report 124, MIT Joint Program on the Science and Policy of Global Change, Cambridge, Massachusetts.

UN, (1992). *Framework Convention on Climate Change*. United Nations, New York.

UN, (2000). *Long-Run World Population Projections: Based on the 1998 Revision*. United Nations, New York.

UN, (2001). *World Population Prospects: The 2000 Revision, Data in digital form*. Population Division, Department of Economic and Social Affairs, United Nations, New York.

UN, (2005). *World Population Prospects: The 2004 Revision, Data in digital form*. Population Division, Department of Economic and Social Affairs, United Nations, New York.

Webster, M., Babiker, M., Mayer, M., Reilly, J., Harnisch, J., Hyman, R., Sarofim, M. & Wang, C. (2003). Uncertainty in emissions projections for climate models. *Atmospheric Environment, 36(22)*, 3659-3670.

Webster, M., Babiker, M., Mayer, M., Reilly, J., Harnisch, J., Hyman, R., Sarofim, M. & Wang, C. (2002). Uncertainty in emissions projections for climate models. *Atmospheric Environment, 36(22)*, 3659-3670.

Webster, M. (2003). Communicating climate change uncertainty to policy-makers and the public. *Climatic Change, 61*, 1-8.

Wigley, T. & Raper, S. (2001). Interpretation of high projections for global-mean warming. *Science, 293*, 451-454.

Wigley, T. & Raper, S. (2002). Reasons for larger warming projections in the ipcc third assessment report. *Journal of Climate, 15*, 2945-2952.

Wigley, T., Richels, R. & Edmonds, J. (1996). Economic and environmental choices in the stabilization of atmospheric co_2 concentrations. *Nature, 379(6562)*, 240-243.

Wigley, T., Smith, S. & Prather, M. (2002). Radiative forcing due to reac- tive gas emissions, *Journal of Climate, 15(18)*, 2690-2696.

End Notes

[1] Although there are many reasons to expect that the three reference scenarios would be different, it is worth noting that the modeling groups met periodically during the development of the scenarios to review progress and to exchange information. Thus, while not adhering to any formal protocol of standardization, the three reference scenarios are not entirely independent.

[2] For simplicity, all hydrogenated halocarbons will be referred to as HFCs in the subsequent text. The greenhouse gas methyl chloroform is often also grouped along with HFCs and HCFCs.

[3] See footnote 1.

[4] It differs from the pure bottom-up approach described in Box 2.1 in that demands for energy are price responsive.

[5] Annex I is defined in the U.N. Framework Convention on Climate Change (FCCC [UN 1992]). However, since the FCCC entered into force, the Soviet Union has broken up. As a consequence, some of the republics of the former Soviet Union are now considered developing nations and do not have the same obligations as the Russian Federation under the FCCC. Thus, strictly speaking, the aggregations employed by the three modeling groups may not precisely align with the present partition of the world's nations. However, the quantitative implications of these differences are small.

[6] The IGSM exceeds the Level 1 target by 0.1 W/m^2, which is a negligible difference that results from the iterative process required to achieve a radiative forcing target.

In: Greenhouse Gas Emission Scenarios For Use In Climate… ISBN: 978-1-61209-610-0
Editor: Jacqueline L. Aigner

Chapter 2

GLOBAL-CHANGE SCENARIOS: THEIR DEVELOPMENT AND USE*

United States Climate Change Science Program

ACKNOWLEDGMENT

This chapter has been peer reviewed in draft form by individuals chosen for their diverse perspectives and technical expertise. The expert review and selection of reviewers followed the OMB's Information Quality Bulletin for Peer Review.The purpose of this independent review is to provide candid and critical comments that will assist the Climate Change Science Program in making this published report as sound as possible and to ensure that the report meets institutional standards. The peer review comments, draft manuscript, and response to the peer review comments are publicly available at: www.climatescience.gov /Library/sap/sap2-1/default.php.

We wish to thank the following individuals for their peer review of this chapter:
Frans Berkhout, Free University of Amsterdam
Garry Brewer, Yale University
Ged Davis, World Economic Forum
Nebojsa Nakicenovic, International Institute for Applied Systems Analysis
Robert Lempert, RAND Corporation
M. Granger Morgan, Carnegie-Mellon University
John Robinson, University of British Columbia
Thomas J. Wilbanks, Oak Ridge National Laboratory

We would also like to thank the numerous individuals who provided their comments during the public comment period. The public review comments, draft manuscript, and

* This is an edited, reformatted and augmented edition of a United States Climate Change Science Program publication, dated July 2007.

response to the public comments are publicly available at:www.climatescience.gov /Library/sap/sap2-1/default.php.

The writing team is grateful for research assistance by Kristen Cleary, Emily Kelly, Paul Porter, Gautam Rao, and NoraVan Horssen.

EXECUTIVE SUMMARY

Ascenario is a description of potential future conditions produced to inform decision-making under uncertainty. Scenarios can help inform decisions that involve high stakes and poorly characterized uncertainty, whichmay thwart other, conventional forms of analysis or decision support. Originally developed to study military and security problems, scenarios are nowwidely used for strategic planning and assessment in businesses and otherorganizations, and increasingly to inform planning, analysis, and decisionmaking for environmental issues, including climate change.

Scenarios can serve many purposes. They can help inform specific decisions, or can provide inputs to assessments, models, or other decision-support activities when these activities need specification of potential future conditions. They can also provide various forms of indirect decision support, such as clarifying an issue's importance, framing a decision agenda, shaking up habitual thinking, stimulating creativity, clarifying points of agreement and disagreement, identifying and engaging needed participants, or providing a structure for analysis of potential future decisions.

Scenarios for Climate Change: Five Types

Developing a scenario exercise involves many design choices, of which the most important involve choosing the few key uncertainties to represent in alternative scenarios. Five types of scenarios have been developed to address different aspects of the climate-change issue; these are distinguished by where they fall along a simple linear causal chain extending from the socioeconomic determinants of greenhouse-gas emissions through the impacts of climate change as shown in Figure ES-1. (This figure does not represent the complete causal structure of the climate issue, which has many linkages and feedbacks. Rather, this simple structure only illustrates how scenarios have been used to fit within the simplest and most prominent causal pathway of the issue.)

Emissions Scenarios for Climate Simulations: Emissions scenarios present future paths of greenhouse-gas emissions or other climate perturbations. A major use of these is to provide needed inputs to climate models. Such scenarios may specify simple arbitrary perturbations of emissions or concentrations (e.g., doubling atmospheric CO_2), or time-paths reflecting specified assumptions for evolution of socioeconomic drivers such as population, economic growth, and technological change.

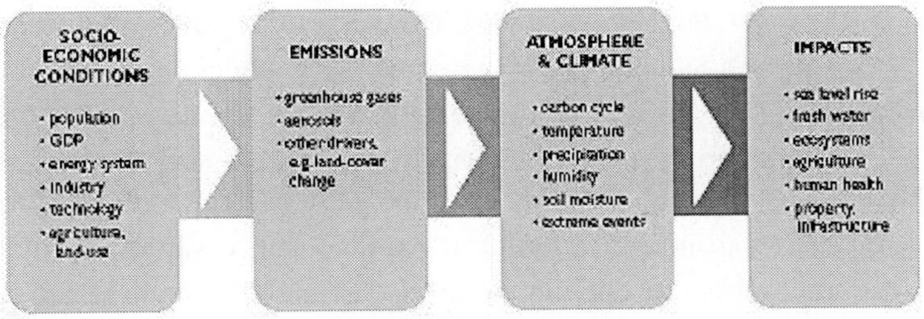

Figure ES-1. Scenarios of anthropogenic climate change: simple linear causal chain.

Emissions Scenarios for Exploring Alternative Energy/Technology Futures: Another use of emissions scenarios involves specifying an environmental or emissions target, arbitrarily or based on normative or political goals, to examine what patterns of socioeconomic change, energy resources, and technology development are consistent with the target and/or what interventions might be needed to meet it. Such scenarios have examined conditions for stabilizing atmospheric CO_2 concentration at various levels and the implications of stabilizing radiative forcing for multigas reduction strategies.

Climate-Change Scenarios: Climate scenarios specify potential future climate conditions to inform assessments of impacts, vulnerabilities, and adaptation options, and inform decisionmaking for adaptation or mitigation. They can be produced by arbitrary perturbation of present conditions, by using climates from elsewhere or the past as a proxy for potential future climate in a given location, or by climate-model simulations driven by some specified scenario of future emissions.

Scenarios of Direct Biophysical Impacts, e.g., Sea Level Rise: Scenarios can specify alternative trajectories for some important form of climate impact that influences many other impacts. For example, scenarios of sea level rise can capture the most important impact pathways in many coastal regions, including the large uncertainties associated with potential loss of continental ice sheets in Greenland and Antarctica.

Multivariate Scenarios for Impact Assessment: Assessing climate-change impacts requires not just considering climate in isolation, but other linked changes and stresses, including both environmental and socio-economic trends. The factors that influence particular impacts and vulnerabilities are likely to be widely variable, and may include demographic, economic, technological, institutional, and cultural characteristics. Consequently, scenarios may have to be generated in an exploratory manner in the context of attempting to assess specific local and regional impacts.

Scenarios for Climate Change: Major Examples

The report reviews four major exercises producing or using scenarios for climate-change applications. The examples include national and international activities, produced by different sets of actors for different purposes.

The Intergovernmental Panel on Climate Change (IPCC) has produced three sets of scenarios of 21st-century greenhouse-gas emissions, of which the most ambitious and important were produced for the *Special Report on Emissions Scenarios* (SRES) between 1997 and 1999. SRES produced four qualitative storylines on which six "marker" scenarios were based – one model quantification of each storyline plus two technological variants of one storyline that stressed fossil-intensive and low-carbon energy supply technologies – each produced by a different energy-economic model. Other models' replications of each other's marker scenarios plus a few additional explorations yielded 40 scenarios in total. These scenarios highlighted several insights, including the ability of alternative paths with similar emissions in 2100 to differ widely in their interim pathways and thus in atmospheric concentrations; the ability of alternative technological assumptions alone to generate as wide a range of emissions futures as substantially divergent socio-economic pathways; and the fact that similar emissions paths can come from widely different combinations of underlying socio-economic factors and so pose distinct mitigation problems. A widely publicized critique of the SRES scenarios alleged over-estimation of future emissions growth due to the metric used to compare incomes in rich and poor nations, but the overestimation was later found to be insignificant. More serious and illuminating challenges associated with these scenarios concerned how to balance and integrate qualitative and quantitative scenarios; how to use and how much to coordinate multiple models to generate the most useful insights; and whether, when, and how it is appropriate to assign explicit probability judgments to alternative scenarios or associated ranges of quantitative variables.

The US National Assessment was a comprehensive assessment of potential impacts of climate change and variability on the United States, focusing on major regions and sectors (agriculture, water, human health, coastal areas and marine resources, and forests). The National Assessment needed scenarios of 21st-century US climate and socio-economic changes. For climate scenarios, it relied principally on climate-model scenarios produced by the UK Hadley Centre and the Canadian Centre for Climate Modeling and Analysis, each driven by a single emissions scenario, with statistical down- scaling based on detailed local conditions and present patterns of fine-scale climate variation. Other proposed types of climate scenario, including historical scenarios and inverse methods to probe for key vulnerabilities, were less used. For socio-economic scenarios, a novel approach was proposed that combined specified scenarios for a few key national-level variables such as population and economic growth, and a common process to elaborate and document additional socio-economic assumptions as needed for specific regional and sector analyses. The National Assessment was criticized for relying on just two climate-model runs and one emissions scenario, although these choices were dictated by time limits and availability of climate-model runs. Limited use was made of the socio-economic approach, principally due to time limits and communication problems.

The UK Climate Impacts Programme (UKCIP) provides common datasets, tools, and support, including scenarios, for climate-impact assessments for UK regions and sectors by researchers and stakeholders. The program produced climate scenarios in 1998 and 2002, all based on the Hadley Centre climate models, and socio-economic scenarios in 2001. The program stresses building a sustained assessment capability by acting as a motivator, resource, and light coordinator with little central authority over separate assessments. The reliance on climate scenarios from just one family of climate models may pose risks of incomplete representation of key uncertainties.

The Millennium Ecosystem Assessment (MEA) examined the status, present trends, and longer-term challenges to the world's ecosystems, including climate change and other stresses. One of the assessment's four working groups constructed scenarios of global ecosystems to 2050 and beyond, largely independently of the group examining current status and trends. All assessment components used a common conceptual framework, which distinguished indirect drivers of ecosystem change (e.g., population and economic growth, technological change, policies and lifestyles), direct drivers (e.g., climate change, air pollution, and land-use and land-cover change), ecosystem indicators, ecosystem services, measures of human wellbeing, and response options. The Scenarios group applied this framework to characterizing potential ecosystem stresses in 2050, with more limited projections to 2100. The four scenarios were based on two dimensions of uncertainty: degree of globalization, and predominance of proactive vs. reactive response to ecosystem stresses. The qualitative storylines underlying these scenarios were more richly developed than in other climate-change scenario exercises. Concerns with these scenarios pertained to the degree of integration and consistency among qualitative and quantitative scenario components; risks of logical circularity within scenarios; and unexplained similarity of projected ecosystem effects among scenarios.

Scenarios for Climate Change: Challenges and Controversies

Scenarios and Decisions

Scenarios can inform climate-change mitigation and adaptation decisions, but most uses so far have had relatively indirect connections to such decisions. Although there is no single global climate-change decision-maker, scenarios can inform the many decision-makers with diverse responsibilities that will affect and be affected by climate change. Three groups of decision-makers with distinct information needs can be distinguished: mitigation policy-makers, who are mostly but not exclusively national officials; impacts and adaptation managers, including national officials and others who are responsible for particular climate-sensitive assets, resources, or interests; and energy resource and technology managers, who include owners, developers, and investors in energy resources and energy-related capital stock and new technologies.

A key issue in creating scenarios for all decision-makers is how to represent decisions within scenarios. In general, decisions by the scenario user should be explicitly examined relative to baseline conditions specified in scenarios, while decisions by others outside their control should be treated like any exogenous uncertainty. The issue is most important in the treatment of mitigation decisions: scenarios to inform mitigation should allow explicit examination of the entire relevant range of mitigation decisions, while scenarios to inform impacts and adaptation should specify the likely range of mitigation efforts – usually yielding a narrower range of emissions futures than is considered in scenarios to inform mitigation.

Scenarios in Assessments and Policy Debates

In climate-change assessments, scenarios can provide required inputs to other parts of the analysis and help to organize multiple components of the assessment. When scenarios are used in a prominent assessment, they may subsequently be adopted in planning or decision-support processes outside the original assessment. Scenarios can also help frame public and policy debate, in part by providing an aggregate metric of the issue's severity. They consequently may gain prominence in contentious policy debates, and so become subject to political attempts to influence their content and political criticism based on their perceived implications for policy action. The unavoidable judgments underlying construction of scenarios provide opportunity for partisan efforts to make scenarios policy prescriptive, and for claims that only certain scenarios are plausible (e.g., high- or low-emissions scenarios, depending on the critic's motivation). These claims are unavoidable, since scenarios represent key uncertainties bearing on high-stakes policy decisions, but such attempts to restrict scenarios should be resisted, principally through prominent communication of the reasoning, assumptions, and treatment of particular uncertainties underlying scenarios.

Scenario Development Process: Expert-Stakeholder Interactions

Scenario developers must decide how and how much to involve scenario users and stakeholders in scenario development. In other fields – where users are clearly identified – relatively few and homogeneous, intensive collaboration between scenario developers and users or their representatives is desirable. Close user involvement is also advantageous in developing scenarios for climate change, but potential users of these scenarios are more numerous and diverse, may not be clearly identified, and may have contending material interests in the scenarios' content or use. This situation calls for delicate decisions about participation and representation to keep scenarios tuned to practical users' needs while keeping the development process small enough to be manageable.

Communication of Scenarios

Climate change scenarios must be communicated to multiple audiences with diverse interests and information needs. In addition to the scenarios' content, sufficient information must be provided about the process and reasoning by which the scenarios were developed, to allow users to scrutinize the underlying data, models, and reasoning; judge their confidence in the scenarios; and have opportunities to critique the scenarios and suggest alternative approaches. Effective communication can help engage a broad user community in updating and improving scenarios. Open communication of the decisions, assumptions, and uncertainties underlying scenarios is likely to both increase users' confidence that the scenarios have reasonably represented current knowledge and key uncertainties, and help them develop alternatives if they are unconvinced.

Consistency and Integration in Scenarios

Scenario developers should strive for internal consistency. At one level, this means avoiding clear contradictions with well-established knowledge and not moving inadvertently outside bounds of historical experience – although such sharp departures from experience may be useful if pursued intentionally to examine low-probability risks or broaden decision-makers' perceptions. Perceptions of internal consistency or coherence in scenarios ultimately rest on subjective judgments, which pose well-known risks of bias if not carefully structured and controlled. Potential inconsistencies grow when scenario exercises use multiple models and attempt to harmonize them, particularly when some key quantities are externally specified for some models and calculated within others. Attempting to avoid such inconsistency by standardizing model outputs, however, can carry more serious risks by obscuring interpretation of results and precluding use of model variation to illuminate uncertainty. Attempts to connect qualitative and quantitative aspects of scenarios have been particularly challenging for pursuit of consistency. Different narrative scenarios often reflect different assumptions about how the world works, which correspond more closely to different model structures than to parameter variation. Better integrating the two approaches will require developing ways to connect narrative scenarios to model structures, rather than merely to target values for a few variables that models are then asked to reproduce.

Treatment of Uncertainty in Scenarios

A scenario exercise can represent a few key uncertainties by variation among scenarios. Extreme economy is required in choosing which uncertainties to represent, what variation (including potential extremes) to represent for each, and how to combine them in a manageable number of scenarios. Complex narrative scenarios pose special problems in representing and communicating uncertainty, usually addressed by seeking underlying structural uncertainties – e.g., deep societal trends such as globalization or values shifts – that are judged to influence many other factors of concern. The most prominent controversy in treatment of uncertainty in scenarios has concerned whether or not to explicitly assign probabilities to scenarios or associated ranges of quantitative outcome variables. The debate rests in part on different views of the typical contents of scenarios, since subjective probabilities can readily be assigned to ranges of one or two quantitative variables. Explicit probability assignment in such simple cases offers clear benefits for assessing alternative choices and avoids the risk of users assigning their own, perhaps less informed, probability judgments. Assigning probabilities to rich multivariate scenarios, particularly if these include narrative elements, is much more problematic, since there is no clearly defined interval "between" such scenarios and their boundaries are not clearly defined.

Conclusions and Recommendations

Use of Scenarios in Climate-Change Decisions

- Scenarios can make valuable contributions to climate-change decision-making. There is a big gap between the use of scenarios in current practice and their potential contributions, but interest in using scenarios is increasing.
- Scenarios of global emissions and resultant climate change are required by many diverse climate-related decision-makers, but beyond these common requirements decision-makers' needs from climate-change scenarios are highly diverse.
- Impacts and adaptation managers include both national officials and others responsible for more specific domains of impact. They need climate-change scenarios, driven by specified global emissions scenarios, to represent potential climate-related stresses on their areas of responsibility, plus other environmental and socio-economic information at appropriate scales. Their combined needs – for centrally produced climate scenario information, associated tools and support, and a capability to develop and apply additional scenario information related to their

responsibilities – suggest the need for a cross-scale organizational structure to provide scenario information.

- Mitigation policy-makers, who are mainly but not exclusively national officials, need scenarios of global and national emissions trends, resultant climate change, and aggregate impacts. In addition, they need scenario information about the potential policy environment for their choices, including alternative scenarios of other nations' mitigation strategies, international mitigation decisions, and implementation and compliance. In some cases, they can usefully employ target-driven scenarios for backcasting analysis. Mitigation decisions require scenario development capacity at the national level.
- Scenarios for mitigation decisions should include a wide range of baseline emissions assumptions and should not prejudge the likely level of mitigation effort, while scenarios for impact and adaptation managers should be based on emissions assumptions that include the range of mitigation interventions they judge likely.
- Energy resource and technology managers, who are mainly private-sector actors, primarily need scenarios that represent alternative policy regimes over the 30- to 50-year time horizons relevant for investment and technology-development decisions. Scenarios of emissions and climate change may provide background, but do not capture the most important uncertainties for these decision-makers.

Use of Scenarios in Climate-Change Assessments

- Large-scale, official assessments are currently the main users of scenarios and will likely remain major users. Scenarios in assessments mostly support further analysis, modeling, and assessment. They can also help frame the climate issue for the public and policy-makers. Presentation of scenarios in assessments leads to additional unforeseen uses.
- Scenarios contain unavoidable elements of judgment in their production and use. This makes them vulnerable both to attempts at bias and to partisan attack. The most productive response lies in transparency about the process, reasoning, and assumptions used to produce scenarios, which can both help limit bias in scenario production and focus subsequent argument on underlying uncertainties.

What Should Centrally Provided Emissions and Climate Scenarios Look Like?

- Centrally provided scenarios of emissions and resultant climate change should be global in scope, with major climate-relevant emissions and other perturbations specified at least for major world regions. They should have a time horizon of a century or longer, with interim results at roughly decadal resolution.
- Centrally provided scenarios of global emissions and climate change can help inform mitigation and adaptation decisions at national and sub-national scales, but such decisions require additional information at these scales.
- Emissions scenarios of several types are needed to serve diverse uses, including alternative baselines, alternative levels of incremental stringency of mitigation effort, and specified future targets to support back-casting and feasibility analysis. Some

emissions scenarios should be coupled to explicit scenarios of wide-ranging alternative socioeconomic futures, but this is not necessary for all uses. Scenarios should reflect various explicit degrees of coordination, including simple fully standardized scenarios for evaluating and comparing downstream models, multi-model scenarios using common input assumptions, and non-standardized scenarios to explore alternative assumptions or meet specific user needs.

- Some scenarios of socio-economic conditions should include qualitative and quantitative elements and sustained analytic efforts to link the two. These elements can provide a vehicle to explore major historical uncertainties with large implications for climate change and vulnerability; provide a logical structure to connect assumed trajectories for multiple variables; and provide guidance to other analysts or users to extend scenarios by elaborating additional detail. Alternative qualitative and narrative elements should be linked not just to alternative parameter values in quantitative models, but also to alternative forms of causal relations and model structures.

Scenario Process: Developer-User Interactions

- There is value in close collaboration between scenario developers and users, particularly at the beginning and ending stages of a scenario exercise.
- The ease of achieving such collaboration and its value are likely to be greater when scenario users are clearly identified, few in number, and similar in their interests and perspectives.

Communication of Scenarios

- Effective communication of scenarios is essential, in forms useful to audiences of diverse interests and technical skills. In addition to scenario contents, communication should include associated documentation, tools, and support.
- Transparency of underlying reasoning, assumptions, and major uncertainties is crucial. Such transparency is necessary to support the credibility of scenarios, to alert potential users to conditions under which they might wish to use or modify them, and to inform criticism and improvement of scenarios.

Consistency and Integration in Scenarios

- Any scenario should be internally consistent in its assumptions and reasoning, to the extent this can be established.
- In scenario exercises that use multiple models to explore potential uncertainties in future conditions, consistency among models should be pursued primarily through coordination of inputs, not outputs, except when coordinated outputs represent common goals for policy evaluation.
- Transparency in reporting scenario and model differences as well as underlying assumptions and reasoning can help mitigate the effects of inconsistencies among scenarios.

Treatment of Uncertainty in Scenarios

- More explicit characterization of probability judgments should be included in some future scenario exercises than has been practiced so far. Means available to express these judgments are of widely varying specificity, ranging from agreed terminology to explicitly quantified probability distributions. All such judgments should include explicit acknowledgement of their inevitable subjective elements and appropriate caveats.
- Explicit probability judgments are easiest to produce and least controversial in scenarios generated using quantitative models of climate change or specific impact domains. These can be conditioned on specific assumptions for socio-economic inputs such as emissions, and can represent explicitly and quantitatively the effects of specified variation in initial conditions or unknown parameter values. These devices are also available, although in less widespread use, in economic models used to project emissions.
- Including explicit probability judgments is likely to be most useful when key variables are few, quantitative outcomes are needed, and potential users are numerous and diverse. It is likely to be least useful when scenarios specify multiple characteristics, including prominent qualitative elements; when the purpose is sensitivity analysis or heuristic exploration; and when potential users are few, similar, and known.
- Because of their large and diverse set of potential users, centrally provided scenarios of global emissions and climate change should attempt to include some explicit probability judgments for ranges of key quantitative outputs, including global emissions and global-average temperature change. These should span a wide range of judged uncertainty on these variables, e.g., 95 to 99 percent. Providing such explicit likelihood statements lets users choose whether to use them or not.
- Scenario exercises should give more attention to low-probability, high-consequence extreme cases, such as loss of a major continental ice sheet or changes in meridional ocean circulation. With these, it is especially crucial to be explicit and transparent about the reasoning and assumptions underlying each scenario, including developers' judgments of relative likelihoods.

Expanding and Sustaining Capacity for Production and Use of Scenarios

- Present scenario capacity is inadequate. To help fulfill these presently unmet needs, the CCSP should establish a program to:
  - Commission scenarios for use in assessments and decision-support activities.
  - Disseminate scenarios with associated documentation, tools, and guidance materials.
  - Commission various groups to evaluate scenarios and their applications, and to develop improved methods.
  - Archive results and documentation related to all these tasks, to provide historical perspective and institutional memory for future scenario-related activities.
- Design and management conditions of this new program should include six elements.

- The program should build and maintain strong connections with outside relevant expertise, and analytic and modeling capability.
- The program should integrate and balance goals and criteria related to scientific and technical quality, and thoserelated to utility and relevance to users.
- The program should be insulated frompolitical control.
- The program should strive for maximumtransparency in its own activities, in addition to demanding it from activities itsupports.
- The program will require the authorityand resources necessary to articulate andpromulgate standards for transparency,consistency, and quality control.
- The program will require adequate sustained resources level of effort.

INTRODUCTION

This chapter examines the development and use of scenarios in global climate change applications. It considers scenarios of various types – including but not limited to emissions scenarios – and reviews how they have been developed, what uses they have served, what consistent challenges they have faced, what controversies they have raised, and how their development and use might be made more effective. The report is Synthesis & Assessment Product 2.1 b of the US Climate Change Science Program. By synthesizing available literature and critically reviewing past experience, the report seeks to assist those who may be conducting, using, or commissioning scenarios related to global climate change.

Scenarios are used to support planning and decision-making when issues have deep or poorly characterized uncertainty and high stakes, often accompanied by long time horizons. These conditions apply to the major decisions about how to respond to global climate change. As scientific research advances our knowledge of the climate's present state and trends, its patterns of variability, and its responses to external forcings, we are gaining an increasingly clear view of risks that may be realized late this century or beyond. These future risks are linked to near-term socio-economic trends and decisions in both public and private sectors. Some near-term decisions – such as investment in long-lived capital equipment, new resources, or new technologies in the energy sector – can influence long-term trends in the emissions contributing to climate change. Other near-term decisions – such as investment in water resources infrastructure or coastal development – can influence how adaptable and how vulnerable future society will be to the impacts of climate change.

Despite pervasiveuncertainties, peoplemust make near-termdecisions related to climate change thathave long-termconsequences, including potentialirreversibilities.

Although such decisions are being made now, making them responsibly requires considering their potential consequences over the longer term, including associated uncertainties. This requires thinking about the future conditions that will shape their consequences, not just next month or next year but 10, 30, 50, or 100 years in the future – longer periods than the horizon of conventional methods of planning or analysis.[2] Attempting to describe potential future conditions over this long time horizon presents a seeming paradox. On the one hand, conditions this far in the future, and the factors and actors that may influence them, are deeply uncertain.[3] On the other hand, we have a great deal of knowledge that can help make informed assumptions about future conditions, even over such long horizons. This includes well-established scientific knowledge about physical, chemical, and biological processes; more weakly, relatively well-established causal mechanisms in economics, sociology, and politics; and more weakly still, certain seemingly robust empirical patterns of historical change in population, economics, and technology. All of these give some guidance to support judgments about future conditions that are more or less likely, virtually certain, or virtually impossible. In some ways we might be highly confident that the future will resemble the present, e.g., in the radiative properties of atmospheric trace gases. In others, we might judge it likely that future conditions will lie within some envelope extrapolated from present and past trends, e.g., in projecting rates of change in fertility, mortality, or labor productivity. Still other areas, such as the development and social consequences of major technological advances, or large-scale political events such as wars, political realignments, or epidemics, may hold more fundamental uncertainties. In some cases, such uncertainties may e adequately represented as wider distributions of recognized uncertain quantities. In others, they may represent events whose character or even possibility we have not yet imagined.

Despite pervasive uncertainties, people must make near-term decisions related to climate change that have long-term consequences, including potential irreversibilities. Scenarios are tools to help inform these decisions by gathering and organizing relevant knowledge, organizing associated uncertainties, and structuring and disciplining associated speculation. This chapter assesses experience to date in developing and using scenarios for global climate change.

Early climate-change debates mainly concerned scientific questions such as whether and how the climate is changing, how much change is caused by human activities, and how sensitive the climate system is. Scenarios did not figure prominently in these debates. But as advancing climate science has increasingly shifted the debate from confirming and describing the climate-change problem toward deciding what to do about it, the need for long-term decision-support tools like scenarios has increased, as have the scrutiny and criticism these have attracted.[4] In a contentious public-policy area like climate change, controversy over scenarios is to be expected: scenarios are a method to structure and communicate the most important uncertainties, and conflicting judgments about uncertainties are a major source of disagreements over what to do. Consequently, we expect the trend of scenarios' increasing prominence and contentiousness to continue – particularly for emissions scenarios, since these are the relevant metric of human environmental burden and the point of most contested proposed intervention.

In this chapter, we try to cast some light on current and coming debates over climate-change scenarios. These debates presently exhibit basic confusion about the definition, purposes, and potential uses of scenarios. We aim to provide clarification and practical advice to two related audiences: those conducting assessments or analyses that develop or use scenarios; and those commissioning, using, and interpreting such assessments or analyses. For the first group, we seek to provide an organized summary of relevant experience in similar efforts, discussion and clarification of key choices and challenges, and – to the extent present knowledge allows – practical guidance about pitfalls, challenges, and opportunities in particular approaches. For the second group, we seek to provide guidance on what to ask for, how and how much to participate in its production, how to interpret the results, and what questions to ask.

Because the charge of this chapter is unlike those of other Synthesis and Assessment products, the approach we have taken to producing it is necessarily different as well. We were tasked with reviewing, interpreting, and evaluating experience with scenario methods in global climate change applications. This is not a narrowly focused question, and there is not a well-developed scientific literature on which we can draw for answers. While we have reviewed the existing literature on scenarios, most of it concerns scenarios in other decision domains than global climate change. In addition, we have examined several major scenario exercises in global-change applications. In this, we have drawn on published materials, both from the exercises themselves and from commentary and criticism, as well as documentary materials and records, interviews with participants and users, and the experience and judgments of team members.

Our review of this experience has not been entirely independent, since members of this writing team were involved as participants, reviewers, and critics in two of the scenario exercises we review, the IPCC SRES process and the US National Assessment. While we have drawn on the experience of these team members, we have attempted to limit the risk of idiosyncratic interpretations and bias by drawing on other sources as well and by engaging all team members in developing our summary and discussions of these exercises. Moreover, our purpose is not to either attack or defend these past exercises, but to seek to understand the decisions made in conducting them, the factors that influenced them, and the constraints under which they operated, in order to assess their experience, identify successes and pitfalls, and to the extent possible, provide guidance to advance scenario methods for climate change and other similar environmental issues. Because the experience we review does not provide a sufficiently large or random sample to support strong scientific inference, the diagnoses, interpretations, and recommendations we present rely on our collective judgment. We have endeavored to follow our own advice, and be as transparent as possible about the foundation and reasoning underlying our conclusions and recommendations.

We aim to provideclarification andpractical advice to tworelated audiences:those conductingassessments oranalyses that usescenarios; and thosecommissioning, using, and interpreting suchassessments oranalyses.

The report is organized as follows. Drawing on the broader literature on scenarios – most of which concerns domains other than climate change – Section 1 introduces scenarios, sharpens their definition, and outlines a few major dimensions of variation and decisions that must be made in developing a scenario exercise. Section 2 focuses specifically on scenarios for global climate change, and outlines the types of decisions that could use scenarios and the main types of scenarios that have been developed for this issue. Section 3 reviews four major experiences in developing and using global-change scenarios. Section 4 discusses several issues that have posed key challenges in climate-change scenarios and that are likely to require particular attention in designing new scenario exercises. In addition to drawing on Section 3, this discussion also makes use of briefer discussions of eight other examples of global-change scenarios that illustrate particular issues or challenges; these examples are presented as short boxes in Section 4. Section 5 provides our conclusions and recommendations for future development and use of global climate-change scenarios.

SECTION 1. SCENARIOS, THEIR CHARACTERISTICS AND USES

1.1. Defining Scenarios

Ascenario is a description of potential future conditions, developed to inform decision-making under uncertainty. The decisions in question can be faced by individuals, groups, organizations, or governments, and may pertain to any subject matter. While many writers on scenarios give no explicit definition, others have offered a wide range of definitions, often substantially more complex and restrictive than this one. The published definitions gathered in Box 1 .1 give a sense of both the broad commonalities among many analysts' conceptions and the significant differences among them.

The historical uses of scenarios for planning and analysis lie in war games, exercises of simulated conflict used for military training, planning, and operational decision-making. The roots of war games extend to antiquity, although the first formalized war games were developed for officer training in 19th-century Prussia.[12] In the 1940s and 1950s, exercises resembling war games began to be applied outside the purely military domain, to study potential international crises that included both high-level political decision-making and the potential for military conflict. These exercises were informed by the then-new field of game theory, which promised new formal insights into situations of conflict and strategic decision-making,[13] motivated by the recognition that the new nuclear age had both raised the stakes of international diplomacy and created profound new uncertainties over how to proceed. In these exercises, principally developed at the Rand Corporation, scenarios were sketches of challenging but plausible situations to which participants had to respond, allowing exploration of associated threats and opportunities. They adopted the term "scenario" from film and theatre, where it denotes a brief sketch of a story that includes only enough detail to convey broad points of plot and character. As in classic war games, scenarios in these exercises served to help organizations and their leaders prepare for novel, complex challenges that they might not anticipate, and which if they did arise – would likely develop too fast to allow adequate reflection or analysis in real time.[14]

Over the past few decades, the use of scenarios has moved outside the realm of military and diplomatic activity. Beginning with strategic planning at the Royal Dutch/Shell oil company,[15] scenarios are now widely used for strategic planning, analysis, and assessment by businesses and other organizations. They have also figured increasingly prominently in planning, analysis, and policy debate for long-term environmental issues, in particular global climate change. Because the total body of experience with scenarios provides useful insights into their use in any particular domain, this section elaborates the meaning, characteristics, and potential uses of scenarios in general. The next section turns to their specific use for global environmental issues.

1.1.1. Distinguishing scenarios from assessments, models, and analyses

Confusion is widespread in discussions of scenarios, because their form and usage are highly diverse, and because writers' uses of the term are often imprecise and occasionally contradictory.

Box 1.1. Scenarios: A Sampling of Published Definitions

A scenario is a coherent, internally consistent, and plausible description of a possible future state of the world.[5]

A scenario is a story that describes a possible future. It identifies some significant events, the main actors and their motivations, and it conveys how the world functions. Building and using scenarios can help people explore what the future might look like and the likely challenges of living in it.[6]

Scenarios are images of the future, or alternative futures. They are neither predictions nor forecasts. Rather, each scenario is one alternative image of how the future might unfold. A set of scenarios assists in the understanding of possible future developments of complex systems. Some systems, those that are well understood and for which complete information is available, can be modeled with some certainty, as is frequently the case in the physical sciences, and their future states predicted. However, many physical and social systems are poorly understood, and information on the relevant variables is so incomplete that they can be appreciated only through intuition and are best communicated by images and stories. Prediction is not possible in such cases.[7]

A climate scenario is a plausible representation of future climate that has been constructed for explicit use in investigating the potential impacts of anthropogenic climate change. Climate scenarios often make use of climate projections (descriptions of the modeled response of the climate system to scenarios of greenhouse gas and aerosol concentrations), by manipulating model outputs and combining them with observed climate data.[8]

[Scenarios] are created as internally consistent and challenging descriptions of possible futures. They are intended to be representative of the ranges of possible future developments and outcomes in the external world. What happens in them is essentially outside our own control.[9]

Scenarios are coherent, internally consistent and plausible descriptions of possible future states of the world, used to inform future trends, potential decisions, or consequences. They can be considered as a convenient way of visioning a range of possible futures, constructing worlds outside the normal timespans and processes covering the public policy environment.[10]

Scenarios are plausible, challenging, and relevant sets of stories about how the future might unfold. They are generally developed to help decision-makers understand the wide range of potential futures, confront critical uncertainties, and understand how decisions made now may play out in the future. They are intended to widen perspectives and illuminate key issues that might otherwise be missed or dismissed. The goal of developing scenarios is often to support more informed and rational decision-making that takes both the known and the unknown into account.[11]

Scenarios must be distinguished, on the one hand, from assessments and various types of decision support activity that often use scenarios; and, on the other hand, from other types of statements about future conditions, such as predictions, projections, or forecasts.

An assessment is any process that reviews and synthesizes scientific or other expert knowledge to provide information of relevance to policy- or decision-makers.[16] The most common methods of assessment are deliberations of expert panels and formal models, but other methods combine human deliberations with formal analysis or modeling, including war games or other simulation games, policy exercises, political-military exercises, constructing future histories, backcasting, and others.[17] These methods may use specifications of potential future conditions – i.e., scenarios – as inputs to or components of their work. Scenarios may even be essential for some of these methods. For example, a war or crisis gaming exercise needs a scenario to specify the nature of the threat or crisis, while a formal model used to represent future development of some issue of concern needs a scenario to specify future values of those inputs not explicitly calculated within the model. But these methods are broader than and distinct from scenarios. For example, models do not need scenario-based inputs when used to reconstruct past conditions or study causal processes.

The distinction between assessments and scenarios is perhaps clearest in conventional assessments based on deliberations of expert panels, such as the IPCC, US National Assessment, or Millennium Ecosystem Assessment (MEA). Such assessments often construct representations of future development of an issue, usually based on formal models. These representations require scenario-based inputs and may produce outputs that are themselves used as scenarios in other activities. But the scenario-related activities are frequently not the central focus of the overall assessment, which may examine many additional things, e.g., the state of knowledge in particular scientific areas, the status of and trends in particular environmental conditions, the evidence attributing particular environmental changes to human inputs, or particular policy-relevant scientific questions. Assessments may also include evaluations of proposed actions or proposed criteria for conducting such evaluations. Scenarios thus may provide required inputs to assessments, but are distinct from them.

Scenarios have figured increasingly prominently in planning, analysis, and policy debate for long-term environmental issues, in particular global climate change.

1.1.2. Distinguishing scenarios from projections, predictions, and forecasts

Scenarios must also be distinguished from other types of statements about the future, such as predictions, projections, and forecasts. All of these satisfy the basic definition above: they are descriptions of potential future conditions whose primary purpose in most cases is to support decisions. How can scenarios be distinguished? Examining the ways scenarios are used and discussed by practitioners and researchers suggests four characteristics that distinguish them from other types of future statements. Although these characteristics are not essential, they are all more likely to be present in scenarios than in other types of future statement, so they help to sharpen and delimit the definition of a scenario.

First, scenarios are multi-dimensional: they describe multiple characteristics that collectively make up a coherent representation of future conditions. To achieve this, scenarios assemble and organize available knowledge, information, and assumptions from diverse bodies of research and expert judgment. The elements of a scenario can be of diverse types: quantitative or qualitative, precise or fuzzy, based on well-established research or informed speculation. Effective scenarios integrate their diverse elements in a way that is coherent, communicates a clear theme or organizing principle, and avoids internal contradiction.

Second, scenarios are schematic: that is, they are multidimensional but not without limit. Scenarios do not seek to describe potential future conditions with complete precision or detail. Rather, they highlight essential characteristics and processes with enough detail that knowledgeable observers perceive them as realistic and relevant, but not so much as to distract from large-scale patterns. Indeed, one potential use of scenarios is to stimulate creative thinking and insights, for which they must leave something to the imagination. How much detail and precision is appropriate is a judgment that depends on the particular application.

Effective scenarios integrate their diverse elements in a way that is coherent, communicates a clear theme or organizing principle, and avoids internal contradiction.

Third, scenarios usually come in groups. To be a useful tool to inform decision-making under uncertainty, scenarios must represent uncertainty. This is most often done by providing multiple scenarios, each presenting an alternative realization of uncertain future conditions.[18] The number of scenarios depends on the application. Scenario exercises usually use between two and seven, depending on the stakes of the issue, the resources invested in the exercise, and the depth of analysis devoted to each scenario. The most frequently proposed numbers are three or four.

Finally, scenarios tend to claim less confidence than other types of future statements. Although different authors' usage is not consistent, "prediction" and "forecast" usually denote statements for which the highest confidence is claimed. "Projection" denotes a less confident statement, which may have some specified confidence level and may be explicitly contingent on specified assumptions about other future conditions. Calling a future statement a "scenario" usually implies still less confidence and more associated contingencies. Any use of a scenario for serious planning or analysis does, however, presume some minimal threshold of likelihood. The situation described must be judged likely enough to merit attention, and to justify expending resources and effort to study its implications and potential responses to it. There may also be a time ordering among these three types of statements – predictions or forecasts tend to describe nearer-term futures and scenarios longer-term futures – but there are exceptions, and the meaning of near term and long term depend on the particular context.

1.2. Creating a Scenario Exercise: Key Characteristics and Choices

Beyond these general characteristics, scenarios vary greatly in their use, production, and contents. Extensive scholarly effort has gone into providing alternative scenario taxonomies.[19] Scenarios can be distinguished, for example, by whether they present a snapshot of a future state or a dynamic account of changes over time to reach that state; by their degree of complexity; by the relative balance of deliberation and intuition versus formal analysis used in producing them; or by their temporal and spatial scale. The set of characteristics on which scenarios could be sorted is long and open-ended, so we make no attempt to provide an exhaustive list. Instead, we summarize the main dimensions of scenario variation in the form of a list of potentially open-ended design choices.

1.2.1. Variation among scenarios: three basic dimensions

Three dimensions of variation, concerned with the purpose of a scenario exercise, have far-reaching implications for its design and use and so merit separate discussion. First, the intended use of a scenario exercise can vary from the more predictive to the more exploratory or heuristic. It is of course a fundamental error to take a scenario's illustrative description of potential future conditions as a confident prediction of what will actually happen – in our terminology, to take a scenario as a projection or even a prediction.[20] Still, a scenario must be judged likely enough to merit the attention of busy people. Exploratory uses of scenarios may presume only this low threshold, yet have great value. For example, scenarios can be used to probe and challenge the mental models, thought habits, and unrecognized assumptions of decision-makers; to clarify points of agreement and disagreement; to identify and engage needed participants; to provide a preliminary structure for advance analysis of potential future decisions; or to seek insights into unrecognized opportunities, risks, causal linkages, or uncertainties.[21] Such insights can arise not just from examination of uncertainties, but also from meticulous critical examination of future factors that are essentially certain (e.g., strongly determined demographic trends such as the aging of industrialized-country populations) or even of present conditions whose significance has been overlooked.[22] Still, the predictive confidence accorded to scenarios is a matter of degree: carefully developed scenarios that are judged to have captured the most important uncertainties may claim some moderate degree of confidence, and reasonable distinctions may be drawn between scenarios that represent conventional versus surprising futures, best and worst cases, etc.

A related dimension of variation among scenario exercises is their proximity to decision-making.[23] Scenario exercises may involve actual decision-makers and seek to directly advise a specific, identified, near-term decision, but more frequently their relationship to concrete decisions and decision-makers is indirect. Scenarios may be used for risk assessment, contingency planning, identification of potential threats or actions to be considered, or early characterization of a poorly understood issue. In such applications, exploratory uses dominate. Scenario exercises that are closer to decisions with significant stakes operate under very different requirements, usually driven by specific user needs. Their uses tend to be more predictive – constrained by limits of available knowledge – so they might be expected, for example, to provide more explicit and complete characterization of major uncertainties. They are also likely to be more integrated with methods to evaluate alternative choices and identify preferred ones.

A third basic dimension of variation concerns whether scenarios are defined primarily normatively, on the basis of their perceived desirability or undesirability, or primarily on the basis of their perceived plausibility or likelihood. Although all scenarios include both positive and normative elements, it is important to keep as clear as possible which elements are included based on perceived likelihood or plausibility and which because of perceived desirability or undesirability. The most frequent use of explicitly normative scenarios involves constructing some hypothetical future state primarily on the basis of its desirability. Such a future might be constructed to embody participants' general intuitions about desirable social trends, or to achieve specific environmental, development, or other goals.[24] The scenario exercise then consists primarily of backcasting – attempting to construct paths that connect present conditions to the specified future conditions, to examine the feasibility of the target, and identify costs, tradeoffs, and conditions associated with meeting it.[25] Similarly, one can posit an undesirable future state and then reason through conditions associated with avoiding it. This approach is sometimes proposed to reduce the risks of hidden bias in construction of scenarios which, like any decision support tool, can be misused to provide support for a decision already made for other reasons, rather than to inform a decision not yet made. By bundling normative assumptions into the future target state or boundary conditions, analysts hope to reduce their penetration into the subsequent instrumental reasoning about actions and conditions to reach the target.[26]

Scenarios can be used to probe and challenge the mental models, thought habits, and unrecognized assumptions of decision-makers; to clarify points of agreement and disagreement; to identify and engage needed participants; to provide a preliminary structure for advanced analysis of potential future decisions; or to seek insights into unrecognized opportunities, risks, causal linkages, or uncertainties.

1.2.2. Developing scenarios: main dimensions of choice

Table 1.1 extends the preceding discussion, summarizing the main areas of variation and choice involved in constructing a scenario exercise. This is a highly simplified representation of a complex set of choices. In any particular scenario exercise some of these choices may be made by default, without explicit consideration, perhaps because the preferred choice is obvious in context. Although we present these choices in simple sequential order, this is not necessary: choices might be made in some other order, or iteratively adjusted. But while the process and sequence of choices may be idealized, the set of choices is not: creating a scenario requires explicit or implicit choices on all these dimensions.

But whether therelationship of ascenario exercise todecisions is near or far, direct or indirect, clearunderstanding of itsfocus and purpose isimportant, andinfrequently achieved.

The most basic choices in developing scenarios, which include the three dimensions of variation called out above, involve identifying the main focus of the exercise: what issues are to be addressed or what decisions informed, for whom? The decision to conduct a scenario-based exercise does not necessarily imply that these matters are clearly understood. The closer a scenario exercise is to concrete decisions, the more likely it is that these definitional issues will be understood clearly, in part through discipline on the process imposed by the involvement of decision-makers. But most often, the coupling of scenarios to decisions is relatively weak.[27] In some applications (e.g., corporate strategic planning, responding to a novel military threat) the relevant decision-makers may be clearly identified at the outset, but the issues to be addressed and relevant choices may not be. In other applications, scenarios may be developed to address some broad issue or concern (e.g., climate change, emerging infectious diseases, or terrorism), but the potential users and decisions to be informed might both be unspecified. But whether the relationship of a scenario exercise to decisions is near or far, direct or indirect, clear understanding of its focus and purpose is important, and infrequently achieved: many scenario exercises muddle through with vagueness, confusion, or disagreement regarding the focus, purpose, and intended user of the exercise. Clarifying the overall focus of a scenario exercise may require broad consultations or scoping workshops involving many potentially interested decision-makers, other stakeholders, and analysts and researchers.

A second basic set of decisions concerns the process by which scenarios are developed. Like the focus of the exercise, decisions about the process of developing scenarios often receive little thought, or are not even explicitly recognized as choices, but are nonetheless highly consequential. What expertise must be included to ensure the scenarios adequately reflect the best available scientific knowledge, data, and models? What decision-makers, stakeholders, or their surrogates must be involved to keep the scenarios relevant, plausible, and credible? For scenario exercises that must integrate knowledge from diverse domains, individual participants' knowledge, flexibility, and imagination can be as important as the disciplines or stakeholder groups they represent. How intensively, for how long, and by what means will participants interact? Will the process be open to outside observers or participants? How and when will feedback on the scenarios be sought, and how will it be used? How and to whom will results be communicated? And crucially, how will the be process be led, and how will disagreements be resolved? With good leadership, resolving differences in a scenario exercise can be less arbitrary and more illuminating than in other group tasks; when disagreements persist after careful examination, they can be treated as important uncertainties to be retained as alternative scenarios, not suppressed by picking a winner, splitting the difference, or retreating to vague language.

Whatever process is chosen, a series of substantive choices must be made about what goes into the scenarios. The most important of these concern what key uncertainties will be explored, and how much richness and detail should be included in scenarios to illuminate these.

What uncertainties are to be explored, and how?Many dimensions of uncertainty may be relevant to the issue being examined, but only a fewcan be examined explicitly in any scenario exercise. Defining these is a crucial choice thatshapes much of what follows in a scenario exercise. For those uncertainties judged most important, alternative outcomes are usuallyrepresented in alternative scenarios. For example, scenarios might present high-and low-growth futures, or alternative forms that acompetitive threat might take. Other uncertainties judged less crucial are usually suppressedby presenting a single "best guess" or "reference case." The few key uncertainties chosencan be represented in the number and characterof scenarios, depending on the intended use. Aparticular uncertainty might be represented byhigh and low values of some quantity, or by areference case supplemented with high and/orlow variants.

Table 1.1. Idealized Sequence of Major Choices in Scenario Development.

Main focus, framing, users, question(s) to be addressed
Process and participation
Key uncertainties to explore: how many, over what range
Narrative, quantitative, or both
Level of complexity (number of quantitative variables, detail of narrative)
Specific variables and factors to specify
Time horizon and spatial extent
Temporal and spatial resolution

If two or more uncertainties interact, they can be represented by scenarios thatcombine different outcomes of each: in the simplest form, two interacting uncertainties can be represented by four scenarios, often illustrated by a two-by-two matrix.[28] Several alternative scenarios might seek to span a plausible range for some key quantitative variable, or present distinct qualitative outcomes for a single uncertainty, e.g., three types of competitive threat, or three political futures for a region in turmoil. Alternatively, scenarios can represent plausible extreme or "worst-case" scenarios, to assess the robustness of decisions or strategies.

How rich and complex should each scenario be? Defining scenarios as we have, as multivariate but synoptic, still leaves a wide range of levels of complexity to choose from. At one extreme, scenarios may specify time-paths for just a few quantitative variables, or even just one. Such scenarios are common, e.g., in applications such as analyzing a firm's profitability under alternative scenarios for oil prices, or projecting tax revenues under alternative scenarios of productivity growth and inflation, often in a standard "high, middle, low" format. A scenario can accommodate more complexity by projecting additional quantitative variables, but as the number of variables increases, so also does the need for an organizing principle or gestalt to tie them together in a non-arbitrary way.

Many dimensions ofuncertainty may berelevant to the issuebeing examined, but onlya few can be examinedexplicitly in any scenario exercise.

At the other extreme, the core of a set of scenarios can be a set of rich, coherent narratives, an approach frequently called the Shell approach.[29] Each narrative, described principally in text, reflects a distinct conception of how the world might develop with a persuasive underlying causal logic. A narrative scenario can stand alone but may also include specifications of important quantitative variables, e.g., of population or economic growth, consistent with the broad causal logic underlying the scenario. The narrative provides the context and explanatory logic that tie together the time-paths of quantitative variables, although particular time-paths are regarded as illustrating, not defining, the scenario – i.e., a different scenario would present substantially different time-paths or relationships among them.

The choice of how rich and complex to make scenarios has far-reaching implications for the process of developing the scenarios, what can be done with them, and the uses they can serve. The two extreme approaches imply large differences in how uncertainty is treated, what aspects of the problem receive attention, and the relationship between scenarios and their users, which we discuss for climate-change scenarios in Section 4. Richer and more complex scenarios require more time and effort to develop, whether based on quantitative models, narratives, or both. Complex, narrative-based scenarios may require many person-months of development to become realistic, relevant, and persuasive, with consistent relationships among scenario elements. In return for the extra effort, this approach allows great flexibility in the way potential futures are described. Narratives can convey different aspects of a future situation with varying degrees of salience or specificity, and they can compactly convey the tone or character of a future situation by allusion, where a precise specification would appear arbitrary or labored. The narrative approach avoids limiting the defining characteristics of a scenario to any particular set of pre-specified variables, but attempts to be alert to a wide range of potentially important characteristics and mechanisms of causal influence. Proponents of this approach argue that a coherent narrative at the core of a scenario is necessary to avoid arbitrariness in specifying multiple variables, and to make the exercise useful to decision-makers: "Most scenarios merely quantify alternative outcomes of obvious uncertainties (for example, the price of oil may be \$20 or \$40 a barrel in 1995). Such scenarios are not helpful to decision-makers".[30]

The remaining substantive choices in specifying a scenario follow from the preceding large-scale choices. They include specifying the time horizon and spatial extent of the scenarios, deciding the particular elements to include, and the temporal and spatial resolution at which scenario outputs are stated. Decisions about temporal resolution (e.g., hourly to multi-decadal) and spatial resolution (e.g., regional, national, continental scales) are particularly important when – as is often the case in global-change applications – scenarios are produced or used by quantitative models. Such models may have very precise requirements for the specification and resolution of inputs and outputs, creating the possibility for serious mismatches between what users need or expect, and what scenario developers feel comfortable and competent providing.

This section has discussed the uses, types, and characteristics of scenarios broadly, in any application area. The next section narrows the focus to climate change and related areas of global environmental change, summarizing the types of scenarios that have been used and proposed, and that might be required, to explore and inform decision-making in this area.

SECTION 2. SCENARIOS IN GLOBAL-CHANGE ANALYSIS AND DECISION SUPPORT

Analysts have tried to develop scenarios to support understanding of and decision-making for global environmental issues, beginning with the global models of the mid-1970s and early assessment of acid rain and stratospheric ozone in the late 1 970s to early 1980s.[31] The reasons for using scenarios in global change are similar to those that apply in other decision domains: high-stakes decisions that must be made under deep uncertainty about the conditions that will determine their consequences, the values at stake, or the relevant set of choices and actors. As in other domains, well-designed scenario exercises can provide a structure for assessing alternative choices and help focus on the nature of the issue, the relevant choices and actors, the values that might be at stake, and the types of research or analysis that might help clarify preferred choices.

For climate-change applications, scenario exercises have been conducted and sponsored by governments, international organizations, non-governmental organizations, and collaborative groups. While these have been diverse in form, details, and purposes, they have tended to focus more on heuristic and exploratory uses than on supporting specific decisions. The boundaries of the climate-change issue are not sharply defined, however: climate change implicates and connects to many other areas of policy, including energy, agriculture, hazard protection, and broad questions of economic development. Consequently, there is substantial uncertainty about what all the relevant decisions, decision-makers, and potentially affected values are. While some decisions are clearly of primary relevance to climate change, many other decisions that appear to be connected have not yet incorporated consideration of climate change or even recognized the connection. Reflecting these fuzzy boundaries of the issue, scenario exercises developed for climate change have overlapped with other exercises primarily focused on ecosystems, energy, and broad issues of world development. The fuzziness of the climate issue's definition increases the challenge of developing useful scenarios, but also increases the potential value of well-crafted and executed scenario exercises.

The decisions most directly related to climate change are conventionally sorted into two categories, mitigation and adaptation.[32] Mitigation consists of actions that reduce the human perturbations of the climate system, by reducing net anthropogenic greenhouse-gas emissions.

Adaptation consists of actions to reduce the harm or increase the benefit from climate change and its impacts. Despite uncertainty about the precise decision agenda, we can identify in general terms the type of information scenarios might provide that would be useful to each type of decision.

Scenarios can help inform adaptation decisions by characterizing the nature and severity of relevant potential impacts; identifying key vulnerabilities, particularly those that might not otherwise have been recognized; identifying research or monitoring priorities that might give advance warning about impacts, particularly acute vulnerabilities; expanding the perceived set of potential responses; and providing a framework for evaluating alternative adaptation measures. They may also help to clarify the time structure of relevant decisions, identifying those near-term decisions that might have important but underrecognized connections to future impacts and vulnerability.

To date, most climate-related uses ofscenarios have notexamined decisionsdirectly, but have beenembedded in largerexercises ofassessment, modeling,or characterization ofthe issue.

Similarly, scenarios can help inform mitigation decisions by characterizing the potential impacts of climate change and their severity, since these provide the motivation for mitigation. But, in addition, mitigation decisions can benefit from information about potential emissions trends, which determine the nature of the challenge of limiting emissions; about potential pathways of the extraction and depletion of current energy resources and development of new ones; and about potential pathways of technological development. Mitigation decisions may also benefit from scenarios representing the potential policy context in which they are made.

To date, most climate-related uses of scenarios have not examined decisions directly, but have been embedded in larger exercises of assessment, modeling, or characterization of the issue. These uses have included formal integrated-assessment models,[33] comprehensive assessments conducted by multidisciplinary expert bodies (e.g., IPCC), and more narrowly focused assessment exercises targeting specific aspects of the climate-change issue. In these uses, scenarios represent components of the climate-change issue that are required inputs to an assessment or model.

The causal logic of the climate-change issue is complex, including multiple two-way causal links and feedbacks among socio-economic, geophysical, and ecological systems. Integrated-assessment models seek to represent many of these linkages and feedbacks explicitly; Figure 2.1 shows a typical example of the "wiring diagrams" that illustrate the increasingly dense linkages and feedbacks represented in these models. But while such diagrams might be taken to indicate that all relationships are represented explicitly within the model – endogenously – this is not the case. All models of the climate-change issue rely on scenarios to specify some future quantities exogenously, and in virtually all cases, scenario-specified inputs are not modified to account for results of the subsequent analysis: i.e., they are truly exogenous, and the causal logic does not close.

When scenarios are used to specify exogenous inputs to a model of some aspect of the climate- change issue, the causal logic of the analysis can be greatly simplified from that shown in Figure 2.1. Instead, the logic can be represented by a simple linear structure that extends from human activities to emissions to climate change to impacts. Figure 2.2 shows this highly simplified structure. This representation is even more suitable for the uses of scenarios in other types of global-change assessments, which have been organized around much simpler causal structures than those that integrated-assessment models seek to represent. Note that we are not claiming this simple logical structure adequately represents the true structure of the climate-change issue: only that it illustrates the ways that scenarios have been used to provide exogenous inputs to global-change models and assessments.

This linear logical structure allows a simple, practical categorization of five types of scenarios that have been developed for the climate-change issue. These types are defined by what quantities they specify and what primary area of analysis they provide input to. Their differences can be represented by where they cut the causal chain in Figure 2.2, with the scenario specifying quantities lying on one side of the cut, and the assessment or other activity using the scenario lying on the other side. The next five subsections discuss these five types of climate-change-related scenarios in turn.

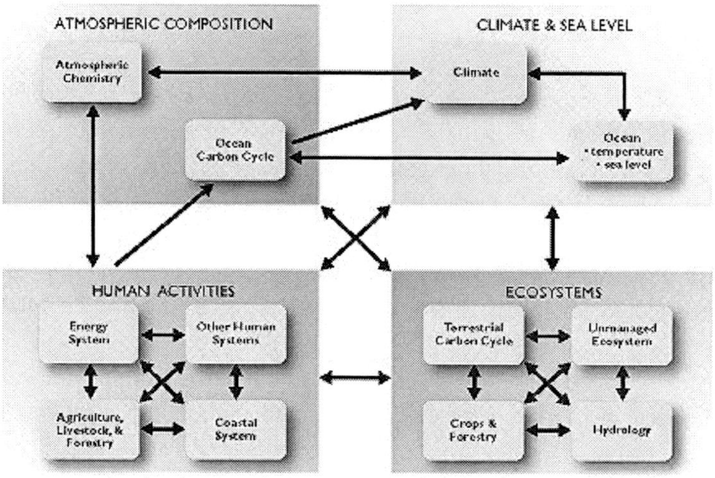

Figure 2.1. Wiring Diagram for Integrated Assessment models of climate change. *(Source: Weyant et al. 1996)*

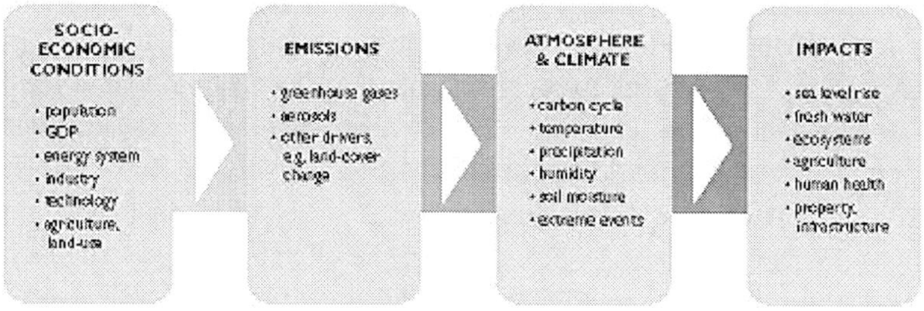

Figure 2.2. Anthropogenic climate change: Simplified linear causal chain.

2.1. Emissions Scenarios for Future Climate Simulations

Scenarios of greenhouse-gas emissions, sometimes including other human perturbations such as land-use change, are the best known type of climate-change related scenario. Emissions scenarios provide required inputs to model calculations of future climate change, as shown in Figure 2.3. As the focus and intended use of climate-model studies have shifted over time, so has the role of emissions scenarios. Early research studies examined the climate system's response to potential (rather than projected) emissions inputs in individual model studies or standardized model comparisons. In such exercises, the purpose of a scenario is to provide a known, consistent perturbation big enough to generate an informative model response. Such scenarios must be standardized, so differences between model runs can be traced to scientific uncertainties and model differences, but they can be simple and arbitrary, making no claim to being realistic. The earliest such scenarios showed a "step-change" increase in atmospheric concentration of CO_2 from its pre-industrial value, to either two or four times that value.[34] Models' equilibrium responses to doubled CO_2 provided a standard benchmark of model responsiveness, which has remained around the range of 1.5 to 4.5°C for more than 20 years. This range of modeled equilibrium responses to a standardized perturbation does not predict actual climate changes under human perturbations, although it has often mistakenly been taken as such.

The next generation of climate-model studies, beginning in the early 1990s, specified a time-path of atmospheric concentrations rather than a one-time perturbation. These studies for the first time allowed comparison of models' transient responses, by examining not just how much the cimate changes, but how fast it changes. They still used a simple, highly idealized standard scenario, most frequently a 1 percent per year increase in atmospheric concentration, expressed as CO_2-equivalent. Only two such transient simulations had been conducted by the first IPCC assessment (1990), but by the time of the second assessment (1996), most modeling groups had produced at least one.[35]

Since the mid-1990s, climate modelers have increasingly sought to produce realistic pictures of how the climate may actually change, requiring a new approach to emissions scenarios. Scenarios must now present well-founded judgments or guesses of actual future emissions trends and their consequences for atmospheric concentrations. The required emissions scenarios have been constructed either by extrapolating recent emissions trends, or, particularly for energy-related CO_2, representing emissions in terms of underlying drivers such as population, economic growth, and technological change and projecting these drivers using some combination of modeling and trend extrapolation. Driven by such scenarios, climate models for the first time can claim to be reasonable estimates of how the climate might actually change. In addition, comparisons using multiple models and emissions scenarios have allowed partitioning of uncertainty in future climate change into roughly equal shares attributed to uncertainty in climate science and models, and in emissions trends.[36] These comparisons have also allowed estimation of the climate-change benefits from specified emissions reductions.

As this shift occurred, advances in climate models – e.g., improved representations of atmospheric aerosols, tropospheric ozone, and atmosphere-surface interactions – produced mismatches between emissions scenarios and the input needs of climate models. For example, climate models now require emissions of several types of aerosols and reactive gases (principally the ozone precursors, hydrocarbons, CO and NO_x), explicit estimates of black carbon and organic carbon, and some disaggregation of different types of volatile organic compound (VOC) emissions. Moreover, because these emissions act locally and regionally rather than globally, they must be specified at the spatial scale of a model gridcell, about 150 sq. km. Models of atmospheric chemistry and transport then use these emissions to generate the concentrations and radiative forcings used by the climate model. Since emissions scenarios often do not provide the required detail, climate modelers meet these input needs through various *ad hoc* approaches.

Changes in standard emissions scenarios pose challenges for maintaining comparability with past model results. For example, the IPCC's IS92 scenarios projected that future SO_2 emissions would roughly double, then stabilize, while the later SRES scenarios projected sharp decreases, giving 2100 emissions about one-quarter the IS92 value. This change caused significant increases in projected warming that were not due to any changed scientific understanding. To help maintain backward comparability, many climate-model groups have continued to run simulations using older standardized scenarios, to provide benchmarks for comparisons both among current models and between current and previous-generation models.

2.2. Emissions Scenarios for Exploring Alternative Energy and Technology Futures

In addition to providing needed inputs to climate models, emissions scenarios have also been produced to examine alternative socio-economic, energy, and technological futures, as shown in Figure 2.4. As in Figure 2.3 the content of the scenario is emissions, but the scenario is now used to examine the socio-economic implications of alternative emission paths, which lie upstream or to the left in the causal chain. A scenario specifying a particular emissions time-path can be used to explore what patterns of demographic and economic change, energy resource availability, and technology development are consistent with that trajectory.

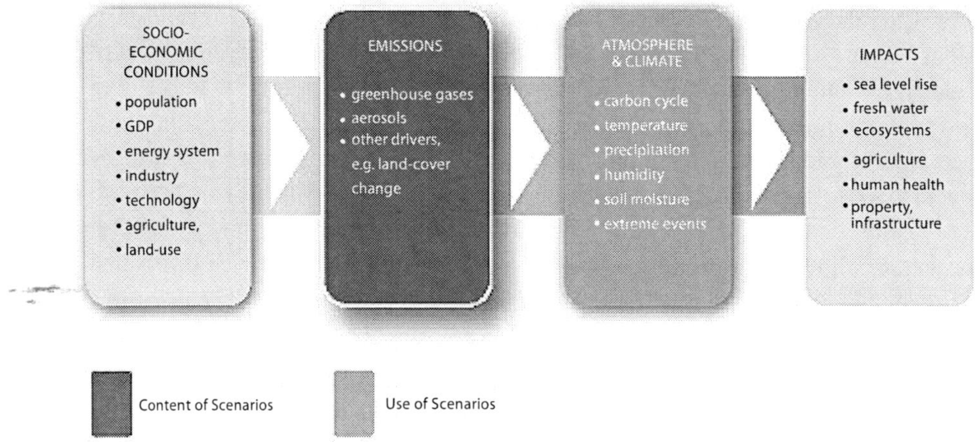

Figure 2.3. Emissions scenarios for climate simulations.

Alternatively, scenarios can be used to examine what changes in policies, technologies, or other factors would be required to shift emissions from some assumed baseline onto a specified lower path, and to estimate the cost of such a shift. To be used in this way, an emissions scenario might be specified arbitrarily, or might specify some environmental target based on normative criteria as discussed in Section 1.2. Such scenarios have been most frequently used to examine emissions trajectories that stabilize atmospheric CO_2 concentrations at specified levels. More recent exercises have instead taken stabilization of radiative forcing as the target, to examine the role of non-CO_2 greenhouse gases in meeting stabilization goals.[37]

Models' equilibrium responses to doubled CO_2 provided a standard benchmark of model responsiveness, which has remained around the range of 1.5 to 4.5°C for more than 20 years.

An important early example is the Wigley, Richels, Edmonds (WRE) scenarios, which presented emissions pathways that stabilized atmospheric CO_2 concentration at five levels, ranging from 450 to 1000 ppm.[38] Developed heuristically from a simple model of the global carbon cycle and two energy-economic models, these scenarios illustrated the large cost savings attainable by approaching stable concentrations through emission paths that initially rise and then decline steeply, rather than by beginning a more gradual decline immediately.

Several other sets of stabilization scenarios have been proposed and used for similar explorations. For example, the Energy Modeling Forum (EMF) has convened several multi-model scenario exercises focusing on emissions, emissions constraints, and their socio-economic effects. These have studied decision-making under uncertainty, international distribution of costs and benefits, the costs and benefits of the Kyoto Protocol, the implications of potential future energy technologies and technological change for emissions, and the implications of including non-CO_2 gases and carbon sequestration in mitigation targets and policies.[39]

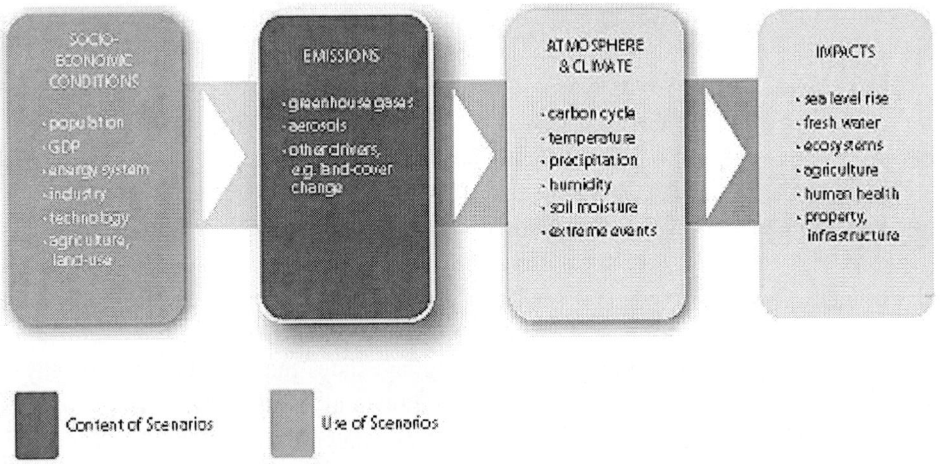

Figure 2.4. Emissions scenarios forenergy/technology futures.

In a recent scenario exercise of this type sponsored by the CCSP, three modeling teams constructed separate reference-case scenarios to examine the implications of stabilizing radiative forcing at levels roughly corresponding to CO_2 concentrations of 450, 550, 650, and 750 ppm. They examined the energy system, land-use, and economic implications of moving to stabilization. This project explored the role of multiple gases and alternative multi-gas control strategies in pursuing atmospheric stabilization. These scenarios may also provide a basis for future analyses by the CCSP, the Climate Change Technology Program (CCTP), or others.[40]

2.3. Climate Change Scenarios

Climate scenarios describe potential future climate conditions (Figure 2.5). They are used to provide inputs to assessments of climate-change impacts, vulnerabilities, and associated options for adaptation, and to inform decision-making related to either adaptation or mitigation. Depending on their specific use, climate scenarios may include multiple variables, such as temperature, precipitation, cloudiness, humidity, and winds. They may describe these at spatial scales ranging from the entire globe, through broad latitude bands, large continental and sub-continental regions, to climate model grid-cells or finer scales. They may project these at time resolutions ranging from annual or seasonal averages to daily or even finer-scale weather.[41]

Three major types of climate scenarios are distinguished by how they are produced: incremental scenarios, analog scenarios, and climate-model scenarios.[42] Incremental scenarios change current conditions by plausible but arbitrary amounts. For example, a region's temperature might be warmed by 2, 3, or 4°C from present conditions, or its precipitation increased or decreased by 5, 10, or 20 percent. Such adjustments can be made to annual or seasonal averages, to finer-period measurements of current conditions, or to the variability of temperature or precipitation over days, months, or years.[43] Like the simple emissions scenarios used for climate-model comparisons, incremental climate scenarios are simple to produce but make no claim to represent actual future conditions. They are used for initial exploratory studies of climate impacts and to test the sensitivity of impact models.

Analog climate scenarios represent potential future climates by the observed climate regime at another place or time. A spatial analog imposes the climate of one location on another, e.g., representing the potential climate of New York in the 2050s by that of Atlanta today or that of Illinois in the 2050s by that of Kansas today.[44] A temporal analog imposes climate conditions observed in the past, in the historical record or earlier paleoclimatic observations, e.g., using the hot, dry period of the 193 0s to study impacts of potential future hot, dry climates.[45] Like incremental scenarios, analog climate scenarios are more useful for exploratory studies of the climate sensitivity of particular resources or systems than for projecting likely impacts. While they represent climate states that are known to be physically possible, they are limited as representations of potential future states since they do not consider the changes in greenhouse-gas concentrations that are the principal driver of climate change.

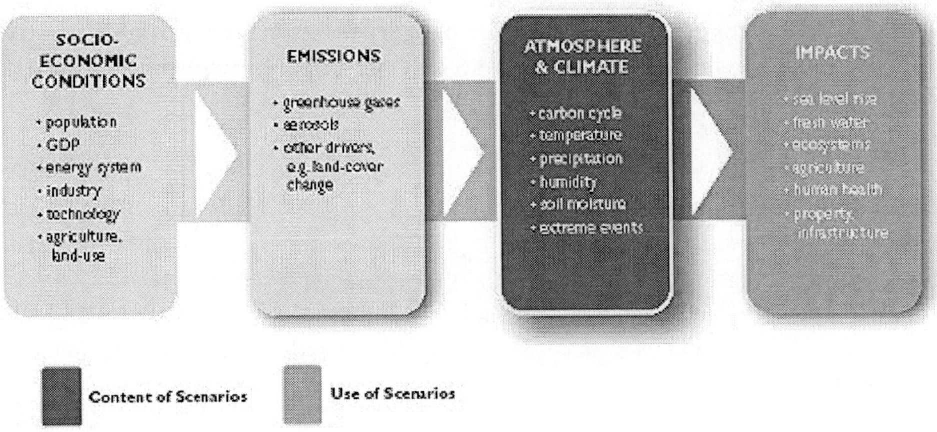

Figure 2.5. Climate-change scenarios.

Climate-model scenarios use computers to produce a physically consistent representation of the movement of air, water, energy, and radiation through the atmosphere. Climate models, also called General Circulation Models or GCMs, approximate this calculation by dividing the atmosphere into thousands of grid-cells, roughly 150 km. square in today's models, with a dozen vertical layers, treating conditions as uniform within each cell and representing finer-scale processes by numerical relationships, called "parameterizations," that are defined at the scale of a grid cell. Climate models are used to study the present climate and its responses to past perturbations like variation in the sun's output or volcanic eruptions, and to construct scenarios of future climate change under any specified scenario of emissions and other disturbances.

Unlike incremental and analog scenarios, climate-model scenarios use emissions scenarios as inputs. Model-based scenarios have a greater claim than the other types to being realistic descriptions of how the climate might actually change, because they are based on specified assumptions of future emissions trends acting on modeled representations of known physical processes. Even with a given emissions scenario, model-based climate scenarios are uncertain. Since climate models are driven by the radiative effects of atmospheric concentrations of relevant species, some of this uncertainty comes from the carbon-cycle and chemical processes by which specified emission paths determine concentrations and radiative forcings. Some of the uncertainty can be seen in the slight differences among different runs of the same climate model, because the models are sensitive to small differences in starting conditions. And some of the uncertainty can be seen in differences between calculations by different models, mainly caused by differences in the computational methods they use to handle errors introduced by finite grid-cells, and the parameterizations they use to represent small-scale processes.

Just as modeling futureclimate changerequires specification of future emissionstrends, assessment offuture climate-change impacts requires specification of future climate change.

Just as modeling future climate change requires specification of future emissions trends, assessments of future climate-change impacts require specification of future climate change. Data from a climate-change scenario might be used as input to impact assessments of freshwater systems, agriculture, forests, or any other climate-sensitive system or activity. Impact studies can use various methods, including quantitative models such as hydrologic and crop models, threshold analyses that examine qualitative disruptions in the behavior of climate-sensitive systems, or expert judgments that integrate various pieces of scientific knowledge.

As with all scenarios, the usefulness of climate scenarios depends on how well they meet users' information needs. The highly specific climate-data needs of impact analyses may not readily be provided by climate-model outputs, or may include results of whose validity climate modelers are not confident. For example, a common need of impact analyses is for data at substantially finer scales than the relative coarse grid of a climate model, which might have only 60 to 100 cells over the continental United States. One advantage of incremental and analog scenarios is that they can typically provide data at substantially finer scales. "Downscaling" techniques seek to combine the benefits of model-based scenarios – physical realism and explicit emissions-scenario drivers – and data at finer scales. The two major approaches are statistical downscaling and nested regional modeling.[46] Statistical downscaling involves estimating statistical relationships between large-scale variables of observed climate, such as regional-average temperature, and local variables such as site-specific temperature and precipitation.[47] These relationships are then assumed to remain constant under global climate change. A nested regional climate model provides an explicit physical representation of climate for a specific region, including local factors such as mountain ranges, complex coastlines, and surface vegetation patterns, with initial and boundary conditions provided by a GCM. Regional climate models can provide projections at scales as small as 10 to 20 km. Although downscaled results are anchored to local features with well-understood climatic effects, downscaling introduces uncertainties beyond those already present in GSM results.[48]

2.4. Scenarios of Direct Biophysical Impacts: Sea Level Rise

Although climate-change scenarios can provide inputs to studies of any impact, scenarios can also be constructed of particularly important forms of impact, such as sea level rise – one of the more costly and certain consequences of climate warming (Figure 2.6).[49] Changes in global mean sea level as the climate warms can be calculated using a GCM with a coupled ocean and atmosphere, which can simulate the transfer of heat to the ocean and the variation of ocean temperature with depth. To construct sea level rise scenarios for particular coastal locations, model-derived projections of global mean sea level rise must be combined with projections of local subsidence or uplift of coastal lands, as well as local tidal variations derived from historical tide-gauge data.

Sea level rise will increase circulation and change salinity regimes in estuaries, threaten coastal wetlands, alter shorelines through increased erosion, and increase the intensity of coastal flooding associated with normal tides and storm surge. Scenarios of sea level rise are consequently needed to assess multiple linked impacts on coastal ecosystems and settlements. In specific locations, these impacts will depend on many characteristics of coastal topography, ecosystems, and land use e.g., coastal elevation and slope, rate of shoreline erosion or accretion, tide range, wave height, local land use and coastal protection, salinity tolerance of coastal plant communities, etc.-in addition to local sea level rise.[50]

In addition to its gradual impacts, sea level rise is subject to large uncertainties from the potential loss of continental ice sheets in Greenland and West Antarctica. The consequences of these events for global sea level rise are well known because they can be calculated quite precisely from the volume of the ice sheets – roughly 7 meters rise from complete loss of the West Antarctic Ice Sheet and 5 meters from Greenland – but the probabilities of these events and their likely speed of occurrence are both highly uncertain. One recent study has suggested a probability of a few per cent that the West Antarctic Ice Sheet will contribute an additional one meter per century beyond that calculated from gradual warming.[51]

There are several reasons to call out sea level rise from other climate-change impacts to be represented in separate scenarios. First, sea level rise is a powerful driver of other forms of climate-change impact, probably the most important driver of impacts in coastal regions. Since it is a direct physical impact of climate change that can be described precisely and compactly, a sea level rise scenario is an efficient way to transmit the most important information about climate change to coastal impact assessments. Moreover, since sea level rise does not depend on socio-economic processes and cannot be significantly influenced by human actions (other than by limiting climate change itself), it may be reasonably treated as exogenous for purposes of impact assessment. For all these reasons, sea level rise is a good proxy for the most important causal routes by which climate change will affect coastal regions.

Finally, because sea level rise is subject to large uncertainties with known consequences but unknown probabilities, it is a useful variable for exploratory analysis of worst-case scenarios in long-range planning. Other forms of climate impact might also merit being called out in separate scenarios: changes in snowpack in mountain regions, seasonal flow regimes in major river basins, or the structure and function of major ecosystem types. Based on present knowledge, however, only sea level rise has shown these characteristics strongly enough to motivate construction of separate scenarios.

2.5. Multivariate Scenarios for Assessing Impacts, Adaptation, and Vulnerability

Many potentially important impacts of climate change cannot be adequately assessed by considering only how the climate might change. These impacts require multivariate scenarios that include climate change and other characteristics likely to influence impacts. This is the case, for different reasons, for both ecosystems and socio-economic systems, although the nature of the multivariate scenarios that are required – i.e., the number and identity of the characteristics that must be specified – will vary widely among particular impacts.

Ecosystems are affected by climate change, but also by many other changes in environmental conditions that are influenced by human activities, such as nitrogen and sulfur deposition, tropospheric ozone and smog, and changes in erosion, runoff, loadings of other pollutants, land use, land cover, and coastal-zone characteristics. Consequently, realistic assessments of potential future impacts on ecosystems require specifying the most important forms of human-driven stresses jointly, not just climate.[52]

In addition, many important forms of climate-change impact depend not just on climate change, its direct biophysical impacts such as sea level rise, and perhaps other forms of environmental stress, but also on the nature of the society on which these climate and other environmental changes are imposed – e.g., how many people there are; where and how they live; how wealthy they are; how they gain their livelihoods; and what types of infrastructure, institutions, and policies they have in place.[53]

Assessment of climate impact on ecosystems that are intensively managed for human use, such as agriculture, managed forests, rangelands, and hydrologic systems, must consider human management as a factor in impacts. The non-climatic factors that influence these management decisions – e.g., changes in market conditions, technologies, or cultural practices – must be considered for inclusion in scenarios if they are sufficiently important in mediating climate impacts.

In other domains, socio-economic factors can mediate climate impacts by influencing vulnerability and adaptive capacity. No general model of the socio-economic determinants of adaptive capacity exists. Important factors are likely to vary across specific types of impact, locations, and cultures, and may include many demographic, economic, technological, institutional, and cultural characteristics.

Some socio-economic characteristics that are likely to be relevant for many impact assessments – e.g., the size and sometimes the age structure of population, the size and sometimes the sectoral mix of GDP – are normally generated in the course of producing emissions scenarios. Consequently, when current emissions scenarios exist for the region for which an impact assessment is being conducted, it makes sense to strive for consistency with them.[54] Even for these variables, however, there may be significant problems of incompatible spatial scale. Impact assessments often examine smaller spatial scales than emissions projections, so they may need these socio-economic data at finer scale than is available. Downscaling future socio-economic projections has proved challenging thus far. There is no generally accepted method for doing so, and several research groups are exploring development of alternative methods.[55]

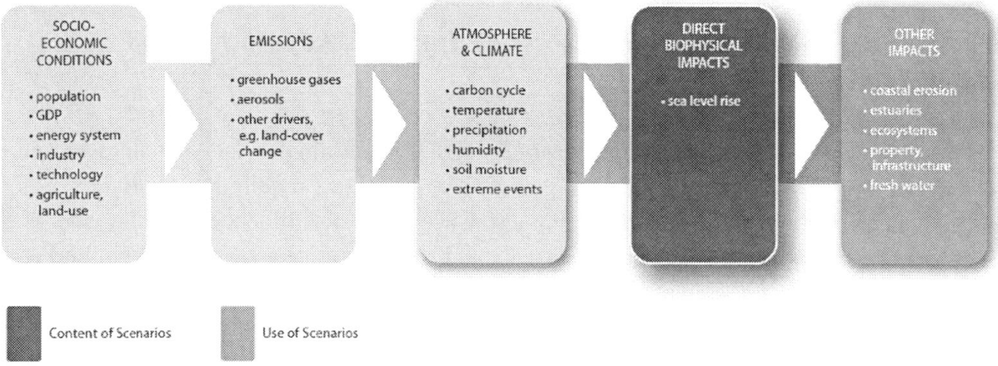

Figure 2.6. Scenarios of direct biophysical impacts: sea level rise.

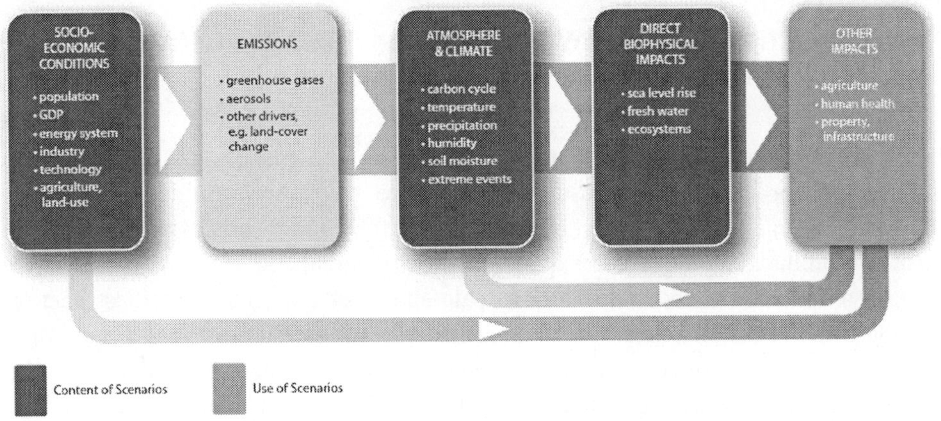

Figure 2.7. Multivariate scenarios for impact assessment.

In contrast to the few clearly identified aggregate characteristics needed to construct emissions scenarios, the socio-economic factors that most strongly shape adaptive capacity and vulnerability for particular impacts may be detailed, subtle, and location-specific. It may not even be clear what characteristics are most important before doing a comprehensive analysis of potential causal pathways shaping impacts. The most important characteristics may interact strongly with each other or with other economic or social trends, or may not be readily quantifiable. All these factors make the development of socio-economic scenarios for impact assessment a much more difficult endeavor than constructing emissions scenarios.

Because scenarios are schematic, not all factors that might be important for impacts can be included. Details are typically not included or treated as merely illustrative. But particular details, which cannot be identified in advance, may be crucial determinants of vulnerability to climate impacts.[56] Impact assessments have responded to this dilemma in two broad ways.[57] First, constructing scenarios of relevant socioeconomic conditions has been delegated to local or regional teams with expertise in the impacts being assessed, subject to constraints to maintain consistency with other assessments. Second, since local or regional scenario groups may not have access to all knowledge relevant to understanding the main determinants of impacts, more open-ended approaches have been employed – e.g., exploratory analyses that iterate between considering particular characteristics that might be important, examining their implications for impacts using the data and models available, then reassessing what variables are most important.

This section has sketched a typology of global-change scenarios and identified major types of decision-makers who might use global-change scenario-based information. The next section turns to current experience with global-change scenarios, summarizing the development, contents, and uses of four major exercises. Informed by these cases plus additional short scenario examples presented in text boxes, Section 4 will summarize and discuss the major challenges for making and using scenarios that are raised by this experience, providing the basis for the conclusions and recommendations presented in Section 5.

SECTION 3. REVIEW OF MAJOR CLIMATE-CHANGE SCENARIO EXERCISES

In this section, we review experience to date developing and using scenarios for global climate change applications. Because little literature on these activities yet exists, our selection of cases has inevitably been both limited by time and resources at our disposal and reliant in part on the knowledge and experience of team members. We discuss four exercises in detail, in an attempt to cover the largest-scale and most important activities. Section 3.1 reviews the IPCC emission scenarios, with particular detail on the most recent and important exercise, the Special Report on Emissions Scenarios (SRES). Section 3.2 considers the US National Assessment, which developed and used scenarios of both climate and socio-economic conditions. Section 3.3 considers the UK Climate Impacts Programme, which has also both developed and used scenarios, following a different approach from the US National Assessment. Section 3.4 reviews the ambitious scenario-generating exercise conducted as part of the Millennium Ecosystem Assessment (MEA), in which climate change was one of several dimensions of stress considered on global ecosystems.

For each exercise, we consider only the development and use of scenarios, rather than examining the larger assessment processes of which the scenarios were part. We consider how the scenarios were developed, including both methods of reasoning and managerial process; how and by whom they were used; and subsequent evaluations, when these are available. General issues and challenges that emerge from these experiences are discussed in Section 4.

To provide more illustrative variation, we also provide shorter summaries of eight additional scenario activities, some of them related to the four we consider in detail. Presented in text boxes throughout Section 4, these are intended to provide additional information to highlight particular issues. We have particularly sought experiences that illuminate potential relationships between scenarios and decision-making.

All these scenario exercises represent early work in an immature field. Our aim is not to criticize particular exercises, but to seek insights from their experience into the general problems of making useful global-change scenarios.

3.1. Ipcc Emissions Scenarios

Since its establishment in 1989, the IPCC has organized three exercises to develop scenarios of greenhouse-gas emissions, of increasing scale and complexity. For its first report, IPCC's Working Group 3 on "Response Strategies" included a sub-group on emissions scenarios. Four scenarios were produced but little used in this assessment because of time limits and because, with one exception, only doubled CO_2 equilibrium climate-model runs were available at the time.[58] The next exercise produced six new scenarios, called IS92a through IS92f.[59] These were the first global emissions scenarios with a full suite of greenhouse gases and at least some explicit calculation underlying each. The IS92a scenario, one of the central scenarios in this group, was used in climate-model comparisons conducted for the 1996 IPCC assessment, along with the simpler transient scenario of 1 percent annual increase in equivalent-CO_2 concentration and further equilibrium runs.[60]

The third and most ambitious IPCC scenario process was established in 1997 and worked for two years to produce the *Special Report on Emissions Scenarios* (SRES).[61] In part, this process was established in response to two widely circulated criticisms of the IS92 scenarios. The first of these criticized the 1992 scenarios for inconsistency with other published scenarios of energy and carbon intensity for major world regions; failing to reflect important recent trends, including the collapse of the Soviet Union and increasing restrictions on sulfur emissions worldwide; relying inappropriately on a single model; and only being useful as climate-model inputs, not for other purposes such as mitigation studies or supporting climate-change negotiations.[62] The second criticized the IS92a scenario for assuming increasing divergence in the per capita emissions of industrialized and developing regions, arguing that this represented a strong bias in favor of already developed regions.[63]

In response, the 1996 plenary session of the IPCC requested a new set of emissions scenarios. These new scenarios were to improve treatment of sulfur aerosols and emissions from land-use change, and were not to rely on a single model or expert team, but instead to draw on the existing literature and invite any group with relevant expertise to participate in an "open process."[64] They were also charged to serve more uses than climate-model inputs, such as supporting impact analyses, but to assume no new climate-policy interventions. Although not explicitly in the terms of reference, it was also clearly understood that the scenarios would address the criticism of the IS92 scenarios by focusing on convergent development paths between North and South.

In January 1997 the IPCC established a writing team, including members of several energy-economic modeling groups and experts in related areas such as population, technological change, and scenario development methods. The process ran under tight time pressure to provide preliminary scenarios by early 1998 for climate-model runs in the IPCC Third Assessment.

Prior scenarios were compiled in a web-based database,[65] and any researcher was invited to submit new ones. By mid-1998 the database contained more than 400 scenarios. Most of these projected only energy-related CO_2 emissions, but they were highly diverse in their coverage and resolution, the variables included, and their methodologies. The usefulness of these scenarios in constructing new ones was limited by several weaknesses, however. Many were incomplete, lacked documentation of inputs, or made inconsistent assumptions. Few included sulfur or land-use emissions, which were specifically requested of the new scenarios. Many were unclear on whether they assumed mitigation efforts, while the new scenarios were instructed to exclude them. Consequently, the development of new scenarios had to proceed largely independently of the collection of existing scenarios through the literature review and open process.

Early on, participants decided to use narrative scenarios in addition to quantitative models, and to include experts in this approach on the writing team. This decision drew on recent successes using such scenarios for energy and environmental applications,[66] and responded to the charge to make the scenarios more integrated and more broadly useful. Participants in an April 1997 workshop chose two key uncertainties to explore in the scenarios: whether world values would mainly stress economic prosperity or balance economic and ecological concerns (labeled "A" vs. "B" scenarios); and whether the organization of economies and institutions would continue shifting toward global integration, or reverse and move toward regional fragmentation (labeled "1" vs. "2" scenarios).[67]

Table 3.1. Target Values for 2100 inInitial Scenario Quantifications

SCENARIO	A1B	A2	B1	B2
Population	7.1	15.1	7.1	10.4
GDP (trillion)	$530	$250	$340	$235
Final Energy (EJ)	~1,700	870	770	950
Model for Marker scenario	AIM	ASF	IMAGE	MESSAGE

Source. Minutes of Laxenburg meeting, July 2-3, 1998.

Combined, these gave four scenarios, which were sketched in preliminary terms at the workshop. In the A1 (economic, global) scenario, economic growth and inter-regional income convergence continue strongly worldwide – all developing countries grow like Japan and Korea from the 1950s to the 1 980s – while world population peaks at 9 billion by 2050. Rapid innovation yields many advanced energy sources, while acid rain and other local and regional environmental problems are aggressively controlled. In contrast, the A2 (economic, regional) scenario has higher population growth, lower economic growth with more continuing regional disparities, slower innovation, and weaker international institutions. B1 (ecological, global) has low population growth, moderate economic growth with strong convergence, and strong reductions in per capita energy use, mostly through higher efficiency, while B2 has intermediate population growth, low economic growth with weaker convergence, and moderate improvements in energy efficiency and development of non-carbon energy sources.[68] The storylines were elaborated in short text descriptions with some preliminary numbers attached in fall 1997.[69]

The SRES interimmarker scenarios wereused to provideemissions scenarios toclimate modelsparticipating in theIPCC third assessment.

Modeling teams were asked to produce initial quantifications of these scenarios in fall 1997, to match specified 2100 target values within 10 percent. In February 1998, the preliminary quantitative targets were reconfirmed and modelers asked to continue work on quantifications, now including a breakdown of economic output into four world regions.[70] In April, one model's quantification was chosen as a "marker scenario" for each of the four scenarios – a particular scenario that would provide the basis for interim reporting to climate modelers, some of whose results other participating models would be asked to replicate. The specifications and models for these marker scenarios are shown in Table 3.1.

These interim marker scenarios were used to provide emissions scenarios to climate models participating in the IPCC third assessment. An IPCC climate modelers' meeting in June 1998 agreed to use SRES scenarios and asked for three cases, central emissions, stabilization, and high emissions.[71] The writing team initially discussed meeting this request by identifying scenarios corresponding to each of these requested cases,[72] but decided to provide only the marker scenarios and recommend that all four be used without identifying any as "central."

The marker scenarios also provided the basis for coordination of subsequent scenario development. Up to this point, there had been substantial discrepancy between different models' quantifications of the same scenarios, particularly at regional level. With the adoption of the markers, other groups were asked to replicate (within 5-10 percent) the marker results on population, GDP, and final energy for the four world regions, for 2100 and several interim years.[73] Achieving this requested replication posed significant challenges for modelers.[74]

The SRES scenarioshave been the most comprehensive, ambitious, and carefully document edemissions scenarios produced to date.

With a further year of work, modeling teams produced a total of 40 scenarios that were retained in the report, of which 26 replicated one of the marker scenarios. Although a few of the 14 non-replicates reflected a model's inability to match the results of a marker scenario, most were produced because a modeling team intentionally sought to explore alternative assumptions. For example, the A1 scenario, which originally balanced fossil and non-fossil energy sources, was augmented by variants with different assumptions about fossil resources and non-fossil technology development, giving widely divergent emissions paths stressing coal, gas, and non-fossil energy technology. Modifications of the scenario set continued until late in the process. For example, it was decided in October 1998 to drop several B variants with explicit mitigation, including one stabilization scenario.[75] At the final IPCC approval meeting, it was decided at the request of the Saudi delegation to reduce the two fossil-intensive variants of A1 to one, a variant of the gas-intensive scenario which was renamed A1FI (for "fossilintensive").[76]

3.1.1. Significance and use

The SRES scenarios have been the most comprehensive, ambitious, and carefully documented emissions scenarios produced to date. They represented a substantial advance from prior scenarios, and contributed to assessments and subsequent research on climate impacts and responses. The SRES scenarios formed the basis for climate-model comparisons in the IPCC Third Assessment (2001) and continuing work in the Fourth Assessment. Most subsequent climate-model work has used only a few of the marker scenarios – typically A2 and B2, sometimes with A1B added. SRES scenarios also provided baselines for analysis of mitigation scenarios in the Third Assessment.[77]

Several significant insights were illuminated by the SRES scenarios.

- Scenarios with similar emissions in 2100 can follow markedly different paths in the interim, giving wide differences in cumulative emissions and concentrations.
- Technology and energy-resource assumptions can strongly perturb future emissions, even with constant socio-economic assumptions. For example, the three A1 variants show that changing these assumptions alone can generate as wide a range of emissions futures as substantial variation of demographic and economic futures.
- Highly distinct combinations of demographic, socio-economic, and energy-market conditions can produce similar emissions trajectories, suggesting that a particular emissions trajectory can pose very different types of mitigation problems, depending on what combination of driving factors underlies the emissions.

3.1.2. Criticisms and controversies

The SRES experience raised issues of great significance for subsequent attempts to develop more useful climate-change scenarios: the desirability of and appropriate methods for characterizing probabilities associated with scenarios; the quantitative representation of the relationship between North and South; methods for developing and using narrative scenarios and integrating them with quantitative model results; alternative modes for coordinating use of multiple models and their implications for the interpretation and use of scenarios; and the relationship between scenario exercises and their users, including the need for clarity about specific intended uses, appropriate methods for engaging users in scenario development, and how to improve utility of scenarios when not all potential user groups are specifically identified. These are discussed in Sections 4 and 5.

The first two of these issues were the subjects of forceful public criticisms. We discuss these, followed by several other issues that have received less attention but which in our view pose more central and instructive challenges for future scenario exercises.

Assigning explicit probabilities

The SRES team decided at the outset to make no probabilistic statements about the scenarios. Their report used great care in its language to avoid any suggestion that one scenario might be more central or more likely than any other.[78] This decision was consistent both with standard practice in developing narrative scenarios, and with the instruction in their terms of reference not to favor any model.[79]

They were sharply criticized for this decision.[80] Critics argued that there were no technical obstacles to assigning probabilities to emissions ranges bounded by the marker scenarios; that scenario developers must have made probabilistic judgments in generating and evaluating the scenario quantifications and that not making these judgments explicit would withhold relevant information; and that if scenario developers decline to assign probabilities, others who are less informed will do so. Indeed, many probabilistic emissions calculations have been produced since the SRES, using various methods such as assigning uniform or other specified distributions over the emissions range of the marker scenarios, counting scenarios lying in specified intervals in the larger SRES set, unbundling and recombining alternative values of the drivers underlying SRES emissions figures, or sampling over parameter distributions within a single model. In response to these criticisms, SRES authors argued that attempting to assign probabilities to scenarios would require assigning joint distributions to the underlying driving factors, and that this would lead to an explosion of combinatoric possibilities over which any attempt to assign probabilities would be spurious and arbitrary.[81]

It may well be unhelpful to assign probabilities to rich, multi-dimensional narrative scenarios, yet still useful to assign interval probabilities when scenarios principally represent uncertainty in one or two quantitative variables

The situation of the SRES scenarios is in fact more nuanced than the arguments of either their authors or critics would suggest. It may well be unhelpful to assign probabilities to rich, multidimensional narrative scenarios, yet still useful to assign interval probabilities when scenarios principally represent uncertainty in one or two quantitative variables. And while the

SRES scenarios began their lives like the former type of storyline scenario, they finished more like the latter. For many users, the scenarios ***are*** their projections of greenhouse-gas emission trends. When they are viewed in this way, a potential user may reasonably ask, how likely are emissions to be higher than this – a distinct and better-posed question than, what is the probability of an A1 world? The uncertainty issue has no clear resolution in this case, and poses hard design problem for scenarios and assessments more broadly. Although the SRES exercise has raised this controversy most explicitly to date, the problem is a general one that any scenario exercise must confront. We discuss it further in section 4.6.

Exchange rates: PPP versus MER

The most prominently publicized criticism of SRES focused on the fact that all but one of the participating models compared GDP across regions using market exchange rates (MER), instead of the more correct purchasing-power parity (PPP) approach. PPP comparisons correct for price differences among countries, providing a more accurate comparison of real incomes. Because lower-income countries have lower price levels, MER-based comparisons overstate the income gap between rich and poor countries.

...while MER over states future income growth in poor countries, it also over states future reductions in energy and emissions intensity.

In a series of letters to the IPCC chairman and several subsequent publications, two critics argued that the use of MER caused SRES scenarios to over-estimate future income growth in developing countries (because they over-estimated the initial income gap), and consequently to over-estimate future emissions growth. Their criticism was widely circulated and repeated by prominent climate-change skeptics.[82]

But, although using MER does overstate future income growth, it does not necessarily follow that future projections of emissions growth are also overstated. MER is universally recognized as a flawed measure of income, whose use in global-change scenarios is only justified by better availability of current and historical data, and the fact that international emissions trades in any future mitigation regime will likely be made at market exchange rates. But changing the measure of income also changes the relationship between income and such physical quantities as energy and food consumption, which determine emissions. Consequently, while MER overstates future income growth in poor countries, it also overstates future reductions in energy and emissions intensity. These opposing errors are likely to be similar in size, in which case any error in emissions projections from using MER will be small.[83]

A related, more serious concern is that all SRES scenarios assumed varying degrees of real income convergence between North and South; this was done in response to criticisms that the IS92 scenarios were biased in favor of the North. But an exercise to construct potential climate-change futures may need to consider less optimistic and less desirable futures in which some currently poor regions fail to solve the development problem. Not considering less fortunate futures, including ones that might challenge the adequacy of current responses, institutions, and decision-making capacity, may limit scenarios' usefulness in supporting longterm risk assessment and planning for the societal response to climate change.

Underdevelopment of narrative scenarios

Although the SRES storylines were produced first and featured prominently in publications, they remained underdeveloped and underused throughout the process. In part due to time pressure, in part due to the predominance of quantitative modelers in the process, little attention was given to further development of the storylines once initial quantifications were established and modeling work began. Nor was significant effort devoted to integration and cross-checking between storylines and quantitative scenarios, although a major purpose of the narratives was to give coherent structure to quantifications.[84] Concerns raised about the storylines included lacking specification of characteristics other than those needed to generate emissions; imbalance between the story- lines, with A1 much more developed than the others and B2, the least developed, likely to be heavily used as the median scenario for emissions; apparent inconsistencies within A2; and lack of clarity regarding the distinctions between A2 and B2 – a serious enough concern that merging them was repeatedly considered until late in the process.[85]

Moreover, participants disagreed over the meaning of some of the scenarios, as indicated by the persistent difficulty they had in agreeing on descriptive names.[86] These names were dropped late in the project, in the context of a broad retreat from attempting to flesh out the storylines. That so little integration of qualitative and quantitative components was achieved in spite of serious and persistent efforts suggests the magnitude of the analytical and methodological challenges involved.

Harmonizing scenarios, interpreting the results

Scenario exercise that use multiple models can coordinate them in several ways: choosing one or a few illustrative scenarios as coordinating devices for subsequent analyses, as was done with the SRES marker scenarios; fixing values of a small set of exogenous inputs to multiple models, to characterize resultant uncertainties and examine their origins through focused model intercomparisons; or fixing key outputs as targets, to reason backwards and examine requirements for achieving them.

Choosing a few quantitative variables as the initial link between storylines and models makes these variables serve as a framework to capture the storylines' basic logical structure. Although these choices are not obvious, the variables chosen here appear reasonable. But the causal structure of a model will not generally mirror the presumed causal logic of a narrative, so a model cannot be expected to calculate values for other variables that flesh out the storyline logic. Moreover, the few key variables so chosen may not be exogenous inputs for every model used in the subsequent quantification. Of the three variables specified in the SRES process, only population was exogenous for all participating models. Because GDP and final energy were endogenous for some or all participating models, matching their specified values required manipulating other internal model characteristics. Once one model run was chosen as the marker for each scenario, subsequent attempts by other models to replicate the results posed the same problem more acutely, since more outputs were specified at this point.

The problems associated with attempting to harmonize model outputs are related to the underdevelopment of narrative scenarios and limited integration of qualitative and quantitative components. The storylines were associated with relatively restrictive numerical targets even though the storylines did not develop the richness or coherence that would carry implications for additional characteristics. The preliminary targets were only slightly modified throughout the project, despite subsequent discovery of significant problems. For example, the United Nations 1998 population projections, with substantial reductions in projected fertility, were completed while the scenario development work was underway but not incorporated.[87]

Clarity about uses, involving users

The SRES scenarios were charged with serving uses beyond driving climate models but given little guidance on what specific additional uses or users to serve, or how the scenarios might best serve them, neither of which is obvious.[88] Providing climate-model inputs remained the most prominent and most clearly specified use, as well as the only use that had an early deadline. But climate modelers were not involved in the scenario development process, and there was substantial divergence between their needs and the outputs and capabilities of the SRES process. A September 1997 briefing identified the principal needs of climate modelers as early availability of scenarios and greater emissions detail.[89] They wanted separate emissions trajectories for major greenhouse gases, not just CO_2-equivalent, including regional detail for some emissions such as sulfur – even suggesting that it would be desirable to have sulfur emissions disaggregated by stack height, to distinguish dispersed emissions from large point sources. Although SRES provided gridded sulfur data by post-processing model outputs, in most cases the emissions included and their spatial detail (not to mention stack height) were limited by the capabilities and structures of participating models.

The US National Assessment was themost comprehensive attempt to date toassess climate impactson the United States over 25-year and 100-year horizons, and to consider both major sub-national regions and sectors.

Other uses received less attention, and representatives of other potential uses were even less involved than climate modelers in the process. Supporting assessment of mitigation strategies was largely deferred to the post-SRES scenarios prepared for the IPCC Third Assessment Report, although ambiguity about the degree of mitigation effort implied by some SRES scenarios complicated that task. Impact and vulnerability assessments depend on diverse, small-scale socio-economic and ecological factors that a global exercise centered on energy-economic models cannot provide.[90] For the population and economic projections that were provided in the course of generating emissions scenarios, the key issue for impacts and adaptation was the degree of spatial detail provided. For consistency among scenarios, and to avoid base-year discrepancies with national and regional datasets, SRES scenario results were reported only for four large world regions. Greater regional detail was available from individual models, but with inconsistent regional boundaries. Providing the greater regional detail desired for impact assessments would generate discrepancies between the global-model results represented in scenarios and the more detailed data and projections available at national and regional levels.[91] Developing valid methods to downscale socio-economic scenario information and integrate it with national and regional datasets remains a key challenge for producing useful scenarios for impact assessment.[92]

3.2. The Us National Assessment

The US National Assessment was the most comprehensive attempt to date to assess climate impacts on the United States over 25-year and 100-year horizons, and to consider both major sub-national regions and sectors.[93] Responding to a requirement in the 1990 Global Change Research Act, the National Assessment was organized by federal agencies participating in the US Global Change Research Program. Work began in 1997, with various components completed between 1999 and 2003. Regional impacts were initially considered in 20 regional workshops, followed by more extended analyses of impacts, leading to published assessments for 12 regions, conducted by university-based teams. Sectoral impacts were examined by national teams focusing on agriculture, water, human health, coastal areas and marine resources, and forests. A federal advisory committee, the National Assessment Synthesis Team (NAST), provided direction for the assessment and synthesized its results in two published reports.[94] Roughly two thousand experts and stakeholders participated.

As an assessment focused on climate impacts, the National Assessment needed both climate scenarios and scenarios of potential future socio-economic conditions over the 21st century, since substantial changes are likely over this period in socio-economic conditions that might influence vulnerability to climate and adaptive capacity.

3.2.1. Emission and climate scenarios

For climate scenarios, the National Assessment relied predominantly on data and model results previously produced. Study teams conducted additional checking, processing, documentation, and dissemination as needed to make these usable. The assessment encouraged the use of three types of climate scenarios: historical scenarios produced by extrapolating observed trends or re-imposing historical climate variability or extremes; an inverse approach using sensitivity analyses to explore the responses of climate-sensitive systems, with particular emphasis on thresholds defining key vulnerabilities; and climate model simulations of future climate conditions.[95]

Of these three approaches, the climate-modelscenarios were the most precisely specified andthe most widely used. The National Assessmentdid not have the resources or time to commission new climate model runs and so had to relyon those completed and published when itbegan its work. A set of criteria was developedby the NAST for the climate model scenarios tobe used in the assessment. Climate-model scenarios used in the assessment should, to thegreatest extent possible:[96]

1. Include comprehensive representations of the atmosphere, oceans, and land surface,and key feedbacks among them
2. Simulate the climate from 1900 to 2100,based on a well-documented emissions scenariothat includes greenhouse gases and aerosols
3. Have the finest practicable spatial and temporal resolution, with grid cells of less than5° latitude x longitude
4. Include the daily cycle of solar radiation, toallow projections of daily maximum andminimum temperatures
5. Be able to represent significant aspects ofclimate variability such as the El Niño-Southern Oscillation (ENSO) cycle

6. Be completed in time to be quality-checked and interpolated to the finer time and spatial scales needed for impact studies
7. Be based on well-documented models participating in the IPCC Third Assessment (forcomparability between US and internationalefforts)
8. Be able to interface results with higher-resolution regional model studies
9. Provide a comprehensive array of resultsopenly over the internet.

To ensure timely dissemination, the National Assessment chose climate-model scenarios to be used in its analyses in mid-1998. At that time, only two groups had completed runs that met most of the key criteria: the UK Hadley Centre (Model Version 2) and the Canadian Centre for Climate Modeling and Analysis (Model Version 1).[97] All participating regional and sector teams were asked to use these scenarios. The climate sensitivity of these models was 2.5°C (UK Hadley) and 3.6°C (Canadian), lying in the middle of the 1.7 to 4.2°C range of sensitivities represented by models participating in the IPCC Third Assessment. [98]

Of the three approaches, the climate-model scenarios were the most precisely specified and the most widely used.

These two models were limited in their ability to reproduce observed patterns of inter-annual and inter-decadal climate variability. But other climate-model runs available at the time failed to meet essential requirements of the ecosystem models that were the basis for an important part of the assessment: availability of documented results, projections to 2100, standard/comparable emissions scenarios, and explicit treatment of the day-night cycle.

For these two climate models, model runs using only one emissions scenario were available, and only one ensemble run was used for each.[99] The emissions scenario was IS92a, which represented the middle of the range of IPCC's 1992 scenarios.[100] In addition to greenhouse gases, the scenario included atmospheric loadings of sulfate aerosols, which were assumed to increase sharply through 2050 and then level off for the rest of the 21st century.[101]

The applicability of these two scenarios was tested by checking the models' ability to replicate broad patterns of US climate change over the 20th century when driven by historical greenhouse-gas forcings. Model results were compared against the VEMAP (Vegetation-Ecosystem Mapping and Analysis Project) dataset, a corrected climatic dataset for the 20th century. This comparison showed reasonable accuracy in reproducing the spatial distribution of average temperatures and century-long temperature trends, but significantly weaker reproduction of observed patterns of precipitation, mainly because the spatial distribution of precipitation depends on topographic detail that is too fine-scale to be captured even by the 0.5-degree VEMAP grid.[102]

With the specified scenario of future emissions, the two climate-model scenarios projected global warming by 2100 of 4.2°C (Canadian) and 2.6°C (Hadley).[103] These projections were at the high end and in the middle, respectively, of the range of warming projected for this emissions scenario by models participating in the IPCC Third Assessment Report.[104] For the continental United States, the two models projected warming by 2100 of 5.0°C (Canadian) and 2.6°C (Hadley), at the high end and below the middle, respectively, of the range of projections in the IPCC Third Assessment.[105] In their projections of precipitation change over the United States, these scenarios both lay at the high end – the Hadley scenario projected the highest precipitation in 2100 and the Canadian the second-highest[106] – but the

Canadian model's greater warming offset the effect of this precipitation increase on soil moisture, which was projected to decrease over most of the continental United States.[107]

Although only the Hadley and Canadian climate-model scenarios were used throughout the assessment, several others that met some or all of the assessment's needs became available during its work. Several region and sector teams were able to use these additional scenarios. In some cases, the additional scenarios allowed groups to strengthen their conclusions. For example, an analysis of future Great Lakes water levels under climate change using eleven climate models found that ten of these showed lower levels and only one higher.[108] In other cases, using multiple models allowed more detailed characterization of uncertainties in future regional changes. For example, the Pacific Northwest team presented distributions of regional temperature and precipitation change in the 2030s and 2090s using seven GCMs.[109]

Despite the National Assessment's aim of exploring future climate using three distinct types of scenario, historical scenarios and sensitivity analyses were less extensively used than GCM scenarios and featured less prominently in the assessment's publications. Two uses of historical climate data – describing observed impacts of climate variability and using observed historical extremes as benchmarks to compare projected future changes – were made by all groups. To support systematic use of historical scenarios, the VEMAP 20th-century dataset was provided to all groups, but no further guidance was provided on how to generate climate scenarios from these historical data, e.g., on what periods to choose or how to use them to assess potential future impacts. Several groups used these historical data to describe the impacts of particular recognized patterns of climate variability, such as ENSO or the Pacific Decadal Oscillation (PDO).[110] Many groups examined past climate extremes, but only in qualitative ways; most did not follow the approach, taken in some previous impact studies, of using historical extreme periods as quantitative proxies for potential future climate.[111]

The third approach, vulnerability analysis, was the least used in the National Assessment. This "inverse" approach involves describing the properties of a climate-sensitive system, specifying some important change or disruption, and asking what climate changes would be required to bring about that disruption and how likely – based on historical data and model calculations – such climate changes appear to be. Given the complex dynamics of climate-sensitive systems and models of these systems, and the multiple dimensions of climate on which these can depend, this approach requires a substantial program of new research, analysis, and methodological development.[112] In part because of the intrinsic difficulty of this task – and in part due to management and resource problems – this approach was not pursued. The NAST proposed it, but more tractable approaches to analyzing climate impacts dominated the assessment's work. This remains an important area for further work in development of assessment and modeling methods.

3.2.2. Socio-economic scenarios

As discussed in Section 2.5 above, assessing impacts of future climate change can require specifying not just scenarios of future climate, but also socio-economic characteristics of the future society that will experience the changed climate. Specifying future socio-economic conditions might be necessary for two reasons. First, socio-economic conditions may influence the demands placed on particular resources that are also sensitive to climate change, the value assigned to them, and the non-climatic stresses imposed on them. For example, future flow regimes in river systems will be influenced by upstream demands for municipal and irrigation water use, in addition to the changes caused by climate. Socio-economic scenarios are also needed to assess climate-change impacts on human communities – e.g., economic impacts and their distribution, human health effects, and vulnerability to extreme events – because socioeconomic characteristics of a community experiencing a changed climate will strongly influence the community's vulnerability to changes and its capacity to adapt.

Socio-economic scenarios are also needed to assess climate-change impacts on human communities.

In contrast to climate scenarios, little prior information or experience was available on constructing scenarios of socio-economic conditions for impact assessment. Consequently, the assessment developed new methods, using an approach that combined centralized and decentralized elements. Centralization was needed because a few variables, such as population, economic growth, and employment, are likely to be important in all regions and sectors. For these, consistent assumptions are required to allow comparison of impacts across regions and sectors, and to aggregate from separate assessments up to overall national impacts. A NAST sub-group developed high, medium, and low-growth scenarios of these variables at the national level. These followed the US Census Bureau high, middle, and low scenarios for fertility and mortality through 2030, but assumed a wider range of values for net immigration to account for possible illegal immigration.[113] Over this period, national population, GDP, and employment were disaggregated among regions and sectors using a commercial regional economic model.[114] Beyond 2030, national projections of these variables followed OECD growth rates in the SRES marker scenarios.[115]

Decentralization was also needed because the particular socio-economic characteristics that most strongly influence climate impacts and vulnerability may differ among regions, activities, and resources. For example, major socioeconomic determinants of climate impacts on Great Plains agriculture may include the crops grown, the extent of irrigation, and the technologies used to provide it, while the main determinants of coastal-zone impacts may be patterns of coastal development, zoning, infrastructure, and local property values. The NAST judged that those assessing regional or sector impacts were likely to know more about such factors than a central body. Consequently, to support decentralized scenario development, the NAST proposed a consistent template for assessment teams to follow in creating their own scenarios. Teams were asked to identify two socio-economic factors they judged most important for their impacts of concern; to identify a range of these factors to represent roughly 90 percent confidence; and to create socio-economic scenarios by combining high and low values of these factors, plus middle or best-guess values if they so chose.

More useful assessments of impacts and vulnerability will require more extensive use of socio-economic scenarios, improved integration of socio-economic with climatic and environmental scenarios, and substantial further investment in development and testing of new methods.

Implementation of socio-economic scenarios in the National Assessment was weak. Few assessment teams used the proposed approach. Many made no socio-economic projections at all, but rather projected only biophysical impacts based on GCM results. One assessment team found the socio-economic scenarios were inconsistent with superior local estimates of current population, and so decided not to use them.[116] The teams that did use the socio-economic scenarios used only aggregate projections of population and economic growth, or in some cases assumed continuation of present conditions in the assessment period. None used the proposed template for identifying and projecting additional important socio-economic characteristics.

Several factors contributed to this limited use of socio-economic scenarios. In addition to various managerial and communication problems, many participants were reluctant to use socioeconomic scenarios, especially the proposed decentralized approach. Some preferred to avoid any socio-economic projections, implicitly presuming either that socio-economic conditions did not matter for impacts, or that those that did matter would remain similar to present conditions. Others objected to specific contents of the scenarios or the methods used to generate them, or judged that their team lacked the expertise required to evaluate them. Still others objected that uncertainties in future socio-economic conditions made any attempt to construct scenarios for more than a few years in the future unacceptably speculative.[117] Consequently, while the assessment attempted to advance scenario methods, weak implementation of these methods limited its ability to identify key vulnerabilities. More useful assessments of impacts and vulnerability will require more extensive use of socio-economic scenarios, improved integration of socio-economic with climatic and environmental scenarios, and substantial further investment in development and testing of new methods.[118]

3.2.3. Criticisms and controversies

The National Assessment was the object of substantial political and scientific controversy. Here, we summarize the major criticisms that pertain to the development and use of scenarios. Criticisms focused predominantly on the climate scenarios, especially those derived from GCMs, probably because these were more precisely defined, widely used in the analyses, and featured in the assessment's publications. Three criticisms of these were advanced.

The first, widely circulated during 2000, was that the use of non-American climate models for climate scenarios was inappropriate and potentially injurious to national interests.[119] While this criticism indicates a dimension of political vulnerability of the assessment, it does not address the assessment's technical quality. Since climate models represent the physics of the global atmosphere, they contain no representation of political or economic factors. The Hadley and Canadian global climate models were extensively documented in peer-reviewed scientific literature – and, moreover, were the only models that met the most critical of the assessment's criteria. That they were developed by scientific groups outside the United States has no significance for their ability to provide scenarios to assess US impacts. Using US models would have avoided this criticism, but at the cost of either weakening the analysis by using scenarios that did not meet the assessment's needs, or delaying the work by one to two years. In deciding to proceed with non-US models, assessment organizers judged that these costs were too high.

The second major criticism was that the two climate-model scenarios used were at the extreme end of available models in their projected climate change. This is partially correct. When temperature and precipitation factors are considered together (because high precipitation in some cases may offset the impacts of high temperature), the Canadian scenario lies at the high-impact end – although not an outlier, as other IPCC model projections lie close to it – while the Hadley lies at or somewhat below the middle for most analyses.

The National Assessment's organizers and its critics agreed that using more models would have been preferable, but the assessment was limited by its schedule and its technical requirements. Given a limit of only two, there can be good reasons to choose one scenario in the middle of current projections and one near the top that provides a plausible upper-bound, but the significance of the results must then be communicated with great care. Some critics suggested that presentation of results based on the relatively high Canadian scenario should be more carefully qualified to highlight its position near the top of current projections.[120] Such qualifications must be crafted very subtly, however, lest they imply these results may safely be ignored, when most analyses suggest the full range of future climate-change uncertainty extends both below the Hadley scenario and – in a long, thin tail – above the Canadian.

A related criticism of the climate scenarios claimed that the emissions scenario driving them was implausibly high. The issues bearing on choice of emission scenarios are similar to those for choice of climate models. It would be preferable to have a wide and relevant range of emissions scenarios driving an impact assessment – at least for the post-2050 period. Using a wide range of emissions scenarios would also allow comparison of projected impacts under high and low emissions futures, and so give insights into what degree of impacts could be avoided by what degree of mitigation effort. Model runs with this emissions scenario were all that were available, however. Moreover, there is no clear basis to reject this particular scenario, since it was the scenario most widely used in climate-model runs at the time and lies near the middle of the range of both the 1992 and the 2001 IPCC scenarios. Finally, there is no support for the claim that this scenario was chosen with the aim of making 21st-century climate change appear as frightening as possible.[121] But, although using just two climate models with one emissions scenario was unavoidable in this assessment, it still represented a serious limitation. With more model simulations testing a range of emission scenarios already available, future assessments will be able to remedy this deficiency.

In contrast with the preceding criticisms that the scenarios used in the assessment understated uncertainty, another criticism focused on the disparities between the two scenarios' projections. Some critics argued that such disparities – e.g., the Canadian scenario projects the Southeastern states becoming much drier than the Hadley model does – show that our limited knowledge of regional climate change makes any attempt to assess future impacts and vulnerabilities irresponsible.[122] This criticism implies that impact assessment should wait until precise, high-confidence regional climate projections are available. Since a major purpose of the assessment was to represent current uncertainty about climate change and its impacts, such discrepancies between model projections served a valuable purpose, as indications of the uncertainty of projections at the regional scale – particularly when the model disparities had a clear origin, such as differences in projected jet-stream location.

Future assessments will need to invest substantial resources in developing the state of underlying knowledge, models, and assessment methods for integrating socio-economic considerations into assessments of climate impacts.

In sum, the National Assessment's use of climate-change scenarios was hampered by the lack of available relevant runs, but reflected an adequate attempt to represent then-understood variation in climate projections for the United States. Future assessments will need to use more climate-model projections – including multiple ensemble runs – informed by a wider range of relevant emissions scenarios. The National Assessment attempted to advance the state of the art in using socio-economic scenarios, but achieved only limited success in implementing its plans. Future assessments will need to invest substantial resources in developing the state of underlying knowledge, models, and assessment methods for integrating socio-economic considerations into assessments of climate impacts. This includes further development of novel approaches to link climate and socio-economic scenarios, such as the proposed "inverse" approach to vulnerability analysis.

The experience of the National Assessment raises three significant issues for future climate-change scenario exercises. First, like several of the experiences reviewed here, it illustrates the difficulty and scale of effort involved in producing scenario-based assessments. Second, the large required start-up effort and time to build the capacity to conduct such an exercise illustrates the great value of sustaining analytic and institutional capacity over time, rather than relying on separate projects. Such continuity of capacity will avoid wasteful repetition of startup efforts, support accumulation of learning and experience, and develop and maintain the required collaborative networks. Finally, the assessment's experience illustrates both the need for consistency in large-scale assessments, and the great specificity of information needs within particular impact and adaptation assessments.

This combination of centralized and decentralized information requirements suggests the need for a cross-scale organizational structure for developing and applying scenarios, including scenarios of both climate and socio-economic conditions.

3.3. The Uk Climate Impacts Programme

The UK Climate Impacts Programme (UKCIP) was established in April 1997 as one element of a broad program of scientific research, assessment, and support for policy-making on climate change. The UKCIP supports research and analysis of impacts for particular regions, sectors, and activities in the UK. The program provides common datasets and tools, as well as ongoing support to university researchers and organized stakeholder groups in all UK regions. As part of its role in stimulating, supporting, and coordinating decentralized and stakeholder-driven impact analyses, the UKCIP has produced and disseminated three sets of scenarios: climate scenarios in 1998 and 2002, and socioeconomic scenarios in 2001.

The 1998 climate scenarios provided information only at the rather coarse scale of the Hadley Centre's HadCM2 climate model, with four grid-cells over the entire UK. Four scenarios, called "high," "medium-high," "medium-low," and "low," combined variation in emissions assumptions with variation in assumed climate sensitivity. The medium-high and medium-low scenarios both used the HadCM2 model, with a sensitivity of 2.5°C.[123] The medium-high scenario was driven by a 1 percent per year equivalent-CO_2 transient scenario, similar to IS92a. The medium-low scenario was driven by a 0.5 percent per year equivalent-CO_2 transient scenario, similar to the lowest IS92 scenario, IS92d. The high and low scenarios used the same two emissions scenarios driving a simpler climate model, whose sensitivity was set at 4.5°C for the high scenario and 1.5°C for the low. These scenarios were used in an initial impact assessment focusing predominantly on direct biophysical impacts.[124] The scenarios did not include any explicit statements of probability, although their documentation suggested that the medium-high and medium-low scenarios "in one sense ... may be seen as being equally likely,"[125] while the high and low scenarios captured part of the tails of the distribution. Nor did they include any potential extreme climate events such as those associated with large changes in the North Atlantic circulation.

The UKCIP's socio-economic scenarios drew on the Foresight Program, a broader exercise of the UK Department of Trade and Industry to develop scenarios for long-range planning in several policy areas, with additional detail in areas relevant to greenhouse-gas emissions and climate impacts.[126] As in several other scenario exercises, developers identified two fundamental uncertainties and combined two alternative outcomes of each to produce four scenarios. The two core uncertainties they chose were similar to those used in the SRES exercise: social and political values, which varied from an increased focus on individual consumption and personal freedom ("consumerism") to a widespread elevation of concern for the common good ("community"); and governance, which varied from authority and power concentrated at the national level ("autonomy"), to power increasingly flowing to global institutions, downward to local ones, and outward to nongovernmental institutions and civil society ("interdependence"). The two dimensions of uncertainty, values and governance, were assumed to be independent of each other. Other major uncertainties such as demographic change, the rate and composition of economic growth, and the rate and direction of technological change, were treated largely as consequences of alternative realizations of the two core dimensions of values and governance.[127]

The four scenarios built around these two dimensions of variation were called "National Enterprise," "World Markets," "Local Stewardship," and "Global Sustainability." Each was initially developed as a qualitative narrative of future conditions in UK society intended to apply broadly to both the 2020s and 2050s. Each scenario specified several dozen socioeconomic characteristics qualitatively, including multiple aspects of economic development, settlement and planning, values and policy, agriculture, water, biodiversity, coastal zone development, and the built environment.[128]

Each scenario was also realized in projections of multiple quantitative variables, at the national scale only. For the 2020s, these provided detail on population, GDP (including the governmental share and the sector split between industry, agriculture, and services); household numbers and average household size; land use and rates of change; total transport and modal split; agricultural production (including such details as chemical and financial inputs, subsidies, yields, and organic area); freshwater supply, demand, and quality; and several indicators of biodiversity and coastal vulnerability. For the 2050s a smaller set of quantitative variables was projected, describing population, GDP, land use, and transport. The plausibility of projections was checked, mainly by comparing projected future rates of change to historical experience. The scenarios were published with a detailed guidance document, which provided suggestions on how to use them together with climate scenarios for impact studies.[129]

The UKCIP's socio-economic scenarios drew on the Foresight Program, a broader exercise of the UK Department of Tradeand Industry todevelop scenarios forlong-range

planning inseveral policy areas,with additional detail inareas relevant to greenhouse-gas emissions.

As of 2005, the socio-economic scenarios had been used in six impact studies.[130] There has been some difficulty applying the national-level scenarios in specific, smaller-scale regions. The most ambitious use has been a preliminary integrated assessment of climate impacts and responses in two regions of England, the Northwest and East Anglia.[131] This study produced four integrated scenarios of regional climate impacts, by pairing each of the four socio-economic scenarios with one climate scenario based on a rough correspondence between the socio-economic scenario and the IPCC emissions scenario underlying the climate scenario[132] Based on these four scenarios, the study elaborated preliminary regional scenarios corresponding to the four national socio-economic scenarios, and conducted an assessment of coastal-zone impacts and responses using these scenarios and a formal land-use model.

Four new climate scenarios were produced in 2002, based on the SRES marker scenarios and new versions of Hadley Center climate models. These new scenarios differed only in their emissions assumptions, not climate sensitivity. The high, medium-high, medium-low, and low scenarios were driven by the A1FI, A2, B2, and B 1 marker scenarios, respectively. These were used to drive the HadCM3 global climate model (with a grid-scale of 250-3 00 km), generating climate-change projections for 30-year future periods centered on the decades of the 2020s, 2050s, and 2080s. For some emissions scenarios and time periods, climate projections were processed through a nested hierarchy of three Hadley Center climate models: the HadCM3 model at global scale, the HadAM3H model at intermediate scale, with a horizontal resolution of about 120 km, and the HadRM3 model for high-resolution climate projections in the United Kingdom and Europe, with a horizontal resolution of about 50 km. This nested processing was done for the baseline period (1960-1990), and for the most distant projection period (2070-2100) to produce three ensemble runs for the medium-high (A2) emissions scenario and one for the medium-low (B2). For the other emissions scenarios and the intervening projection periods, results of the global-scale model were downscaled using statistical patterns of fine spatial-scale climate variation derived from full runs using scenario A2. These scenarios were widely distributed and supported through a web-based interface, including map-based graphical display of projected changes in more than a dozen climate indicators on a fine-scale (50 km) grid of the United Kingdom.

The UK program's experience highlights some of the same issues for future scenario exercises as the US National Assessment, in particular the importance of continuity of institutional and analytic capability and the desirability of developing and supporting scenarios using an organizational structure that combines centralized and decentralized elements.

Several analyses are continuing to use the 2002 climate scenarios in conjunction with the socioeconomic scenarios. For example, a 2004 integrated analysis of flood risk and erosion control over a 30-100 year time horizon produced a threat assessment, a set of scenarios of floodrisk, and a set of policy recommendations. Anevaluation of this study's effects one year laterfound that it was being used by several publicand private actors to inform decision-making.[133]

Table 3.2. Millennium Ecosystem Assessment Scenarios

Ecosystem Management	World Development	
	Global	Regional
Reactive	Global Orchestration	Order from Strength
Proactive	TechnoGarden	Adapting Mosaic

The UKCIP, in contrast to the US National Assessment, has built a sustained assessment capability. In addition, the central program has less authority over the separate assessments, instead acting more as motivator, resource, and light coordinator. Access to scenarios is to licensed users, of whom there are about 130 – roughly half in universities, the rest about equally split among private sector and all levels of government. Most active users have been national officials responsible for climate-sensitive resources, with less participation from the private sector and local governments.[134]

The program has invested in generating, disseminating, and documenting useful climate scenarios for impacts users. The jury appears to still be out on whether the level of effort and success is similar for socio-economic scenarios, which have not yet been either downscaled or repeated. Getting scenarios used is a slow process, but the scenarios produced by this program are starting to be used by decision-makers in support of their practical responsibilities. A significant limitation of the program, however, is its exclusive reliance on just one family of climate models. This may pose risks of under-estimating future climate uncertainty and over-confidence in assessments of potential climate impacts and responses. Although the UK program followed a substantially different organizational model from the US National Assessment, its experience highlights some of the same issues for future scenario exercises, in particular the importance of continuity of institutional and analytic capacity and the desirability of developing and supporting scenarios using an organizational structure that combines centralized and decentralized elements.

3.4. The Millennium Ecosystem Assessment

The Millennium Ecosystem Assessment (MEA) was a large, United Nations (UN)-sponsored assessment of the current status, present trends, and longer-term challenges to the world's ecosystems, including climate change and other sources of stress. Conducted between 2001 and 2005, the MEA sought to assess changes in ecosystems in terms of the services they provide to people and the effects of ecosystem change on human well-being. It also sought to identify and assess methods to mitigate and respond to ecosystem change, for various private and public-sector decision-makers, including those responsible for the several international treaties that deal with ecosystems.[135] More than 1350 authors from 95 countries participated in the global assessment's four working groups, and hundreds more in about 30 associated sub-global assessments. The assessment's goals were broad, ranging from providing a benchmark for future assessments and guiding future research to identifying priorities for action.[136]

Results of the global assessment were presented in a March 2005 synthesis report, and in additional volumes presenting the output of the assessment's four working groups, "Current State and Trends," "Scenarios," "Policy Responses," and "Multi-Scale Assessments." The current state and trends group examined ecosystem trends over the past 50 years and projections to 2015; the scenarios group took a longer view to 2050 and beyond. Because of time limitations, the work of these two groups proceeded largely independently.

All components of the assessment used a common large-scale conceptual framework, which distinguished indirect drivers of ecosystem change, direct drivers, ecosystem indicators, ecosystem services, measures of human wellbeing, and response options. Direct drivers included direct human perturbations of the environment such as climate change, air pollution, land-use and land-cover change, resource consumption, and external inputs to ecosystems such as irrigation and synthetic fertilizer use. Indirect drivers included underlying socio-economic factors such as population, economic growth, technological change, policies, attitudes, and lifestyles.[137]

The MEA sought to assess changes in ecosystems in terms of the services they provide to people and the effects of ecosystem change on human well-being.

The scenarios working group sought to apply this conceptual framework to long-term trends in ecosystems, looking ahead to 2050 with more limited projections to 2100. They developed the structure of the scenarios in an iterative process, including consultations with potential scenario users and experts in a wide range of decision- making positions around the world.[138] Like several other major scenario exercises, they initially sought to identify two basic dimensions of uncertainty in long-term ecosystem stresses, which together would produce four scenarios.[139] For the first dimension, like SRES they chose globalization: continuation and acceleration of present global integration trends, versus reversal of these trends to increasing separation and isolation of nations and regions. For the second dimension, in contrast to the broad value-based uncertainties used in the SRES and UKCIP scenarios, they chose one more specifically related to ecosystems: whether responses to increasing ecosystem stresses are predominantly reactive – waiting until evidence of deterioration and loss of services is clear – or predominantly proactive, taking protective measures in advance of their clear need. The combination of two polar values of each of these uncertainties yielded four scenarios, summarized in Table 3.2.

The Global Orchestration (global, reactive) scenario presented a globally integrated world with low population growth, high economic growth, and strong efforts to reduce poverty and invest in public goods such as education. In this scenario, society focuses on liberal economic values, follows an energy-intensive lifestyle with no explicit greenhouse-gas mitigation policy, and takes a reactive approach to ecosystem problems.[140] In Order from Strength (regional, reactive) the reactive approach to ecosystem problems takes place in a fragmented world preoccupied with security and less attentive to public goods.[141] This scenario exhibits the highest population growth and lowest economic growth. Economic growth is particularly low in the developing countries, and it decreases over time. In Adapting Mosaic (regional, proactive), political and economic activity are concentrated at the regional ecosystem scale. Societies invest heavily in protection and management of ecosystems in locally organized and diverse efforts. Population growth is nearly as high as in Order from Strength, and economic growth is initially slow but increases after 2020. Finally, TechnoGarden (global, proactive) presents a world that is both focused on ecosystem management and globally connected, with strong development of environmentally friendly technology. Population growth is moderate, and economic growth is relatively high and increasing.[142]

Each scenario was initially constructed as a qualitative description. Population and GDP were specified quantitatively, while all other indirect drivers – including social, political, and cultural factors – were qualitative. Population scenarios were derived from the International Institute for Applied Systems Analysis' (IIASA's) 2001 probabilistic projections, capturing the middle 50-60 percent of the distribution, with world population in 2050 ranging from 8.1 billion (Global Orchestration) to 9.6 billion (Order from Strength).[143] No statements of probability or likelihood were made about the scenarios.

From the indirect drivers, a more specific and quantified set of direct drivers was developed, using formal models where possible. Species introduction and removal was the only unquantified direct driver.[144] Separate pre-existing models were used of the world energy-economy, greenhouse-gas emissions and climate change, air pollution, land-use change, freshwater, terrestrial ecosystems, biodiversity, and marine and freshwater fisheries. To the extent possible, these quantitative models were used to reason from indirect and direct drivers to ecosystem effects, changes in ecosystem services, and effects on human well-being.[145] In some cases this was achieved by soft-linking models, using outputs from one as inputs to another, but this was limited by different variable definitions, spatial and temporal resolution, and other model incompatibilities.[146] Not all scenario elements could be modeled quantitatively, so expert judgments were also extensively used. The qualitative scenario process proceeded in parallel with quantitative modeling – elaborating aspects of the scenarios that were not amenable to modeling, filling gaps, and stipulating feedbacks between ecosystem services and human well-being and behavior.[147]

The groups attempted to check for consistency between quantitative and qualitative scenario elements through periodic consultations, particularly for feedbacks that could not be modeled analytically. Some of these were interactions between direct drivers and ecosystems, but the most difficult occurred in scenarios that assumed strong socio-economic feedbacks and regulating mechanisms. Adapting Mosaic, for example, assumed strong feedbacks from new ecosystem observations and knowledge to changes in human behavior that could not be incorporated into the models used. Representing these required allowing qualitative scenario logic to override both the quantitative results and the structure of models. Unfortunately, time limits prevented this consistency checking from being done thoroughly, so unexamined disparities between qualitative and quantitative aspects of the scenarios remained a significant weakness.

Many of the conclusions developed from the scenarios are common to all four scenarios, while in others Order from Strength is the exception. For example, one major conclusion is that rapid conversion of ecosystems for use in agriculture, cities, and infrastructure will continue, and that habitat loss will continue to contribute to biodiversity loss. However, if ecosystem services increase as projected, some ecosystem services – although not biodiversity – may be decoupled from ecosystem stresses. Food security is projected to remain out of reach for many people. Extreme, spatially diverse changes are projected for freshwater resources, with general deterioration in developing countries under both "reactive" scenarios. Increasing demands for fishery products are projected to increase risks of regional marine fishery collapses.[148] In sum, ecosystem services show mixes of improving and worsening trends in all scenarios except Order from Strength, in which nearly all ecosystem services are projected to be more impaired in 2050 than in 2000. The same three scenarios also suggest that significant changes in policies, institutions, and practices can mitigate some negative consequences.[149]

In sum, the MEA scenarios project invested substantially more effort in developing rich qualitative and narrative scenarios than the SRES, but also fell short on integrating qualitative and quantitative components. In part because of the greater elaboration of the qualitative components, this limited coordination resulted in significant inconsistencies and requirements to resolve conflicts between the two components. These inconsistencies arose even with just one model used for several components of the assessment, so the challenges of harmonization among models – and the associated possibility to explore model-structure uncertainty – did not arise. A related problem was that for many factors it was difficult to generate the desired level of variation between scenarios.[150] This raises issues of potential methodological interest, such as how to distinguish robust results from inadvertent convergence of assumptions or model structures, which remain to be investigated. Finally, the great breadth of conditions represented in the scenarios, as well as possible concerns with logical circularity between their presumptions and results,[151] make interpreting the significance of the results difficult.

The experience of this scenario exercise provides a different perspective on some of the same key challenges for future scenarios highlighted by the other activities reviewed. The quite distinct difficulties faced here in attempting to combine quantitative and qualitative scenarios highlight the central importance and the difficulty of developing new methods to integrate these two approaches. In addition, this experience highlights the value of clarity about the intended uses of scenarios, including clarity about whether they are intended to address specific questions, guide decisions, or explore longterm conditions. The risk of scenarios becoming less useful due to breadth and vagueness may be particularly acute for scenarios that attempt to capture multiple stresses on some system – even though such multi-stress assessment is repeatedly advocated for climate-change and other forms of environmental assessment.[152]

SECTION 4. CHALLENGES AND CONTROVERSIES IN SCENARIOS FOR CLIMATE CHANGE

This section discusses several issues that have arisen in multiple scenario exercises related to climate change, issues that pose challenges for expanding the usefulness of scenarios to climate change analysis, assessment, and decision support. Section 4.1 examines the type of information needs of specific types of decisions related to climate change and considers the requirements and challenges of crafting scenarios to serve these needs. Section 4.2 considers the use of scenarios that has been more common thus far, in structuring climate-change assessments and framing broad policy debates, and identifies the distinct challenges in enhancing the value of scenarios in these purposes. The remaining sub-sections examine particular design challenges in crafting scenario exercises: how to structure interactions between experts and stakeholders in developing scenarios; how to communicate scenarios to potential users not involved in their creation; how to pursue the two, not perfectly aligned goals of consistency and integration in scenarios; and how to represent and interpret uncertainty in scenarios. Throughout this section, we present illustrative examples of scenario activities in text boxes. These examples shed additional light on various challenges, especially relating to scenarios' use in decision-making.

4.1. Scenarios and Decisions

As discussed in Section 1, the general purpose of scenarios is to inform decisions, but their connection to specific, identified decisions can be more or less close and direct. In interpreting and evaluating present experience with scenarios and identifying key challenges in making them more useful, it is important to distinguish scenario exercises by their major characteristics, including their specificity, their proximity to decisions, the degree of normative presumptions embedded in them, and where they lie in the causal chain outlined in Section 2. To consider how scenarios can help inform climate-change decisions, we must first specify the relevant decisions and decision-makers more sharply. This section considers the major concrete decisions that comprise a response to climate change. Decisions related to assessment, modeling, and research are considered in Section 4.2. This discussion must be somewhat hypothetical, extending from rather thin current practice to reasonable speculation about future decisions and likely information needs.

Because the dynamics of climate change operate on multiple spatial scales from the local to the global, there is no single global climate-change decision-maker. Rather, many distinct decision-makers with diverse responsibilities will affect and be affected by climate change. Because of climate's recent appearance on policy agendas and its dense connections to other issues, many of these decision-makers' primary responsibilities are defined as something other than climate change. Some of them are already considering how climate change might affect their responsibilities, but many are not.

Section 2 described climate-change decisions using the conventional dichotomy of mitigation versus adaptation. To consider potential contributions of scenarios in more detail, we discuss three types of decision-maker: *national officials*, *impacts and adaptation managers*, and *energy resource and technology managers*. These can often be identified as particular programs, divisions, agencies, organizations, or individuals, each with different responsibilities and types of information they might consider in making their decisions. All three groups face decisions under uncertainty with long-term consequences related to climate change, and so might benefit from scenarios providing structured information and assumptions about the values at stake, the available choices, and their consequences under alternative climate-change futures.

All three groups face decisions under uncertainty with long-term consequences related to climate change, and so might benefit from scenarios providing structured information and assumptions about the values at stake, the available choices, and their consequences under alternative climate-change futures

National officials' responsibilities are the broadest and the most likely to be explicitly related to climate change. They develop national policies that target greenhouse-gas emissions and motivate investment in technologies that will influence future emissions trends. They negotiate policies internationally with officials from other nations, and with sub-national officials who may share mitigation responsibilities or undertake mitigation measures at their own initiative. They also have responsibilities to anticipate and respond to climate-change impacts in their nations. Their climate-change responsibilities are open-ended, not limited to mitigation and adaptation: these decision-makes will determine the extent to which other responses such as geoengineering are considered, and the design of systems and institutions for assessment. They are also responsible for overall national welfare, including not just the environmental effects of their decisions but also other linked national interests such as economic prosperity and security.

Impacts and adaptation managers have responsibility for particular assets, resources, or interests that might be sensitive to climate change. They must decide how to anticipate, prepare for, and respond to the threat, minimize its harm, and maximize any associated benefit. They may be private or public actors – e.g., owners or managers of long-lived assets such as ports or water-management facilities; managers of lands, forests, or protected areas; emergency preparedness or public health officials; officials making zoning or coastal development policy; or firms in insurance or financial markets who may bear secondary risks from impacts or seek to develop new instruments to exchange these risks. Unlike national officials, these actors' decisions are purely *responses* to climate change, realized or anticipated: they have little influence over how the climate will change. Their responsibilities will often connect with the impacts-related responsibilities of national officials, but are narrower in scale or scope. Impacts and adaptation managers would be concerned not with aggregate climate-change impacts on the United States, but with more specific impacts such as those on seasonal flows and water-management operations on the Upper Mississippi.

Energy resource and technology managers include developers and operators of fossil or non-fossil energy resources, investors in long-lived energy-dependent capital stock such as electrical utilities, and researchers, innovators, and investors in new energy-related technologies. These decision-makers are mostly but not exclusively in the private sector. Their decisions may have consequences that interact with various processes operating over multiple time-scales, from short-term market responses, to decadal-scale processes of investment, resource development and depletion, and penetration of new technologies, to century-scale processes of climate change.[153] These actors' decisions will strongly influence society's ability to control greenhouse-gas emissions. This group also includes energy consumers such as firms or public agencies considering mitigation actions in their own operations. While their areas of responsibility may be vulnerable to climate change and its impacts, the largest climate-related risks for this group are likely to come not from climate change itself, but from climate-change policies: national mitigation policies, and other market and regulatory decisions that shape the outcomes of private mitigation activities.

At greatly varying levels of precision and specificity, scenarios can present two types of information to support decisions by these three types of actors. Scenarios can represent potential future developments that may threaten decision-makers' interests or values, call for decisions, or challenge conventional thinking and practices. And they can provide a structure to assess potential consequences of alternative decisions for things that matter to the decision-maker. Beyond this generalization, the three types of decision-makers will differ substantially in the specific types of information they need, the time horizons of their decisions, and the type and extent of causal connections between their decisions and the conditions specified in scenarios.

4.1.1. Scenario needs: national officials

As national officials have the broadest responsibilities related to climate change, they are also likely to have the broadest information needs. In their responsibilities to build national adaptation capacity and manage key vulnerabilities, their needs are similar to those of impacts and adaptation managers: scenarios of potential future climate change under specified emissions assumptions and resultant impacts on particular resources and communities in their nation, with particular focus on areas of greatest vulnerability. They will likely have less need for fine spatial and sectoral detail in potential impacts, but more need for consistent scenarios that allow comparison and aggregation across sub-national regions and sectors. These will help to prioritize, identify key areas of vulnerability, and estimate aggregate costs for planning purposes.

In their responsibilities for national mitigation policy, national decision-makers will also need information about the aggregate impacts of climate change, since the more severe climate impacts are likely to be, the greater the justification and likely political support for mitigation measures. But mitigation decisions also require additional information – including projections of future emissions in the absence of explicit mitigation efforts, and the consequences of alternative mitigation policies, in their effects on emissions, their cost, and their implications for other dimensions of national interest.

These needs introduce a dimension of complexity into mitigation scenarios, sometimes called "reflexivity." Because mitigation policies seek to reduce future emissions by altering the socio-economic drivers of emissions growth, the analysis of mitigation policies and their consequences must be coupled to the causal logic of emissions scenarios. Whereas climate scenarios can be treated as exogenous when assessing adaptation decisions, emissions scenarios cannot be treated as exogenous in assessing mitigation decisions. Any emissions scenario embeds some assumptions about mitigation policies, assumptions that may have to be changed to assess particular mitigation policies. This effect will be strongest when emissions projections and mitigation options are being considered at the same spatial scale, e.g., national mitigation policies are being assessed relative to national emissions projections. The effect of national mitigation strategies on global emissions will be weaker: no nation controls global emissions trends, and the effects of small nations' mitigation strategies on global trends can be very small.

Any emissions scenario embeds some assumptions about mitigation policies, assumptions that may have to be changed to assess particular mitigation policies.

Scenarios to inform mitigation decisions are also likely to require considering alternative assumptions about the policy context in which these decisions are made. The effects of national mitigation strategies – including how much they reduce national emissions, as well as their costs and other consequences – will depend on the economic, technological, and policy context, including related decisions by other major nations, individually and through international coordination. Assumptions about the policy context will be less important in scenarios to inform international mitigation decisions, since when decisions are globally coordinated there is no "elsewhere" – but alternative assumptions about nations' degrees of compliance and form of implementation of international commitments may still be needed.

Scenarios of emissions, climate change, and impacts inform mitigation decisions by helping to characterize the potential severity of climate change and therefore how important it is to control emissions. This support is indirect, serving primarily to elevate or moderate the general level of concern on the issue. More focused work on mitigation has been done using constructed scenarios of limited emissions, often aiming at stabilizing atmospheric concentrations or radiative forcing at various levels, and examining the configurations of technology, energy resources, and economic and population growth that are consistent with the specified scenario. Some studies have used quantitative models to estimate costs of such scenarios, relative to an assumed baseline emissions scenario.[154]

4.1.2. Scenario needs: impacts and adaptation managers

To assess the threats and opportunities they face and evaluate responses, impacts and adaptation managers need scenarios of potential future climate change, its impacts in their areas of responsibility, and the factors that influence vulnerabilities. With few exceptions, these actors' decisions will have no effect on the climate change to which they must respond, so scenarios of climate-change stresses can be constructed independently of the assessment of potential adaptation decisions, without concern for feedbacks that may modify the conditions specified in the scenario.

Impacts and adaptation managers need scenarios of potential future climate change, its impacts in their areas of responsibility, and the factors that influence vulnerabilities. Energy and technology managers will most benefit from scenarios that explore alternative policy regimes and their consequences for thevalue of energy and technology assets and investments.

Particular decision-makers' needs will be highly specific in the variables they require, and their time and spatial scale and resolution. A planner of water-management infrastructure may need monthly or finer-scale rain and snow projections over a watershed; a designer of coastal infrastructure may need probabilistic projections of sea level, storm intensity and frequency, storm surge, or saltwater intrusion. But in their climatic elements, these information needs all rest on a common core of scenarios of global climate change and emissions drivers. This dual structure of information – highly particular climate variables, based on a set of common "core scenarios" – suggests a cross-scale organizational structure for providing scenario information: commonly produced scenarios of climate change and other components requiring consistency, specialized expertise, or high-cost resources; development of decentralized capabilities in impact assessments to adapt these core scenario elements and develop assessment-specific extensions; and close communication between these groups to ensure that useful variables are generated and saved, and that data and documentation are transferred accurately.

This is the area of climate-related decisions for which the provision of information from climate-change scenarios is most advanced. Still, further progress is needed in the development and use of scenarios of socio-economic conditions, and in creation of methods and tools to augment centrally provided scenario information with information tailored to specific impact assessments. In addition, many impacts-related decisions will require scenarios of climate change in the context of other linked stresses and changes.

4.1.3. Scenario needs: energy resource and technology managers

Energy and technology managers will most benefit from scenarios that explore alternative policy regimes and their consequences for the value of energy and technology assets and investments. For some, the predominant concern may be overall policy stringency, perhaps summarized as alternative emissions-price trajectories over time; for others, specific details of policy design and implementation may need to be considered. Scenarios of emissions, climate change, and impacts only matter for decisions via their likely influence on policy stringency, and so do not need to be explicitly represented in scenarios. These actors may have some influence on policy, but probably not such strong influence that climate-policy scenarios would have to incorporate feedbacks from their own advocacy efforts.

Unlike the other two types of decision-makers, these actors are likely to compete with each other. If, for example, they are investors allocating research effort between higher and lower-emitting energy sources, those who better anticipate future policy will benefit relative to those who do worse. If they use scenarios, they may consequently choose to produce them privately, perhaps coupled with other analyses to generate practical guidelines for investments.[155] As for the other types of decision-makers, these specialized scenarios could be based on general scenarios of global emissions and climate change. Published scenarios produced to date on the climate-change issue, however, have not considered mitigation policies with the specificity necessary to inform these actors' decisions.

4.1.4. Representing decisions in scenarios

A major challenge to developing scenarios to support decisions is reflexivity, that is, how to represent decisions within scenarios without making scenarios either circular or contradictory. In meeting this challenge, the most basic distinction to draw is between decisions by the scenario's targeted users and decisions by other actors. From the users' perspective, decisions by others over which they have no influence are indistinguishable from non-choice events. If the factors influencing these decisions are confidently understood, they might be represented deterministically, like well-understood biophysical or economic processes. In the more frequent case that others' choices cannot be confidently predicted, they might be represented as uncertainties – again, just like uncertain biophysical or economic processes. As with all uncertainties, how to treat them depends on their judged importance for the users' decisions: if it is high, they can be represented in alternative scenarios; if not, they can be fixed at some best-guess value for all scenarios. In either case, these decisions are treated exogenously.

Representing decisions by the scenario users is fundamentally different. Since the scenarios are intended to inform these decisions, alternative choices should not be represented as exogenous uncertainties but be stipulated independently from the scenarios. Users can then explore their implications under conditions imposed by scenarios, including representation of major uncertainties. Various degrees of coupling can be required between the logic of scenarios and the analysis of consequences of the users' decisions. In scenarios for impacts, these can usually be separate; in scenarios for mitigation, they may have to be closely coupled, since emissions scenarios may change under alternative assumptions about mitigation decisions.

In scenarios to inform global climate-change decisions, the sharpest question is how to represent mitigation decisions within scenarios.

Following the general reasoning above, how these are treated should depend on what type of decision is being informed. In climate scenarios to inform impact assessments and related decisions, the scenario users are not considering mitigation decisions and have little influence over them, so emissions scenarios should include assumptions about the likely or plausible range of mitigation efforts. The range of future climate change considered may thus be narrowed to reflect the possibility of negative social and political feedbacks: sustained rapid emissions growth may generate pressure for aggressive mitigation, due to increasing signs of climate change, alarming projections of future change, or other environmental harms from rapid expansion of coal or synthetic fuels.

In climate scenarios to inform impact assessments and related decisions ...emissions scenarios should include assumptions about the likely or plausible range of mitigation efforts.

Such a negative-feedback mechanism may not be effective, of course. Many factors could intervene: mitigation measures may not gain enough support to be adopted, socio-political capacity to enact stringent policies may be diminished, policies adopted may be ineffective, or early technology or policy decisions may lock in high-emitting future paths. But to the extent that such a negative-feedback mechanism does operate, persistence of the highest emissions paths beyond a few decades would become unlikely.

Parallel reasoning may apply to extremely low emission paths, if sustaining such low emissions requires continued costly mitigation efforts that come to be seen as unnecessary. This negative-feedback mechanism would likely be weaker than that operating at the high end of the emissions distribution, however, because long time-constants mean that increasing signs of climate change are likely to continue through most of the 21st century even if we follow a low-emissions path. If impacts assessors and managers judge these negative feedbacks to make extreme emissions paths sufficiently unlikely, particularly high ones, they may reasonably decide not to consider these extreme emissions futures in their planning for adaptation.

For scenarios to inform mitigation decisions, particularly at the international level, the situation is different. Informing these choices requires information about potential emissions paths and their consequences under all levels of mitigation effort that decision-makers might plausibly consider – including no additional measures, or even reversal of previous measures if this is on the agenda. Consequently, in contrast to scenarios for impacts, extreme emissions futures should not be excluded when assessing mitigation decisions. For example, if scenarios that truncate high-emissions futures by assuming stringent mitigation are used to support a decision that stringent mitigation is not necessary, the result is contradictory: a conclusion supporting a decision is based on the presumption of the contrary decision. To avoid such contradictions, scenarios to inform mitigation decisions must consider alternative mitigation choices explicitly, not embed them implicitly in the underlying logic of the scenario.

Moreover, national officials act only for their own nations in the near term, even when they negotiate global mitigation. They may make choices for long-term planning and institutional design for future mitigation as well, but it is their successors who will decide whether to continue, strengthen, or otherwise change measures adopted today. From the perspective of current national officials, mitigation decisions by other nations and in the future fall between the two cases discussed above: they are not controlled by the scenario user, but can be influenced to some degree. For policy choices by other nations, national officials may need to be advised in two modes, reflecting their dual responsibilities to make national policy and to negotiate international agreements. In the latter capacity, alternative approaches to global mitigation strategy should be represented as choices. But when they consider national decisions separate from globally coordinated strategy, relevant decisions of other major nations should be represented as uncertainties. This may require use of two distinct types of scenarios to advise development of different aspects of national mitigation policy.

...in contrast to scenarios for impacts, extreme emissions futures should not be excluded when assessing mitigation decisions. ... Scenarios to inform mitigation decisions must consider alternative mitigation choices explicitly, not embed them implicitly in theunderlying logic of the scenario.

How to represent future mitigation decisions poses a still harder dilemma. On the one hand, it appears risky or even irresponsible to assume that the bulk of mitigation efforts can be left to future decision-makers, even if we assume this will be easier for them because of greater wealth or technological prowess. On the other hand, assuming that future decision-makers cannot be relied on to act responsibly at all can easily lead to decisions that incur excessive costs, by trying to achieve rapid mitigation immediately or tie future decision-makers' hands.

Two approaches appear promising for integrating future mitigation decisions into scenarios to inform current decisions. Scenarios could presume that today's decision-makers choose the future path of mitigation, allowing them to assess and contribute to a trajectory of effort that considers the welfare of both current and future citizens. Alternatively, scenarios could treat future large-scale mitigation choices as uncertainties represented in alternative scenarios, while also considering how current choices can seek to influence the opportunities and incentives faced by future decision-makers.

In sum, the importance of connecting scenarios to actual decisions is widely recognized, but there is a large gap between, on the one hand, the value scenarios could provide to climate-change decisions and the aspirations of scenario producers to provide that value, and current practice on the other hand. There has been little use of scenarios to directly inform climate-change related decisions, although there appears to be a sharp increase in the interest of decision-makers and early attempts. The rapid increase in interest is particularly evident for informing decisions related to climate-change impacts and adaptation. There are fewer indications of similarly direct use of scenarios to inform mitigation decisions, perhaps in part because nearly all current mitigation decisions have been near-term.

Mitigation decisions at the national and international level have taken scenarios into account indirectly. Most scenarios have been constructed to provide inputs to assessments, models, or other analyses. This has included serving as inputs to the production of other types of scenarios, which then describe other potential future conditions that depend on those specified in the scenario, as for example a model-based climate scenario depends on inputs from an emissions scenario. While these uses can be characterized as supporting decisions (i.e., decisions about assessments, modeling, and research), their connection to concrete decisions of mitigation and adaptation is indirect, achieved through contributions such as supporting strategic planning and risk assessment, providing advance analysis for potential future decisions, exploring plausible extreme cases, helping to characterize and prioritize key uncertainties, or educating decision-makers or the public. This description applies to the major scenario exercises discussed in this chapter, including the IPCC emissions and climate scenarios, the US and UK assessments of climate impacts, and the MEA scenarios.

Box 4.1. Scenarios for Climate-Change Adaptation in the New York Metropolitan Region

Three linked activities – the Metropolitan East Coast (MEC) assessment of the US National Assessment, the New York Climate and Health project (NYCHP), and the New York City Department of Environmental Protection (NYCDEP) Task Force on Climate Change – have used or are using scenarios to assess impacts of climate change on the New York Metropolitan Region, identify areas of vulnerability, and inform regional planning and decision-making.[156]

The MEC assessment began with a regional workshop in April 1998, involving about 150 participants, including public agencies at all levels of government as well as climate researchers. The subsequent assessment was conducted by sector teams of researchers and officials from agencies responsible for the study sectors. Teams developed regional scenarios of climate change and sea level rise based on the downscaled climate-model scenarios provided by the US National Assessment, plus two additional scenarios based on extrapolation of recent regional climate trends and historical extremes. The scenarios were used to project climate-change impacts on beach nourishment, 100 and 500-year flood heights, wetland aggregation and loss, adequacy of the water supply system under droughts and floods, illnesses from acute air-pollution episodes, and peak energy loads. These impact projections were used for preliminary assessment of adaptation strategies and policies.

Following the MEC Assessment, the NYCHP created updated regional climate scenarios in consultation with an expert-stakeholder Advisory Board. This study further analyzed public health impacts, focusing on air quality and extreme heat events. The updated climate scenarios used the IPCC A2 and B2 emissions scenarios driving global and regional climate models to create downscaled scenarios for the region. These were augmented with newly developed scenarios of future regional land use and population growth based on the IPCC A2 and B2 storylines.

In response to the widespread public attention received by the MEC Assessment Report, the Commissioner of the NYCDEP established the Climate Change Task Force, a collaboration among regional researchers and the agency that manages the water system. The Task Force uses the latest climate-model simulations from the IPCC Fourth Assessment Report, as well as additional global and regional climate models, to develop new regional scenarios. These will include probability distributions of average and extreme temperature and precipitation change, as well as sea level rise. The Task Force is also developing qualitative scenarios of extreme sea level rise in the region. DEP is using these results to develop a comprehensive adaptation strategy for the New York City water system, including assessment of many specific adaptation options, that considers both uncertainties in future climate change and managerial factors such as the time horizon of different adaptation responses and capital turnover cycles.

This is a successful example of scenario-based assessment of climate impacts and adaptation options. The scenarios are connected with the concrete responsibilities and concerns of stakeholders, who were involved in their design from the outset. Although officials have found the wide range of uncertainty in climate scenarios difficult to incorporate into infrastructure design specifications, particularly for precipitation, the exercise has effectively conveyed the challenges posed by future climate uncertainty to current decisions of planning and infrastructure design. Stakeholders' willingness to support and participate in three separate phases of these activities and NYCDEP's incorporation of them into a strategic planning exercise provides clear evidence of the practical utility of the exercises.

Box 4.2. Scenarios of Sea Level Rise along the Gulf Coast

Sea-level rise is one of several factors that contributed to the decline of coastal ecosystems along the US Gulf of Mexico coast in the 20th century (Figure 4.1).[157] In southeastern Louisiana, where the local rate of land surface subsidence is as high as 2.5 cm per year, rise in local "relative sea level" may be the most important factor in the rapid loss of coastal zone wetlands over the past several decades.[158]

Despite the importance of sea level rise in historical losses of coastal lands, planning projections of future changes in coastal Louisiana used by both federal and state agencies prior to the devastating impact of Hurricanes Katrina and Rita in 2005 were based on just one scenario: no change in the rate of sea level rise. No alternative sea level scenario was considered in the plans then being developed to restore and protect the Louisiana coastal zone.[159] This assumption sharply contrasts with the IPCC projections, which state that the global average rate of sea level rise in the 21st century may increase 2- to 4-fold over that of the 20th. Such increases will exacerbate wetland losses throughout the Gulf Coast region and obstruct restoration plans that do not take account of likely increases in water levels and salinity.

The ecosystem modeling team working for the State of Louisiana and the US Army Corps of Engineers is presently integrating accelerated sea level rise scenarios into planning exercises that will aid federal and state agencies in evaluating restoration alternatives.[160] The State of Louisiana is consulting with the Rand Corporation to obtain probability estimates for various scenarios of sea level change to help guide engineering decisions and the design of projects aimed at restoring levees and coastal landforms that protect coastal communities.[161] Sea level rise scenarios are also being used to assess the impacts of climate change and variability on the Gulf Coast transportation sector. To assess transportation impacts, a sea level rise simulation model developed by the US Geological Survey generates scenarios of sea level change using over a dozen GCMs and six SRES emission scenarios.

Sea level rise scenarios are important not just in regions like Louisiana. The Big Bend region of the Florida panhandle is experiencing very little vertical movement of the land surface, so sea level there has been rising at approximately the global average rate of 1 to 2 mm per year. But even here, coastal wetlands positioned on flat limestone surfaces may be subject to highly nonlinear effects as sea level reaches a threshold at which large areas are subject to increased salinity or inundation.[162]

Regional scenarios of potential sea level rise are needed to support coastal management and protection activities, as well as plans for wetland restoration and post-hurricane reconstruction. Absent consideration of such scenarios, restoration and rebuilding programs are likely to lock in errors that result in wasted resources and avoidable increases in future vulnerability.

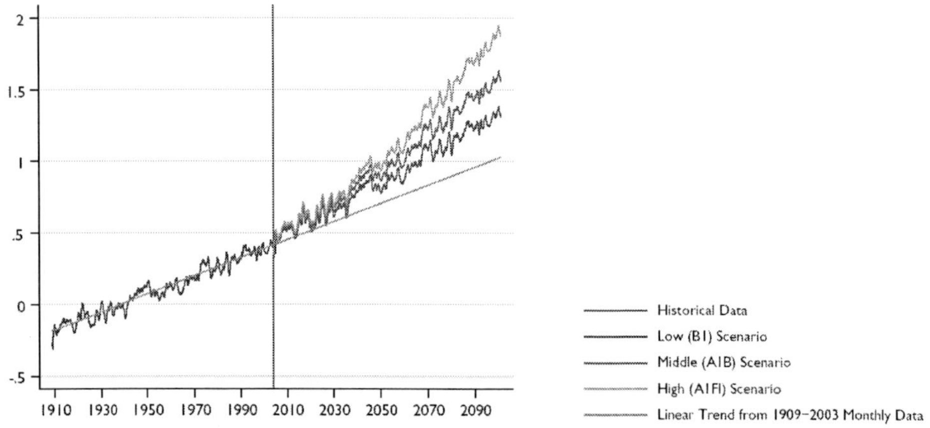

Historical sea level change and projected sea level rise under three greenhouse-gas emissions scenarios, in meters, are shown for Galveston, Texas. Both historical data and future projections are smoothed from monthly data using a 12-month moving average. *(Source:ThomasW. Doyle, National Wetlands Research Center, United States Geological Survey.)*

Figure 4.1. Output from a Gulf Coast sea level rise scenario tool

4.2. Scenarios in Assessments and Policy Debates

Within large-scale assessments of climate change or other environmental issues, scenarios can serve several roles. Most straightforwardly, they can provide required inputs to other parts of the analysis, as the IPCC emissions scenarios support the controlled comparison of climate-model runs. They can also serve as devices to organize and coordinate the multiple components of a large-scale assessment, particularly when much of the assessment is forward-looking. In the IPCC assessments, for example, emissions scenarios have not just been used to drive coordinated climate-model projections, but have also increasingly been followed through to coordinate characterization of climate impacts and adaptation opportunities, and used in a more preliminary way to organize assessments of the economic and technological implications of alternative mitigation strategies.

BOX 4.1.3. SCENARIOS IN THE CALIFORNIA WATER PLAN

The California Department of Water Resources (DWR) produces an updated California Water Plan every five years. The plan projects water supplies and demands, and evaluates current and proposed demand-management programs and supply investments, to "provide a framework for water managers, legislators, and the public to consider options and make decisions regarding California's water future."[163]

In contrast to prior plans that constructed only one future scenario, the 2005 plan explicitly considered uncertainty in supply and demand projections. Three alternative scenarios of supply and demand conditions were constructed through 2030: one extending current trends in population and economic growth, agricultural production, environmental restrictions on water use, and water conservation occurring without policy initiatives (e.g., through equipment replacement, technological change, and revised building codes); and two presenting higher and lower increases in demands. The report of the 2005 plan discusses global climate change and the potential challenges it poses to water supply and demand in California, but climate change is not explicitly represented in the plan's three scenarios.

In addition to adopting these scenarios, the State of California is developing data and analytic capacity to enrich the treatment of uncertainty and climate change in future plans. In parallel with development of the three principal scenarios in this plan update, DWR sponsored development of several analytic tools to strengthen the treatment of uncertainty in future plans. In addition, the California Climate Change Research Center with co-sponsorship from DWR is developing fine-scale regional climate-model scenarios to support analysis of climate-change impacts on water resources.[164] The DWR plans to incorporate these climate-change scenarios explicitly in the next plan update in 2010.

Similarly, the US National Assessment and UK Climate Impacts Programme have both attempted to identify a small set of climate and socio-economic scenarios, to coordinate and gain comparability across multiple studies and allow aggregate assessment of impacts and vulnerabilities at the national level.

In a broad assessment including many teams considering separate questions of climate-change, impacts, mitigation, and adaptation, simple coordinating devices are needed to make teams' work comparable and allow synthesis to produce aggregate conclusions. Emissions scenarios are natural devices to provide such coordination, both because emissions hold the clearest near-term opportunities for intervention, and because they have clear and recognized connections both directions in the causal chain, to every aspect of the climate-change issue. However, in part due to management issues, these efforts to use scenarios as broad co-ordinating devices have not been wholly satisfactory in practice. To serve as coordinating devices, scenarios must be developed and disseminated early in the process, preferably before the work of assessment teams even begins. Moreover, they must be documented with detailed information about the process and reasoning used to generate them, including explicit identification of underlying assumptions and supporting data, models, and arguments. In practice, timely, detailed, and transparent dissemination of scenario information has rarely been achieved.

Scenarios used in large-scale assessments can also make other contributions that are related to the prominent dissemination a major assessment receives. They may, for example, be used as inputs to planning or decision-support processes that were not part of the original assessment. In such use, they may gain a more direct connection to decision-making than they had in their original production or use. Scenarios of global emissions and the model-based climate scenarios based on them especially lend themselves to such derivative uses, informing many different decisions by diverse actors.

Scenarios inevitablybecome politicalobjects, in two senses.They are subject topolitical forces thatseek to influence theirdevelopment, andpolitical reactions tothem once developed.

Scenarios in prominent assessments can also contribute to the framing of public and policy debates. In this role, scenarios inevitably become political objects, in two senses. They are subject to political forces that seek to influence their development, and political reactions to them once developed. These pressures pose challenges and risks that differ quite markedly from those that apply in using scenarios to inform decision-making, where we tend to assume a greater degree of commonality of knowledge, perspective, and interest in the process among participants and some group of relatively well-defined users.

Within scenario exercises, various actors may seek to bias scenarios' content to help advance their policy preferences or their broader political objectives, by limiting consideration to futures they judge desirable or showing some problem in an acute state that would appear to demand a response. While it is not possible to eliminate biases in scenarios, unacknowledged normative biases in scenarios can pose the risk of excluding consideration of futures that are judged undesirable or that pose sharp decision-making challenges. Such biases can be difficult both to detect and to correct. Beyond exhorting developers to scrutinize scenarios critically to avoid bias, the best protection against such biases lies in transparency about the assumptions and information underlying scenarios and associated judgments of likelihood.

Other political pressures come onto scenarios in the broader use, debate, and criticism that they encounter after release. For impartial support of policy decisions, scenarios should fairly present knowledge and uncertainty about potential variation on important dimensions. This typically requires consideration of a wide range of potential futures – often a wider range than relevant decision-makers might initially think plausible, due to well-known habits of conventional thinking and excessive confidence.

Sometimes a scenario's implications for decisions may be obvious. For example, a scenario might represent developments so severe that most people would judge it to demand intervention. Another might represent developments that most people would judge inconsequential or beneficial, so not meriting any intervention. A wide-ranging set of scenarios may include examples of both such extremes. Consequently, such a wide range of potential futures in a set of scenarios – even if this is faithful representation of present knowledge and uncertainty – provides opportunity for partisan distortion and efforts to make scenarios policy-prescriptive.

In global change scenarios, conflicts and opportunities for bias arise most acutely over emissions scenarios. Since much of the uncertainty about climate change beyond 2050 comes from uncertainty in future emissions trends, actors with strong policy preferences can highlight emissions scenarios that lend support to their views. Those who advocate aggressive mitigation may highlight the highest-emissions scenarios to emphasize the elevated risk of climate change that would follow. Those who oppose mitigation may highlight the lowestemission scenarios to suggest that no action to limit emissions is warranted. Because scenarios are used when knowledge of causal processes is weak, it is easy to make any scenario appear salient and likely, even if it is extreme. It is equally easy to probe inside the details of any scenario to find inconsistent or implausible implications, particularly when a scenario is rich in detail.

But, although political actors may have legitimate reasons to highlight one extreme scenario, it is not appropriate for any scenario to dominate assessment or consideration of decisions. A claim that only a single scenario is plausible – especially one near the top or bottom of the present range – is a claim to predict the future, which can be readily dismissed. Claims that a particular scenario is *implausible* cannot be so easily dismissed, however, since scenarios represent only the imperfect judgment of the team that produced them. Leaving aside scenarios that violate clear principles of science (e.g., one whose energy assumptions violate the laws of thermodynamics) or economics (e.g., one that presumes a large new capital stock in a few decades without the investments needed to create it), it is possible to construct pictures of the next century so extreme or unprecedented that most observers would agree they do not merit serious consideration. But short of such an extreme – which describes no global-change scenario discussed here or known to us – claims that a broad class of potential futures is implausible should have to pass a high hurdle. Identifying specific extreme or implausible elements within a scenario does not suffice to make this case, since virtually any scenario will be found to contain these if scrutinized closely enough. Nor does identifying ways that a scenario of future change diverges from some established trend or pattern, since established trends can and do change.

Historical studies of forecasting exercises such as energy forecasts have repeatedly found them too confident that the future will extend recent trends.[165] The threshold any single scenario must pass is to appear sufficiently plausible or instructive to merit consideration in planning and analysis, and this is a judgment to be made by developers and users – with enough transparency about underlying assumptions and reasoning that users can make an informed judgment. A set of scenarios should be constructed so that the range of conditions they represent encompasses present knowledge and relevant uncertainties that might influence mitigation or adaptation decisions. Since subjective judgments cannot be avoided in constructing scenarios, the range provided should err on the side of being broad rather than narrow, at least initially. Identifying problems with one scenario does not necessarily impugn the credibility even of that one scenario, certainly not the whole set, because scenarios cannot be consistent in every underlying detail.

In subsequent revisions as knowledge advances, scenarios can continue to play their role coordinating assessments and framing policy debates with more focus and less arbitrariness. Continuing research and analysis might come to identify some scenarios as severe in their consequences and others as inconsequential, or might revise the initial characterization of the determinants and feasibility of particular scenarios, including suggesting that some are too unlikely to merit serious consideration. These judgments can feed into decisions about continuing analysis of scenarios: which ones can be dropped and what new ones should be added. One major basis for updates in scenarios will be policies adopted, which can set a baseline to focus further deliberations. Perfect attainment of targets and success of policies should not be assumed, but scenarios can focus subsequent debate by posing such questions as "What if we just meet this target; what if we fall short by this much; and what if we exceed it by this much, or adopt these additional measures?"

A claim that only asingle scenario is plausible – especially one near the top or bottom of the present range – is a claim to predict the future, which can be readily dismissed.

Box 4.4. Scenarios of Ozone Depletion in International Policy-making[166]

Emission scenarios of chlorofluorocarbons (CFCs) and other ozone-depleting chemicalssubstantially influenced policy debates over control of these chemicals to protect theozone layer. Until the early 1980s, these policy debates used a convention to projectozone losses that originally served as a simplifying research assumption: constant emis-sions forever. This convention has obvious research benefits, like the simple doubled-CO_2 equilibrium scenarios used in climate models. It standardized input assumptions,allowing exploration of scientific and modeling uncertainties without the confounding effect of different emissions assumptions. Moreover, this convention made no claim to realism, and so avoided distracting arguments over whether one emissions projection oranother was more realistic. Nevertheless, the resultant calculations were frequentlymistaken for projections of realistic future trends.

The question of what future emissions trends were likely only became prominent inpolicy debates around 1983. World CFC production had dropped nearly a third in thelate 1970s due to both regulatory and market-driven reductions in their largest use,aerosol spray propellants, and declined further in the early 1980s recession. It waswidely argued that further restrictions were unnecessary—CFCs' major markets weresaturated and further growth was highly unlikely. The resumption of sharp growth in1983 undermined this claim, making it clear for the first time that managing the ozonerisk required considering scenarios of CFC growth as well as steady-state and decline.How much emissions might grow and what that would mean for the atmosphere remained highly controversial, however.

Emissions of other chemicals complicated the picture. Advances in stratospheric chemistry showed that future ozone loss depended not just on CFCs, but also on emissionsof several other gases, including carbon dioxide, methane, and nitrous oxide. But theknowledge and computing capacity to credibly model interactions among all these pollutants only began to appear in the early 1980s. In 1984, a major scientific assessmentconducted the first standardized comparison of multiple stratospheric models using afew simple scenarios of emissions trends for CFCs and other chemicals. This exercise had the striking result that under a wide range of trends in other emissions, constant CFC emissions would lead to only very small ozone losses, while CFC growth above about 1 percent per year would lead to large losses.

This result, together with resumed growth in CFC production, had a powerful effect inbreaking the deadlock in international negotiations that had persisted since the mid-1970s. Although not the only factor that mattered, this result was crucial in persuadinglong-standing opponents of CFC controls to accept limits on their future growth. This decisively shifted the agenda for the subsequent negotiations that in 1987 yielded agreement on the Montreal Protocol, which cut CFCs by 50 percent.

In this debate, scenarios used in model-based projections of ozone loss identified divergent trends in future risk that were robust to a wide range of assumptions abouttrends in other emissions over which there was disagreement. By parsing projected futures into high-risk and low-risk cases, scenarios served to coordinate and simplify apolicy debate and so help to focus an agenda for collective decision-making.

BOX4.2.2. CLIMATE-CHANGE SCENARIOS FOR THE INSURANCE INDUSTRY

The insurance and reinsurance industries face large financial risks from climate change.These are present in many business lines, but the clearest risk is in insurance for property damage from weather-related events, especially windstorms and floods.

In the past two decades, weather-related insurance losses have increased rapidly. Bysome estimates losses have doubled, even controlling for population and insured value– a much faster increase than that in losses from non-weather events. Climate changeis likely to increase insurance risks in multiple ways, increasing the frequency and severity of loss events and also their correlation. Historically based pricing, which is often required by regulations or market conditions, can compound insurers' vulnerability bypreventing them from anticipating and adapting to a changed risk environment.

Insurance companies do not use scenarios of future climate change in pricing decisions,because property and casualty contracts are written for short periods, usually one year.Since 1992's Hurricane Andrew, these have mostly been priced using historically basedCatastrophic Event Risk Models (Cat models). These estimate losses using a simulateddistribution of storm conditions based on historical experience, together with estimatesof the durability of insured property. While future climate change poses no risk forthese short-term pricing decisions, insurers are concerned that climate change may already have invalidated the historical distributions on which these models are based, byincreasing either the probability of severe events or the correlation among them.

Two published exercises have used climate-change scenarios to explore longer-termrisks to the insurance industry. The first, conducted for the Association of British Insurers in June 2005, examined potential impacts of climate change on the costs of extreme weather events (both insured and total economic costs) under the six SRESmarker scenarios, as well as IS92a and CO^2 stabilization at 550 ppm. The analysis calculated changes in losses due to US hurricanes, Japanese typhoons, and Europeanwindstorms.

The second scenario exercise, conducted by Harvard Medical School researchers withsponsorship by Swiss Re and the United Nations Development Program, used two scenarios of 21st-century climate change to examine potential impacts on human andecosystem health, and associated economic costs, not limited to the insurance industry.The two climate scenarios both assumed CO_2 doubling by approximately mid-century, one with continued incremental climate changes and one with hypothesized nonlinearimpacts and abrupt events. The exercise examined potential changes in infectious andwater-borne diseases, asthma, agricultural productivity, marine ecosystems, freshwateravailability, and natural disasters (including heat waves and floods). The analysis was based primarily on qualitative judgments.

The first scenario showed increased property losses and business interruptions relativeto recent trends, emergence of new types of health-related losses, and increasing difficulty in underwriting. The second scenario was qualitatively similar but more severe, withsubstantial increases in both average losses and variability leading to large premium increases and withdrawal of insurers from many markets, particularly along coastlines. Asmany insurance firms succumbed to mounting losses, those remaining established strictlimits on coverage, shifting more exposure back to individuals and businesses.

Neither of these exercises was connected to any specific, near-term business decisionfaced by insurance companies. Both could serve longer-term decision-making, however,including planning for reserve accumulation, providing supporting analysis for advocating mitigation and adaptation policies, and supporting changed regulations to allow moreflexible pricing of risks experiencing long-term increases. Such exercises can also serveto inform firms' long-term risk-avoidance strategies, including decisions to exit certainareas of business.

BOX 4.6. SCENARIOS OF CLIMATE IMPACTS IN THE COLUMBIA RIVER BASIN

In conjunction with the US National Assessment, researchers at the University of Washington studied climate impacts on the Columbia River system, which is the primarysource of energy and irrigation water for the Northwest and one of the most intensivelymanaged river systems in the world.[167] The project examined the response of annual andseasonal flows both to existing patterns of climate variability –the El Niño Southern Oscillation (ENSO) and the Pacific Decodal Oscillation (PDO) –and to projected 21st century climate change.

The study projected climate change through 2050 using eight climate models driven byone emissions scenario (1 percent per year CO_2 concentration increase), which on average projected 2.3°C regional warming by the 2040s with precipitation increases of 10percent in winter and a few percent in summer. In the Columbia, these changes areprojected to increase flows in winter (because there is more winter precipitation andmore of it falls as rain) and to decrease flows in summer (because there is less snow-pack and it melts earlier in the spring). The impact of summer decreases is likely to besubstantially more serious than that of winter increases. Because the Columbia is a snowmelt-dominated system, winter flows could double or even triple and remain belowthe present spring peak.

Assessing the impacts of these flow changes requires assumptions about water demands and system management. The study used a reservoir operations model that calculatedthe combined effects of flow changes and alternative system-operation rules on the reliability of different water-management objectives, such as electrical generation, flood control, irrigation supply, and preserving flows for salmon. Under historical climate variability, all objectives can achieve high reliability in high-flow years (i.e., in the cool phaseof ENSO or PDO), but conflict between them occurs in low-flow (warm) years, whenonly one top-priority objective can be maintained at or near 100 percent reliability andother uses suffer substantial risks of shortfall. Different operating rules distribute thisrisk among uses.

Under the projected climate and flows of the 2040s, this model showed a pattern ofcompetition between uses similar but additional to what already prevails in low-flowyears, suggesting increases in already sharp conflict among uses over flow allocations.One objective could be maintained near full reliability, but other uses suffered reliabilitylosses up to 10 percent from the changed climate, in addition to effects of continued climate variability.

In this analysis, scenarios helped to illustrate interactions between management decisionsand climate change and variability, and to explore opportunities and limits for adaptationthrough management changes alone, with no change in infrastructure or larger-scalepolicies. This analysis has not been incorporated into any operational decisions, but hasbeen integrated into the Fifth Conservation Plan issued by the Northwest Power andConservation Council.[168] More detailed assessment of climate-change impacts would require extending this analysis to include projected changes in water demands, boththrough direct climate effects and through scenarios of regional economic and population growth, allowing a more realistic assessment of potential effects of new water-man-agement investments and changes in large-scale policies to alter water demand, balancecompeting uses, or improve coordination among the multiple organizations involved inmanaging the river system.

4.3. The Process of Developing Scenarios: Expert-Stakeholder Interactions

Scenario exercises are collaborative activities that need to be managed. As Section 1 discussed, managing a scenario exercise includes deciding who participates, what jobs they are assigned and how these jobs fit together, how disagreements are resolved, and how much time and money are dedicated to the exercise. These matters can be decisive for the success of an exercise. For some of them, the nature of challenges and tradeoffs they pose are fairly obvious. For example, scenario exercises need enough time to build a team, research scenario components, consult with users, and disseminate results, but often too little time is available, so various compromises must be made. Adding participants expands the expertise and the range of views represented, but increases the time needed for team building and internal communication. Delegating parts of the exercise to smaller groups can overcome this tradeoff, but can introduce coordination problems and inconsistencies among groups. Accepting external direction on scenario exercises increases the chance that decision-makers will take the scenarios seriously, but also increases the risk that they are seen as biased or simply reflect conventional wisdom. These issues pose various challenges, but the challenges are not unique to scenario exercises.

The more central process problems for scenarios concern the relationship between experts and stakeholders in the design, creation, evaluation, and application of scenarios. There has been substantial experience and research in processes for involving stakeholders in environmental decisions, in the United States and other regions.[169] In the most well-established areas of scenario use – e.g., strategic planning for corporations or other organizations, and military and security planning – it is widely understood that there should be close, intensive collaboration between developers and users in the production, revision, and application of scenarios. While high-level decision-makers are not usually involved in the detailed work of scenario construction, they or their surrogates may be intensively involved in problem definition, identification and elaboration of key uncertainties, large-scale scenario design, evaluation and criticism of scenario outputs, and deliberation over lessons and implications. Their level of involvement must be high for results to be useful, particularly if a major purpose of the exercise is to challenge decision-makers' assumptions and promote creative thinking.

Managing a scenario exercise includes deciding who participates, what jobs they are assigned and how these jobs fit together, how disagreements are resolved, and how much time and money are devoted to the exercise. These matter scan be decisive for the success of an exercise.

In these areas, scenarios typically serve a clearly identified, relatively small and homogeneous set of users who have some degree of agreement on what values they are trying to advance, what issues are relevant, and what choices are feasible, acceptable, and within their power and authority. This is most clearly the case when scenarios are developed for a single organization, but also applies to scenario exercises for larger groups that are sufficiently homogeneous in their interests and perspectives, e.g., scenarios for property and casualty insurers, for organized labor in the United States, or for European environmental groups. In such context, the problems of deciding participation are likely to be manageable.

Intensive user involvement has also been advocated in developing scenarios for climate change. This is obviously correct when climate-change scenario exercises serve specific, clearly identified user groups. The strongest examples are scenarios to support narrowly targeted assessments of impacts and adaptation in particular industries, resources, or regions, e.g., scenarios for coastal managers considering the establishment or revision of setback lines for coastal-zone construction as sea level rises,[170] for rangeland managers considering the purchase of conservation lands or easements for the purpose of providing migration corridors, or for insurance and reinsurance firms examining the nature of climate-change risks they may face and potential responses. In such cases, intensive participation of users is relatively easy to achieve and provides access to valuable expertise and assurance of practical utility.

But climate-change scenarios typically serve larger and more diverse sets of users and stakeholders. This is especially true for scenarios produced in large-scale, official assessments such as the IPCC or US National Assessment. Climate-change stakeholders – defined by the CCSP as "individuals or groups whose interests (financial, cultural, value-based, or other) are affected by climate variability, climate change, or options for adapting to or mitigating these phenomena"[171] – are an enormous group, diverse in their interests and responsibilities.

Climate-change stakeholders are an enormous group, diverse in their interests and responsibilities.

Even when the set of all potential users is numerous and diverse, there may be some types of users who are clearly identified – e.g., climate modelers who need input from emissions scenarios or impact assessors who need input from climate-change scenarios – and who have highly specific scenario needs, including such prosaic factors as data format and resolution. Close consultation with such users is clearly important, especially when their desires exceed what scenario developers can confidently provide, e.g., when climate modelers need emissions data at fine spatial resolution and for specific gases or aerosols, which are not readily available from the energy-economic models used for emissions scenarios. These situations call for particularly close and sustained consultation, so the two sides can understand each other's needs and capabilities in enough detail to develop workable resolutions.

Other users, however, may be numerous, diverse in their disciplinary foundations, methods, and tools, and not clearly identified. Their information needs may have some commonalities but substantial differences. They may even have points of conflicting interest in the construction and use of scenarios. The general case for stakeholder involvement remains strong with such diverse users, especially in the initial design of a scenario exercise, and in the evaluation and refinement of scenarios for relevance, practicality, and utility. In principle, the required approach is to involve a reasonably diverse and representative group of users and stakeholders, as well as an appropriate range of disciplinary and modeling experts, while keeping the size of the scenario team manageable. But the judgments about participation and representation needed to carry out this approach in any particular scenario exercise will be complex and challenging.

Can a scenario process be completely open? In political settings, some insulation from users may be needed to insure consistency across participating models and analyses. Whatever approach to stakeholder participation is adopted, numbers must be kept manageable. Despite recent progress in scenario methods allowing a substantial increase in the number of participants, there are still practical limits. Although requirements for expertise external to the core scenario team increase with scenario complexity, a scenario process is unlikely to work with a hundred people in the room. A few scenario processes have engaged much larger numbers of participants, but these have greatly reduced the complexity of the scenario-creation process by limiting it to specifying inputs to a single interactive model, or have involved large numbers of participants in independent, parallel sessions interacting with a computer-based model or scenario construction system.[172] These tensions between representational realism, participation, and managerial feasibility pose challenges for design of processes of representation and consultation in scenario development, on which further progress is needed.

Box 4.7. Scenarios in Acid-Rain Assessments: Two Approaches

Two programs, one in the United States and one in Europe, developed scenarios inintegrated-assessment models of acid rain to inform policy decisions over sulfur emissions. Amongmany other differences, the two programs differed strongly in their approaches to involving stake-holders and in their effectiveness at informing decision-making.

The US National Acid Precipitation Assessment Program (NAPAP) was created in 1980 as a 10-yearprogram to study all aspects of acid deposition: emissions, transport and deposition, impacts, andcontrol strategies.[173] Managed by a committee of six government agencies and supported by a full-time staff office, the program involved roughly 2,000 researchers.[174] Although charged to conductboth scientific research and assessment, NAPAP strongly emphasized research. Its assessment report was opaque on the origin and interpretation of its scenarios, and did not use the scenarios tointegrate across the issue or examine implications of alternative policies. Overall, NAPAP is regarded as having succeeded as a research program, but fallen critically short in providing useful information for decision-making.[175]

An alternative approach to acid-rain assessment was taken in Europe as part of the policy debatesunder the Convention on Long-Range Transboundary Air Pollution (LRTAP). The core of this assessment was a cooperative program to monitor and model acid emissions, transport, deposition,and impacts. In contrast to NAPAP, this program focused more on assessment than research, beingspecifically established to inform the policy process.[176] Scientific models of components of the acid-rain issue were chosen to contribute to a simplified integration of the problem; scenarios ofemissions and controls were chosen in consultation with officials, in an attempt to replicate thepolicy alternatives under consideration.

The culmination of this pursuit of simple, accessible, and policy-relevant analyses was the RAINSmodel, developed at the International Institute for Applied Systems Analysis (IIASA) in Austria. Asa result of its flexibility, ease of use, and relevance to policies under consideration, the RAINS modelwas used extensively by policy-makers in the negotiation of sulfur-control agreements under theConvention, and had substantial influence over the controls adopted.[177]

The contrast in approach and outcome between these two programs suggests the potential valueof close interaction between experts and stakeholders in producing scenarios, at least when thestakeholders are relatively expert officials responsible for a specific set of decisions. In the Europeancase, such close interaction helped to ensure the credibility of baseline emissions scenarios and therelevance of proposed control scenarios, despite the diverse and sometimes contending interestsof the participating officials. The contrast between the two programs also suggests that there canbe significant tradeoffs between scientific and assessment objectives in programs that seek to integrate the two activities.

4.4. Communication of Scenarios

Scenarios related to climate change must be communicated to multiple audiences, with diverse interests and information needs. Involving users in scenario development can aid subsequent communication in various ways – e.g., by ensuring that scenarios are understand-able and practically oriented, and helping to disseminating scenarios to their constituencies.

But, in all likelihood, most users to whom scenarios must be communicated did not participate directly in scenario development.

Although specific needs will vary from case to case, any communication of scenario-based information to a large, diverse public audience is likely to require a few common elements. First, in addition to the scenarios' content, information should be provided about the process and reasoning by which the scenarios were developed. This allows users and stakeholders to understand and critique scenarios, and to determine their own levels of confidence. Second, scenario developers should identify the uncertainties considered. A particularly important distinction to communicate clearly is between scientific uncertainty and scenario uncertainty, including explicit statements of when and how scenarios change (e.g., the reduced SO_2 projections in the IPCC SRES scenarios), and clear explanations of the effects of such changes. Third, related to uncertainty, developers should acknowledge the unavoidable elements of subjective judgment in developing scenarios, and be prepared to explain and defend the judgments they made. Fourth, when particular scenarios were constructed to have specific meanings – e.g., a reference case, a plausible worst-case, or the exploration of a particular causal process taken to its extreme – these should be clearly conveyed. Fifth, if scenario developers have articulated any indicators of the confidence they place on scenarios or distributions of associated variables, this information and any supporting reasoning should also be made available.

Developers should acknowledge the unavoidable elements of subjective judgment in developing scenarios, and be prepared to explain and defend the judgments they made.

A communication strategy should attempt to steer users away from certain common pitfalls, such as choosing one scenario and treating it as a highly confident prediction, or taking the range spanned by a set of scenarios as encompassing all that can possibly happen. An effective strategy of communicating scenarios and their underlying reasoning can help to engage users in the process of updating and improving scenarios. Providing transparency rather than claiming authoritative status for scenarios is likely to increase users' confidence that the scenarios have reasonably represented current knowledge and key uncertainties. It also provides users with the tools to develop alternative representations if they are unconvinced.

In large and complex assessments such as theIPCC and US National Assessment, communication of scenarios and underlying informationboth to various groups within the assessmentand to potential outside users poses representational and managerial challenges. Scenario developers have experimented with various visualtechniques for conveying complex informationin vivid and understandable form, includinglandscape representations, maps, and pictures, as well as various graphical and tabular formats.[178] In the US National Assessment, climate scenarios and other related information were provided to participating assessment teams in several formats (e.g., tabular summaries, models, graphic representations), through web-sites backed up with workshop presentations. In the IPCC, the Task Group on Data and Scenario Support for Impact and Climate Analysis (TGICA) was established in 1997 to facilitate distribution of climate scenario data, model results, and baseline and scenario information on other environmental and socio-economic conditions, for use in climate impact and adaptation assessments. Data, scenarios, and supporting information are distributed over the internet by the IPCC Data Distribution Center (DDC).[179]

To compactly communicate uncertainty in climate scenarios, the TGICA and several national scenario efforts have developed various graphical methods, including scattergrams showing the range of projected temperature and precipitation changes generated by several climate models using four SRES marker scenarios, and comparing these projected changes to estimates of natural variability.[180] In Figure 4.2, each data point represents one climate-model projection associated with a given SRES emissions scenario. Efforts to develop similarly compact representations of the distribution of scenarios for extremes as well as annual and seasonal averages are underway.

To help users select climate scenarios for impact assessments, an alternative to summarizing climate-model scenarios in such scatter-grams is to combine various climate-model results using statistical methods to construct explicit probability distributions for important climate variables.[181] Figure 4.3 shows one such method, which assigns weights to model results based on their bias in simulating the current climate (smaller biases are assigned higher weight) and their correspondence with other model results (outliers are assigned lower weights). This method compactly communicates multiple model results, clearly conveying which ones fall at the top and bottom of the distribution ("unlikely to be higher/lower than this"), and which fall in the middle of the range.

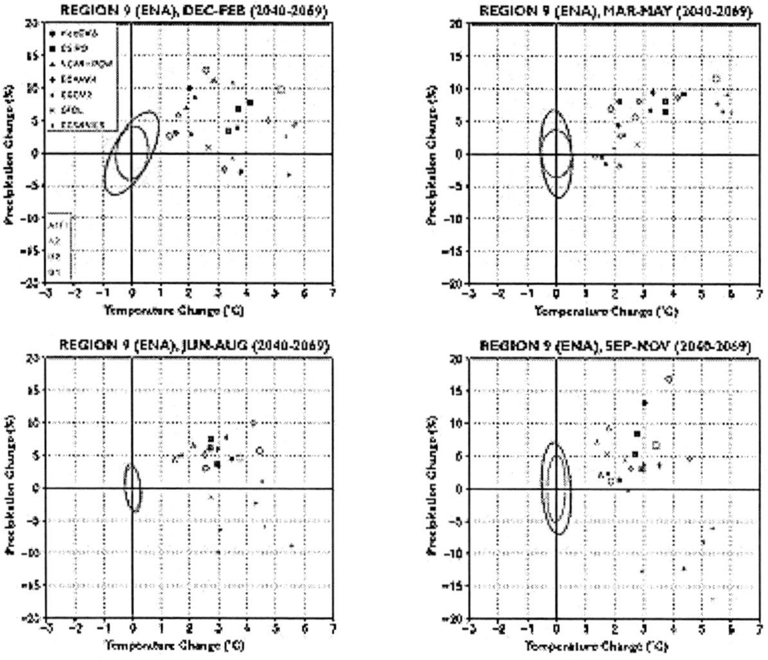

The x-axis shows temperature changes in °C, the y-axis precipitation changes in percent. Each point shows one model's projection under one emissions scenario. A point's color denotes the corresponding emissions scenario, its shape the corresponding model (per legends in upper left figure). Ovals show 95 percent confidence bounds for natural 30-year climate variability, calculated from unforced 1000-year runs of the models CGCM2 (orange) and HadCM3 (blue). Points outside the ellipses indicated projected climate change significantly outside the range of natural variability, most frequently due to changes in temperature rather than precipitation. *(Source: Ruosteenoja et al. 2003.)*

Figure 4.2. Regional scattergram for eastern North America, 2040-2069.

This current focus on collections and intercomparisons of model-based projections with various emission scenarios represents a new approach for communicating scenario-driven model output to users engaged in assessment and adaptation activities. It has enabled users to consider a broader range of emission scenarios and climate models than was feasible in the US National Assessment and previous IPCC assessments. It allows users to consider all available model and scenario combinations to span the literature, or to select only scenarios that exceed some threshold of interest or fall within some specified probability range. Future assessments should benefit from this type of multi-model, multi-scenario approach, which allows users more effective and informed choice over scenarios to consider.

The current focus on collections and intercomparisons of model-based projections with various emissions scenarios represents a new approach for communicating scenario-driven model output to users engaged in assessment and adaptation activities.

4.5. Consistency and Integration in Scenarios

One of the most often stated requirements for scenarios is that they be "coherent" or "internally consistent." This is clearly an important goal. Scenarios usually specify multiple characteristics of an assumed future, whether as multiple elements of a narrative or multiple quantitative variables, so these elements should fit together. Difficulties arise in the pursuit of such consistency, however, and in some scenario exercises the pursuit of consistency, together with the goal that scenarios integrate many components of a broad issue such as climate change, may jeopardize the validity and usefulness of the scenarios.

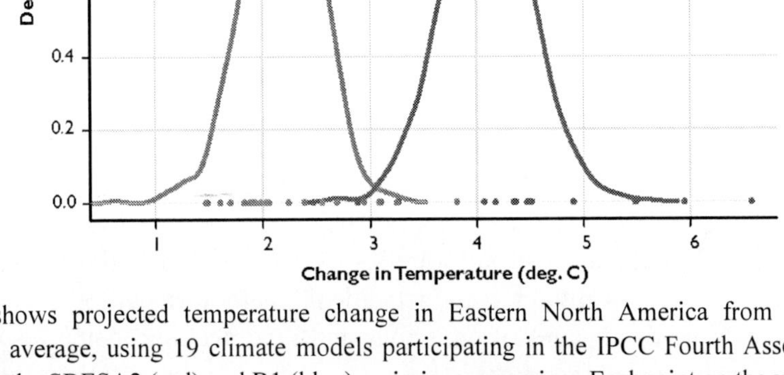

The x axis shows projected temperature change in Eastern North America from the 1980-1999 historical average, using 19 climate models participating in the IPCC Fourth Assessment Report driven by the SRESA2 (red) and B1 (blue) emissions scenarios. Eachpoint on the x axis shows the result from one model. The curves above the axis showprobability distributions constructed from thes eindividual model results.(Source: Tebaldi et al. 2005.)

Figure 4.3. Constructed probability distributions of model-simulated temperature change in 2080-2099.

Certain elements of internal consistency in scenarios are unproblematic, such as avoiding gross contradictions with well known principles of behavior of biophysical or socio-economic systems, and not inadvertently moving far outside the bounds of historical experience. Inadvertently implausible assumptions can arise, for example, when multiple elements of a scenario are specified without cross-checking; e.g., end-year specifications of a region's population and GDP without checking the implied growth rate in GDP per capita, or specifying energy-related emissions trajectories without checking what they imply for resource availability. Avoiding these pitfalls requires thorough cross-comparisons of related values with each other, of terminal values with implied time-trends in the intervening period, and of values within and among regions. Scenario developers should not always and necessarily avoid extreme or unprecedented outcomes, however. Presenting extreme or seemingly implausible future conditions *intentionally*, with an explanation of how they could in fact arise, can contribute to several of the major purposes of scenarios, e.g., shaking up habitual thinking and broadening expectations of what future developments are plausible.

But statements about internal consistency in scenarios usually claim much more than the mere absence of gross contradictions and inadvertently implausible values. They tend to claim that the multiple elements of a scenario are related in a way that reflects reasonable, well-informed judgments about causal relations, suggesting that some events or trends are more likely to occur together, some less. Expressing the goal as "coherence" rather than "internal consistency" suggests a higher level of perceived affinity among scenario elements, evoking normative or even aesthetic aspects.

Expressed in probabilistic terms, statements about internal consistency may be interpreted as claims that alignments of factors similar to those in the scenario are more likely than other, dissimilar alignments. One might, for example, claim that a scenario with rapid growth in both the economy and energy use is more internally consistent than one in which the economy grew rapidly but energy use did not. But where do these perceptions of greater or lesser likelihood come from, and how valid are they? In some cases a well-founded theory or model might say that certain outcomes tend to be related. Alternatively, explicit analyses might connect the claim to underlying assumptions that are open to scrutiny and criticism. But in the absence of such transparent foundations for judgments of what scenario conditions are consistent and what are not, these claims can only rest on more diffuse judgments by scenario developers, refined and tested through various deliberative processes – e.g., arguing about the claims, working through their implications relative to those of alternative specifications, and identifying additional bodies of research and scholarship that can be brought to bear.

These difficulties can be compounded when, in addition to consistency, a goal of scenario "in tegration" is also pursued (although the precise meaning of "integrated" can be difficult to ascertain). The integration of a scenario is a function of its complexity or breadth, which is related to the number of characteristics it jointly specifies. In global climate-change scenario applications, integration typically refers to including all major elements of the causal chain of the issue, i.e., multiple dimensions of emissions and their socio-economic drivers, climate, impacts of climate change, and responses.

Asking a scenario to be integrated in this way imposes on the scenario the burden of capturing all relevant elements of the future. Such an expansive scenario may occasionally be needed – e.g., for preliminary assessment of a threat for which no relevant data or current research exists. However, the risks of error, bias, and arbitrariness in such a scenario are greatly increased, because so much of reality (with whatever unknown causal processes by which it actually operates) is being stuffed into the scenario.

More likely, an integrated scenario would be constructed by combining exogenous assumptions about some elements with model-calculated values for others. This approach does not avoid increasing risks of inconsistency and contradiction as a scenario is expanded, particularly when multiple models are used. Since models embody specific, quantitative causal relations among variables, they do not require – or indeed allow – all variables to be specified. Scenarios provide only those exogenous inputs that the model does not produce. These scenario-based inputs should be consistent with each other, but to a lesser extent than the precise standard that defines consistency in a scenario. These exogenous inputs, together with model results, canjointly comprise a scenario that is generated forsome alternative use.

Consistency problems grow when scenario exercises involve multiple models and attempts are made to achieve model harmonization. When scenarios are constructed partly out of exogenous inputs provided by a scenario (made consistent as much as possible through qualitative or intuitive causal reasoning) and partly out of models, multiple models are often used. Using multiple models in parallel can allow for more extensive exploration of causal relations, and helps to characterize uncertainty in scenarios since different models embody different representations of causal processes. It may also enhance the credibility of the process. But models of the same broad set of phenomena – e.g., models of the economy and energy sector – frequently differ in which variables they require as exogenous inputs and which ones they calculate endogenously. In this case, some variables must be specified exogenously for some models, but are calculated endogenously by others.

In the absence of transparent foundations for judgments of what scenario conditions are consistent and what are not, claims can only rest on more diffuse judgments by scenario developers, refined and tested through various deliberative processes.

This creates various problems for consistency. In general, when scenario exercises are conducted in this way, some elements are assumed and others are model-calculated. Attempting to avoid this poses even more serious problems, however. It is not usually possible to arbitrarily perturb the exogenous input variables so all inputs and outputs match across all models, since such perturbations will influence other variables in the model. Consequently, avoiding these inconsistencies will require manipulating internal relationships within models to make their outputs match the specified values, given the common inputs. But such reverse-engineering of internal model relationships to match specified outputs, in addition to being exceedingly cumbersome and arbitrary, can corrupt the internal logic of models, obscure the interpretation and significance of results, and make it impossible to use model variation to illuminate uncertainty.

For example, in an exercise to generate non-intervention scenarios of potential future emissions, little insight is likely to be gained from defining scenarios in terms of the resulting emissions and forcing the different models to generate these emissions targets.[182] It may be equally fruitless to define scenarios in terms of GDP and energy consumption trajectories and to force multiple models to reproduce these. For this reason, multi-model exercises such as the Energy Modeling Forum usually seek to harmonize only a few of the most essential and commonly used inputs.[183] When multiple models are used to generate scenarios, the most useful way to pursue consistency may be to develop common assumptions for the variables furthest back in the causal chain, but the wide variation of model structure can make even this approach to harmonization challenging.

When multiple models are used to generate scenarios, the most useful way to pursue consistency may be to develop common assumptions for the variables furthest back in the causal chain, but the wide variation of model structure can make even this approach to harmonization challenging.

In addition to consistency within a scenario, consistency across scenarios within an exercise also requires attention. Ideally, factors not explicitly recognized as the basis for inter-scenario differences should be consistent across scenarios. Or alternatively, all bases for differences between scenarios should be explicitly recognized and stated.

When models are used in a scenario exercise, significant variation in model structures suggests less mature underlying knowledge, or at least greater recognition of knowledge gaps, than when model structures converge and all remaining uncertainty is over exogenous input parameters. For scenarios to provide faithful representation of present knowledge and uncertainty, this variation should not be suppressed or concealed. Consequently, when scenarios are defined over variables that include outputs of some participating models as well as inputs, it is crucial not to pursue false consistency by forcing models to match the target outputs through manipulation of their internal causal processes. This is suppressing model uncertainty.

One preferable alternative would be for the results of scenario exercises involving both exogenous inputs and multiple models to explicitly distinguish between three classes of variables: (1) a minimal set, exogenous to all; (2) those specified exogenously for some models, but generated by others; (3) model outputs, whose variation reflects partly model and partly parameter uncertainty.

An alternative way to use multiple models is to let each model produce one scenario, as was done in the selection of the SRES marker scenarios. With this approach, each scenario represents a particular realization of uncertainty over both exogenous inputs and model structure. But this approach confounds model uncertainty with parameter uncertainty. It may be preferable to cross exogenous inputs with models to produce a larger number of scenarios from which subsets can be extracted as needed, perhaps organizing these as a nested hierarchy of scenarios similar to the SRES: six marker scenarios, 40 SRES scenarios in total, and hundreds of scenarios in the literature review.

There are good reasons to combine narrative with quantitative approaches, as scenario exercises have increasingly sought to do. But the connection between qualitative and quantitative aspects of global-change scenarios has been inadequate, diminishing the usefulness of the exercises due to inconsistencies within each type of scenario and between the two types. This problem has partly been due to limited time and resources, but has also reflected substantive difficulties in linking the two types of scenario, difficulties that have not been understood or managed well. Narrative scenarios typically specify deep structural characteristics like social values and the nature of institutions, which are associated with structural characteristics of models such as the determinants of fertility trends, labor-force participation, savings and investment decisions, and substitutability in the economy. Consequently, the differences among alternative narrative scenarios, reflecting different basic assumptions about how the world works, correspond more closely to variation of model structure than to variation of parameters. Better integrating the two approaches will require developing ways to connect narrative scenarios to model structures, rather than merely to target values for a few variables that models are then asked to reproduce. This has not happened because scenario exercises have not had the capability or resources to direct new model development, or to induce modelers to undertake substantial structural changes to their models. This would require substantial efforts, including getting modelers to interact with scenario exercises in a new way, but might hold more promise for allowing scenarios to usefully inform discussions about large-scale policy choices for mitigation and adaptation.

4.6. Treatment of Uncertainty In Scenarios

Representing and communicating uncertainty is perhaps the most fundamental purpose of scenarios. This section discusses how scenarios represent uncertainties, how these methods connect scenario exercises to simpler formal exercises in the analysis of decisions under uncertainty, and what challenges are posed in how uncertainty is represented. It also addresses several important debates in the treatment of uncertainties.

In most scenario exercises, uncertainty is represented not in a single scenario, but in variation across several scenarios considered together.[184] The choices to be made in deciding how to represent uncertainty include the following:

- What characteristics are varied
- By how much these characteristics are varied, separately and together (e.g., should extreme values of multiple characteristics be combined, or extremes of some combined with the middle cases of others?)
- How many scenarios to create and consider together
- What description, documentation, or other information is attached – including whether and how specifically measures of likelihood are assigned.

4.6.1. Uncertainty in simple quantitative projections: basic approaches

How these choices are made and their implications for scenario use and effectiveness are closely related to the large-scale decisions in designing a scenario exercise outlined in Section 2.1. In particular, the role of uncertainty in a scenario exercise is strongly linked to scenario complexity, richness, and use. In the simplest case, a scenario exercise may be dominated by a single quantitative variable, so all uncertainty could be represented by alternative future levels or time-paths of that variable. This case is so simple that many scholars and practitioners argue it should not be considered a scenario at all.[185] Still, even this simple and extreme case raises significant issues. We begin here and then move to more complex cases.

Representing and communicating uncertainty is perhaps the most fundamental purpose of scenarios.

If we also assume that the probability distribution is known, the situation reduces to a formal exercise in analysis of decision-making under uncertainty. Given a known set of choices and outcomes of each choice under each uncertain outcome, alternative choices can be evaluated by formal methods such as seeking the best outcome on average or under some risk-averse valuation scheme, or seeking robust strategies. This decision-analytic approach can be extended to situations of a few uncertain variables with a known joint distribution, multiple decision-makers who evaluate outcomes differently, or (with somewhat more difficulty) decision makers with different probability distributions.

Further relaxation of these simplifying assumptions moves us toward activities that are more widely recognized as scenario exercises. First, if a scenario exercise is addressed to more than just a few decision-makers with known choice sets and outcome valuations, scenarios can no longer simply be inputs to an analytic exercise, but rather become descriptions of potential future states that must be communicated directly or indirectly to decision-makers for their reflection and deliberation. Second, if distributions of important quantities are unknown, it is necessary to exercise judgment regarding how to draw on relevant knowledge to construct and describe alternative future values of the quantities, and how to represent these values to users with a manageable number of scenarios.

Probabilistic statements about future conditions always incorporate elements of subjective judgment. Many forms of current knowledge – including data, models, and expert judgments –are relevant to forming these judgments about future conditions.

Since scenarios describe future conditions, the distributions of quantities in scenarios cannot be known in the same sense that the distribution of current characteristics – e.g., the November daily high temperature at O'Hare Airport – can be known through repeated observations. Probabilistic statements about future conditions always incorporate elements of subjective judgment. Many forms of current knowledge – including data, models, and expert judgments – are relevant to forming these judgments about future conditions. In constructing scenarios of population growth, for example, the distribution of observed past growth rates can be used to construct a range or distribution of plausible future values. But while scenarios can draw on, and be made conditional on, such knowledge, this does not overcome their unavoidable reliance on subjective judgments as well.

Scenarios can also be based on model representations of knowledge of causal processes. For example, instead of simply extrapolating past population growth rates, one could use a demographic model that represents trends in fertility rates, lifespan, and migration to calculate a resultant population trend. Formal modeling can represent the structural relationships transparently, reducing the risk of generating inconsistent projections. Structural models can possibly also perform better in extrapolating to conditions beyond the observed range of behavior. Because models represent causal relationships among multiple variables, these models can extend the range of current and historical data that are relevant to projections, although this may result in an expansion of data needs. Models can also help characterize uncertainty in future quantities of interest, by allowing the uncertainty to be attributed to input parameters – explored through sensitivity analysis or simulation techniques such as Monte Carlo – or to model structure.

Estimating output distributions based on assumed distributions of uncertain input parameters does not capture all uncertainty of importance for assessment and decision-making. The input probability distributions are not known with certainty, nor are the structural assumptions that determine the mapping of inputs onto outputs within any particular model. Uncertainty analysis can embrace this additional level of uncertainty, sometimes called "meta-uncertainty," by stepping up one more level of abstraction – considering not just uncertain quantities, but uncertainty about their uncertainty, or alternatively, probability distributions over probability distributions of unknown quantities. Methods to represent and process such meta-uncertainty mirror those used for first-order uncertainty. This is an active area of research, but its importance for assessment methods and their application is unclear. This level of abstraction increases the difficulty of communicating scenarios and their underlying reasoning transparently and comprehensibly to non-specialists. Moreover, since any step of analysis represents an act of potentially fallible judgment, taking the step to meta-uncertainty still does not capture all possible uncertainty. It is not clear whether, for purposes of constructing and using scenarios, the explicit separation of uncertainty in outcomes from uncertainty in probability distributions brings more benefit than could be gained from simple heuristic guidance to assume distributions are wider than initially seems necessary.

A major risk in all scenarios is subjective bias, which can be reduced but not eliminated through use of existing data and formal modeling. Judgment is an essential element in constructing scenarios, both to apply relevant data and models when these are available, and to build future descriptions using less formal methods when they are not. The expert judgments supporting such less formal projections may be better founded than mere uninformed speculation, since there is typically much relevant knowledge available beyond what is explicitly captured in present datasets and models.

Approaches to developing expert-judgment based projections vary widely in their structure and formality, from simply asking one or more experts to state their best estimate of some unknown quantity, to highly structured elicitation exercises that provide multiple cross-checked estimates of the same quantity.[186] Such methods must attend to risks of overconfidence and bias, which are well documented in experts as well as laypeople. Carefully designed elicitation protocols can reduce the effects of such biases, e.g., by prompting experts to broaden their estimates of uncertain quantities, but cannot eliminate them.[187] An additional challenge to these methods is that there is no generally accepted method for selecting or aggregating estimates from multiple experts.

4.6.2. How many scenarios, over what range?

In communications of scenarios, limited time, resources, and attention usually require that only a few discrete values or time-paths are specified, not a complete distribution. Scenario developers must decide how many scenarios to provide and how to space them.

How many scenarios to provide rests on a judgment of the value provided by each additional point from the underlying distribution relative to the burden of producing and using each new scenario. If the use made of each scenario is expensive – e.g., consuming large quantities of time of busy senior people, or running a large model – then the number of scenarios that can be adequately treated may be very few. The 1992 IPCC scenario exercise provided six separate scenarios, of which nearly all subsequent analyses used just one or two. Of the 40 scenarios produced by the SRES process, only 6 (initially 4) were highlighted as "marker" scenarios, while most subsequent analyses used just 2 or 3.[188]

Deciding how many scenarios to provide also involves some element of attempting to avoid predictable errors in their use. While the most obvious and frequent choice in providing scenarios of a quantitative variable has been to provide three – one high, one low, and one in the middle – it has been widely noted that this practice runs the risk that users will ignore the top and bottom, pick the middle, and treat it as a highly confident projection, suppressing the uncertainty that scenario developers tried to communicate by providing, and carefully spacing, three scenarios. The same risk applies to any odd number of scenarios, leading many developers of quantitative scenarios to the informal guideline that the number provided should always be even, so there is no "middle" scenario that users can inappropriately fix on.

Similarly for global environmental change, low-probability risks might need to be considered if their consequences or their effects on preferred decisions are large enough.

More specific guidance on the appropriate number and range of scenarios must reflect both scenario developers' sense of the underlying distribution from which scenarios are drawn, and their intended use. One must consider whether departures in both directions from the middle are of similar importance, or whether only departures in one direction need be represented. For example, one might judge that in an assessment of impacts of climate change a scenario drawn from the lower tail of potential climate change is likely to provide little substantive insight, since in most cases the impacts of a small-change scenario are predictably small.

One must also consider how far a set of scenarios should extend toward including extreme or unlikely futures. In estimating unknown quantities, many fields of empirical research draw intervals to capture from 90 percent to 99 percent probability, but in constructing scenarios to inform decisions there may be good reasons to consider more extreme and less likely possibilities, whether these likelihood judgments are expressed quantitatively or qualitatively. Assessments and policies in both regulation of health and safety risks and national security, for example, routinely focus on high-consequence risks that are judged much less than 1 percent likely. Similarly for global environmental change, low-probability risks might need to be considered if their consequences or their effects on preferred decisions are large enough.

It is often suggested that a set of scenarios should "span the literature" of prior scenarios or projections of the same quantities. However, there may be good reasons for a wider or different range, or even a narrower range – although developers should be cautious about a set of scenarios that spans a much narrower range than published estimates of the same quantities. A published scenario may have been constructed to serve various purposes other than providing an independent new estimate of a quantity of interest. Previous scenarios developed to serve some particular purpose may or may not be relevant to a new scenario exercise, depending on the relationship between their intended purposes. Moreover, previously published scenarios can be highly self-referential, since many published analyses use prominent pre-existing scenarios as inputs to a new study, or examine a new model by forcing it to reproduce some pre-existing scenario. For all these reasons, previously published scenarios are better regarded as one input to the judgment of developers of new scenarios than an authoritative picture of present knowledge that new scenarios must follow.

Previously published scenarios are better regarded as one input to the judgment of developers of new scenarios than an authoritative picture of present knowledge that new scenarios must follow.

4.6.3. Bifurcations and major state changes

While many uncertainties may be treated as a continuous range of possible values, some uncertainties may capture large-scale bifurcations or abrupt changes. For climate change, potential abrupt changes include melting of major continental ice sheets or shifts to some new mode of ocean circulation.[189] Large-scale bifurcations may also arise from breakthroughs in energy technology. Such possibilities are typically not captured in either historical data (because they are by assumption novel), or models (because they would represent a change in the causal structure represented in models).

Abrupt changes can pose particular challenges for deciding the number and range of scenarios to include in an assessment or decision-support exercise, either because their consequences are so extreme or because they would fundamentally change our understanding of how the system operates. The decision of whether and how to consider these uncertainties consequently turns on the balance between their probability – which is believed to be low but not well characterized – and their high consequences, which must be evaluated relative to the scenarios' intended use. This will be a particularly difficult choice when only a few scenarios are being generated. For example, in a coastal impacts assessment the enormous consequences of the difference between a half-meter and a five-meter sea level rise over this century – and the well-identified mechanism by which such a large rise could occur – may suggest the importance of explicitly considering a scenario involving loss of one of the major continental ice masses. But including such a scenario runs the risk that users will assign it a much higher probability than is appropriate, either because of its vividness and extremity or because they presume that developers' decision to include the scenario meant that they assigned high probability to it. When such a scenario is included, scenario developers have a serious responsibility to communicate, loudly and consistently, its different status.

A further challenge in representing large-scale or discrete changes in scenarios is that there might be many such possibilities, all of them high-consequence but believed to be unlikely. Including any particular one may mislead both by exaggerating its likelihood and by strengthening users' tendency to ignore others, when these all represent "unknown unknowns" that should receive some consideration. The more there are, the more the right approach might be to shift all scenarios further out to reflect the various mechanisms by which conventional understanding may under-represent the tail of the distribution, rather than highlighting any particular abrupt-change mechanism by giving it a scenario of its own.

4.6.4. Uncertainty in multivariate or qualitative scenarios

As the characterization of future conditions within scenarios grows more complex, so does the process of representing uncertainty within them. While many of the issues discussed above in the simplified context of scenarios on a single variable also apply to multi-dimensional scenarios, several additional issues arise.

The most basic of these is that with multiple dimensions of variation in scenarios, it is necessary to decide which uncertainties are represented. Even when scenarios include only multiple quantitative variables, it is no longer possible for a few scenarios to span all corners of the joint distribution of these variables. Rather, they must combine variations in ways that are most illuminating and important for the purpose at hand, massively reducing the dimensionality of the problem to make it intelligible for users. In addition, increasingly detailed and realistic scenarios often specify characteristics that are qualitative, or described less precisely than cardinal variables. For example, alternative scenarios might specify that current trends of globalization increase, stagnate, or reverse, or that decision-making capacity on climate change increases or decreases. Such characteristics may be judged crucial to include because they may be among the most important drivers of preferred choices or consequences of concern.

Scenarios of this kind pose substantial further challenges in representing uncertainty and interpreting its meaning. Relative to the simple quantitative scenarios we have considered up to this point, these lie in a much higher dimensionality space of future possibilities; they may not lie in any ordinal relationship to each other; and they may include characteristics whose definitional boundaries are not precisely specified. Defining a small set of scenarios to reasonably span the most important uncertainties is consequently even more difficult than for simple quantitative scenarios.

The approach most widely proposed to represent key uncertainties in such scenarios is to seek underlying structural uncertainties that satisfy two conditions: they appear to be most important in influencing outcomes of concern or relevant decisions; and they are linked with variation in many other factors. These underlying uncertainties can be simple discrete states such as peace or war, prosperity or stagnation; or, as in several major global environmental scenarios, they can be deeper societal trends, such as more or less globalization or shifts in societal values toward greater environmental concern, from which variation in many factors is assumed to follow.

This approach, formalized in the Shell scenarios method,[190] involves two steps: first identifying a small number of fundamental uncertainties and a small set of alternative realizations of each; and then elaborating additional future characteristics associated with each realization through both qualitative reasoning to fill in a narrative, and assembly of data and model results to build a parallel quantitative description to the extent this is judged useful. Repeated, critical iteration between the qualitative and quantitative elements is conducted, to bring additional relevant knowledge and expertise to bear and to check for consistency.

Even rich narrative multivariate scenarios must imply certain claims of likelihood. Every scenario included must be deemed likely enough to merit the resources and attention spent on developing and analyzing it. This applies even to extreme-event scenarios that are intentionally constructed to capture the low-probability tail of the distribution, since even they must be perceived likely enough to merit time and attention given their severity. Since users would reject any scenario that they persistently judged too implausible to consider, when decision-makers find a scenario exercise useful, it validates developers' judgment that each scenario was likely enough to consider.

The approach most widely proposed to represent key uncertainties in scenarios is to seek underlying structural uncertainties that satisfy two conditions:they appear to be most important ininfluencing outcomes of concern, and they are linked with variation in many other factors.

In a purely mathematical sense, any one specific rich multivariate scenario must be arbitrary and of vanishingly small probability. There are, however, ways in which it may be reasonable to assign non-zero probabilities to multivariate scenarios. First, if scenario designers in fact succeed at identifying a few deep structural uncertainties that strongly condition outcomes on many other characteristics in a scenario, then the richness of a scenario description need not imply that it is vanishingly unlikely. Whether this is true or not is a judgment to be made by scenario developers and users in each application. If they are sufficiently careful in their development and critical examination of scenarios, their judgment may well be correct. But there will often be no way to further test these judgments, so it is of course possible that the proliferation of additional detail in scenarios – even detail that developers and users recognize is crucial for determining valued outcomes and preferred choices – is arbitrary or erroneous.

A second way in which rich, detailed scenarios may be judged sufficiently likely to consider concerns the precision with which scenario characteristics are specified. In rich multivariate scenarios, many characteristics are often specified diffusely: economic growth may be merely "high" or "low," rather than being stated as a particular value. Even when a characteristic is stated quantitatively, its particular value may be treated as merely illustrative of a range of similar values; e.g., annual GDP growth might be set at 4 percent because a user needs a numerical model input, but it is understood to represent a broad range of similar values that all count as "high" growth. Interpreted in this way, a multivariate description may remain likely enough to merit examination – and indeed, a modest number of scenarios may exhaust the set of potential futures that matter for the issue at hand. Here one is not assigning likelihood to the precise numerical assumptions used to flesh out the details of a scenario, but rather to a thick slice of future conditions that resemble that scenario more than the other scenarios in the set.

The simpler the contents of scenarios, the more readily they lend themselves to explicit quantification of probabilities.

4.6.5. The debate over quantifying probabilities

A major debate in the use of global-change scenarios has concerned whether or not to specify quantitative probabilities associated with scenarios. This debate is central to the meaning and use of scenarios, and has been sharpest over the IPCC's SRES scenarios. Developers of the SRES scenarios decided at the outset of their process that they would make no attempt to assign probabilities to scenarios, in part because they were adopting the Shell approach of developing scenarios from storylines, in which quantitative probabilities are usually avoided. After the scenarios were published, several critics argued that since the most prominent and important outputs were the projections of emissions under the six marker scenarios, it was natural – and essential for development of rational climate-change policy – to describe the distribution of emissions in probabilistic terms. For example, how likely are 2100 emissions to lie above the 30 GtC of scenario A2 or below the 5.2 GtC of B1? Should the range spanned by all 40 SRES scenarios be understood to comprise 90 percent of all probability? 99 percent? All of it?

Developers of the SRES scenarios stood by their initial decision not to quantify probabilities. Since the controversy only became promi nent long after the decision had been made by a writing team no longer in operation, it would have been virtually impossible for the group to retrospectively assign such probabilities. But rather than rely on this argument of managerial infeasibility alone, SRES organizers offered a vigorous substantive defense of their initial decision. This defense relied in part on the statement that the six marker scenarios were all "equally sound," without providing any guidance regarding what this meant other than explicitly denying that it meant "equally likely." Describing each of the six marker scenarios as "equally sound" represents the entirely reasonable case that in the developers' judgment these all needed to be considered seriously – without making any further judgment as to their likelihood. While clearly frustrating to those wanting to use the scenarios as a basis for policy, the result is entirely consistent with the IPCC mandate to do assessment, but not to reach policy conclusions.

However, this debate will continue; it rests in part on different conceptions of the meaning and typical contents of a scenario. The simpler the contents of scenarios, the more readily they lend themselves to explicit quantification of probabilities. When scenarios consist only of alternative time-paths of a single quantitative variable, or one such variable is of predominant importance, it is straightforward and sensible to understand the intervals between those time-paths to have probabilities associated with them and there are several strong arguments for being explicit about these probabilities. First, stating probabilities allows comparative risk assessment between scenarios and explicit exploration of risk-reducing strategies.[191] Second, sophisticated decision-makers whose choices depend on uncertainty in these variables need probability information about possible values, not just a set of alternative values, to evaluate choices – whether their approach to decision-making is based on expected values, risk-aversion, seeking robust strategies, or some other approach. Finally, when such scenarios are presented without probability judgments, users may attach their own, often via simple heuristic devices that may misrepresent the developers' understanding. For example, many subsequent users of the SRES emissions scenarios have simply assumed the probabilities they needed to conduct further assessments, using such simple devices as counting scenarios or assuming a uniform distribution over the entire range.

Opponents of explicit quantification of probabilities do not dispute that such probabilities can coherently be assigned to simple scenarios in one or two quantitative variables. Rather, they raise principled objections to the appropriateness of attempting to quantify probabilities for more complex scenarios, particularly those involving socio-economic conditions, as well as practical objections to the use of probabilities even in the case of simple quantitative scenarios.

Many researchers are less comfortable using probabilities for complex scenarios that include explicit socio-economic elements than for uncertainties that are purely bio-physical, such as probabilities of different rates of climate change, conditional on a particular emissions scenario. Four main arguments are advanced against the use of probabilities for such scenarios.

First, some argue that the large multivariate space of possibilities from which such scenarios are drawn, and the vague and qualitative way that some scenario characteristics are specified, make it impossible to coherently define the boundaries of the outcome space to which probabilities are being assigned. There is no way to clearly define the interval "between" one scenario and another; and if probability is attributed to a slice of possibilities around each scenario rather than to the intervals between them, is it not possible to define clearly the boundaries of the slice to which the probability is assigned. To the extent that scenarios describe different types of worlds, which are distinguished from each other by alternative resolution of a few key uncertainties – e.g., high or low growth, high or low globalization – where the location of the boundary is not precisely specified, it may be difficult to create a shared understanding of these boundaries between users and creators. But if assigning a precise numerical probability is judged too difficult in these cases, less precise descriptions such as "highly likely," "more likely," "less likely," or "roughly equal" could be assigned. In some applications where scenarios are intended to capture all the uncertainty of concern to the decision-maker – i.e., scenarios are intended to be mutually exclusive and collectively exhaustive – there may even be a reasonable basis for numerical probability.

The second argument for rejecting probabilistic description of socio-economic conditions is based on "reflexivity" – the proposition that scenarios may influence the behavior or decisions driving the scenarios, so probability judgments about scenarios could reflect back on themselves, becoming either "self-fulfilling" or (more plausibly) "self-denying" prophecies. Section 4.1 addresses this issue in some detail, in particular in the distinction between how to treat mitigation decisions in scenarios to inform mitigation decisions and impacts or adaptation-related decisions. We might only add here that for scenarios of global emissions, reflexivity could only operate if both the influence of scenario judgments on their users' behavior and the influence of their users' behavior on global emissions were extremely strong. Moreover, it is not evident why scenarios with explicit likelihood judgments should raise this concern, while scenarios presented without such judgments – which also presume some claims of plausibility or likelihood – should not. Concern about reflexivity appears more serious for scenarios prepared in close consultation with national mitigation policy-makers, and it is for this reason (among others) that we judge explicit attempts to assign probabilities less valuable for scenarios prepared in such settings.

Concern about reflexivity appears more serious for scenarios prepared in close consultation with national mitigation policy-makers, and it is for this reason (among others) that we judge explicit attempts to assign probabilities less valuable for scenarios prepared insuch settings.

Third, some argue that it should not be scenario developers or experts who make judgments about likelihood of alternative scenarios, but users – particularly when scenarios are used to inform high-stakes public decisions. But this depends on the details of the content and use of scenarios. For some scenario elements in some settings, particularly use of scenarios to advise specific policy decisions, the scenario users may be as expert as the developers in associated uncertainties and risks, or more so. But in such settings, the use of scenarios normally highlights critical examination of these assumptions, and users have the knowledge and assertiveness to probe, critique, modify, or reject scenario elements that they find weak, including probability judgments. When scenarios are produced to serve many diverse users and consequently cannot rely on intensive interplay with representative, well-informed, and challenging users, scenario developers frequently have the best access to available knowledge relevant to forming probability judgments. Not making these judgments explicit is withholding information that users may need to understand and interpret the scenarios. If scenarios and their underlying reasoning and assumptions are presented clearly enough, users can make informed choices whether or not to use probability judgments that are provided.

Overall, we find the arguments in favor of quantifying probabilities to be strongest for scenarios whose major outputs are projections of one quantitative variable (or very few), and weakest for complex multivariate scenarios with substantial qualitative or narrative elements.

Finally, some argue that probabilities cannot be known, or even sensibly estimated, for socio-economic futures – perhaps because socio-economic processes and mechanisms are intrinsically less knowable than biophysical ones, perhaps due to the unpredictable effects of human creativity and leadership, and perhaps because causation does not operate in the human domain as it does in the bio-physical domain. Although these arguments raise deep philosophical questions, as a practical matter probabilistic projections are routinely done in some socio-economic domains, including projections of population and economic growth, but not, or not well, in others, such as projecting technological innovation. Provided the basic concept of subjective probability is accepted, weaker knowledge and deeper uncertainties can be accommodated by broadening the relevant uncertainties rather than declining to make probabilistic judgments, but the question remains of whether the resultant broad uncertainty ranges are meaningful or operationally useful.

Several practical objections have also been raised to associating explicit likelihood judgments with scenarios. These include the difficulty of developing probability estimates from multiple information sources that can achieve sufficient agreement from diverse experts, and the non-intuitive nature of probability distributions in using scenarios to communicate with non-expert users. These are both valid concerns, although active areas of research and development in expert elicitation techniques and in simple intuitive devices to communicate uncertainty are making some progress in mitigating them.

An additional practical argument against quantifying probabilities is that attempting to do so may represent a distraction that uses time, generates conflicts, and is of little value to scenario users. Whether this is indeed the case, however, is in part a judgment to be made by scenario users, not developers. Opponents of quantified probability argue that users typically only need scenarios to pass some probability threshold. Beyond this threshold, they will seek robust choices that yield acceptable outcomes under all possibilities, so further refinement of probability serves no purpose. This argument has merit, but only to the extent that it accurately describes how these scenarios will be used. Quantitative assignment of probabilities to scenarios when high-stakes decisions are implicated is clearly difficult and contentious, as the SRES controversy illustrates. Even if this argument correctly characterizes how scenarios are used, users might still be able to profitably exploit more detailed probability information if it were available – although one must also consider the risk that non-technical users might somehow be more likely to misunderstand scenarios with explicit probability judgments attached (perhaps by taking a stated probability distribution as the "true" distribution) than to misunderstand a simple collection of scenarios presented with no such probability information (perhaps by taking the range presented to embrace the totality of all possibilities). It is also possible that engaging scenario users in an attempt to assign probabilities, even only illustratively, could both draw on relevant knowledge of uncertainties that they possess more than scenario developers, and provide a valuable device to probe and sharpen their understanding of the situation. Any argument based on the information needs of specific users becomes less persuasive as the set of potential uses and users, and the likely diversity of their information needs, grow larger.

Overall, we find the arguments in favor of quantifying probabilities to be strongest for scenarios whose major outputs are projections of one quantitative variable (or very few), and weakest for complex multivariate scenarios with substantial qualitative or narrative elements. The controversy over probabilities in SRES reflected in part different perceptions of what type of scenarios these were. SRES initially followed a storyline-based process and rejected quantification of probabilities on that basis. Subsequent efforts, however, consisted predominantly of developing quantitative emissions projections and neglected further development of the storylines. Moreover, with a few significant exceptions, subsequent applications of the scenarios have principally used their emissions figures, sometimes together with population and GDP, and made little or no use of the underdeveloped storylines that lay behind them. The controversy over quantitative probability in this case suggests that when quantitative projections are a major output of a scenario exercise, developers may have a responsibility to go further in characterizing the likelihood of the resultant emissions intervals than would be appropriate for the more complex underlying storylines.

Moreover, even for rich narrative scenarios, the arguments against rendering probability judgments are strongest when the exercise is produced for a small number of users with similar responsibilities and concerns. In such a setting, intensive interaction between scenario developers and users can provide whatever additional detail about, or confidence in, the scenarios that users may require to benefit from the scenarios. When scenarios serve potential users who are more numerous and diverse, however, such intensive interaction is not possible. As a result, the value of explicit likelihood judgments increases. To the extent that future global-change exercises continue to strengthen their qualitative aspects and the integration between qualitative and quantitative – valuable directions for future efforts – they should still seek to move further toward explicit characterization of likelihood than has been done thus far.

BOX 4.8. THE GLOBAL BUSINESS NETWORK ABRUPT CLIMATE CHANGE EXERCISE

In 2002, the Office of Net Assessments (ONA), a small strategic planning office in the Office of the US Secretary of Defense, asked the Global Business Network (GBN), a consulting firm expert in scenario methods, to develop a scenario of potential national-security implications of abrupt climate change. This request was stimulated by widespread scientific interest at the time in abrupt climate change, particularly shifts in North Atlantic circulation, including a 2002 report by the National Academy of Sciences.[192] In addition, several scientific papers had reported changes in Atlantic circulation and salinity that some scientists thought might indicate impending larger disruption, as well as new evidence of rapid climate shifts in the past.[193]

GBN staff developed the scenario by reviewing scientific literature and informally consulting with climate and ocean scientists.[194] They reviewed three past climate events of diverse severity and decided to base their scenario on the one in the middle, the century-long period of strong cooling 8,200 years ago. Coming after an extended warm period, this event brought cooling of about 5 °F over Greenland, with cold and dry conditions extending around the North Atlantic basin and substantial drying in mid-continental regions of North America, Eurasia, and Africa.[195]

For their future abrupt-change scenario, the authors constructed a path of climate change to reach these conditions by 2020. The pathway involved rapid warming through 2010, as high as 4 – 5 °F per decade in some regions,[196] followed by a rapid turn to cooling around 2010, as melting in Greenland freshens the North Atlantic and substantially shuts down the thermohaline circulation. By 2020, hypothesized conditions have approached those of the 8,200-year event – cooling of 5 °F in Asia and North America and 6 °F in Europe, with widespread drying in major agricultural regions and intensification of winter storm winds. The authors acknowledge that the scenario pushes the boundaries of what is plausible, both in the rapidity of changes and in the simultaneous occurrence of extreme changes in multiple world regions. They contend that this is defensible and useful, however, for an exercise focused on sketching the nature of challenges posed by a plausible worst case.[197]

The socio-economic and security implications of the climate scenario were developed judgmentally, in consultation with ONA. Incremental changes are projected for the first 10 years, with general increase in environmental stresses and approximate maintenance of present disparities between rich and poor countries. After 2010, catastrophic cooling in Europe and drying of major agricultural regions worldwide brings widespread shortages of food, due to decreased agricultural production; of water, due to shifted precipitation patterns; and of energy, due to shipping disruptions from increased sea ice and storminess. These shortages produce 400 million migrants over the period 2010-2020, as desperate scarcity generates violent conflict in Europe, Asia, and the Americas. Extending their speculation on security implications into the 2020s, the authors hypothesize widespread southward migration of Europeans and near-collapse of the European Union, sustained conflict in East and Southeast Asia including struggles between China and Japan over access to Russian energy supplies, and increasing political integration of a fortress North America to manage security risks and refugee flows.

Controversy and Criticism

The project was completed in October 2003, its report published in February 2004 and reported in Fortune Magazine the same month.[198] A few weeks later, the *London Observer* claimed to have obtained the report secretly and used the scenario to criticize US refusal to join the Kyoto Protocol.[199] Subsequent news coverage took up the theme that the report was secret or suppressed, and suggested the reason was that the scenario called for more urgent action on climate change.[200] In the resultant controversy, ONA stated correctly that the report did not represent US policy, but was merely a speculative consultant's study. Although the controversy subsided after a few weeks, interest and concern about the possibility of abrupt climate change, although not of this precise character, have continued to grow.[201]

This scenario is a sketch of an abrupt climate-change event, with little fine-scale detail about the hypothesized changes or underlying reasoning and no attempt to suggest how likely or unlikely such an event is. Rather, it seeks a preliminary answer to the question, what might the worst case look like? Such questions are more often posed in security studies than other fields, because of the unique nature of responsibilities of military organizations – responding to diverse, novel, unknown threats with extremely high cost of failure. Many climate-change decision-makers could likely benefit from such upper-bound scenarios too, but this exercise is the only example of a worst-case scenario produced for climate change. Major official assessments have focused overwhelmingly on average or best-guess projections.

But the response to this chapter vividly illustrates the risks of worst-case or extreme scenarios. Produced in consultation with a sophisticated user – and in this case, one closely connected to senior decision-makers – who thoroughly understands the outer-bound nature of the underlying assumptions, they can be valuable devices for preliminary risk assessment and threat identification. But in a wider and polarized policy debate they are hard to explain and may be misunderstood or misrepresented. Attempting to manage the process through secrecy appears counterproductive, foregoing the potential value such analyses could provide to multiple decision-makers. More promising might be to integrate extreme-case scenarios explicitly into analyses that also present multiple midrange scenarios.

SECTION 5. CONCLUSIONS AND RECOMMENDATIONS

This section presents our conclusions regarding the present state of development and use of scenarios for climate-change applications, and some recommendations for specific changes or initiatives to advance current practice to make scenarios more useful.

Before doing so, we briefly reprise some key definitional points, because uses of the term scenarios are so divergent. We have defined scenarios as descriptions of future conditions produced to inform decision-making under uncertainty. This definition distinguishes scenarios from assessments, models, decision analyses, and other decision-support activities. Scenarios may be developed and used in conjunction with these – for example, scenarios can provide descriptions of potential future conditions used as inputs to such activities – but are not identical to these, and not alternatives to them.

We have also distinguished scenarios from other types of future statements intended to inform decisions, such as projections, predictions, and forecasts. Relative to these, scenarios tend to be more multivariate (but still schematic), tend to be developed in groups, and tend to presume lower predictive confidence. The last condition is the case in part because scenarios tend to be used in situations where the basis for forecasting is less established because of deeper uncertainties, or for situations that pertain to further in the future beyond the range for which there is high confidence in specific projections, even contingent ones.

Having distinguished scenarios from these related activities, we consider a broad set of scenarios of diverse characteristics and uses, including simple and complex scenarios, quantitative and qualitative scenarios as well as various combinations of the two, and scenarios whose primary use and interpretation is positive or normative. Where we intend our conclusions and recommendations to apply to only certain types or uses of scenarios, we state this explicitly. Unless stated otherwise, they pertain to all types of global-change scenarios we are considering.

5.1. Use of Scenarios in Climate-Change Decisions

Scenarios can make valuable contributions to climate-change decision-making. Many of the decisions that will comprise society's response to climate change – whether mitigation, adaptation, or other responses – involve high stakes, deep uncertainties, and long time horizons. Scenarios can help inform these decisions by structuring present knowledge and uncertainty, prompting critical examination of present assumptions and practices, stimulating new insights, identifying key pitfalls and opportunities, or providing a framework for the assessment of particular decisions. For some decisions, which involve irreversible near-term commitment to choices whose consequences extend over a horizon involving substantial uncertainties, some form of scenario-based reasoning may be essential.

There is a big gap between the use of scenarios in current practice and their potential contributions. Despite their evident value and capability, many climate-related decisions that could benefit from scenarios (e.g., many decisions regarding long-term management and investments in climate-sensitive areas such as freshwater systems or coastal zones) are not using them. Indeed, many such decisions are still being made without considering climate change at all. Conversely, many climate-change scenarios have only weak and indirect connections to practical decisions related to climate-change mitigation or adaptation.

Interest in considering and using climate-change scenarios is sharply increasing. There is increasing interest in considering climate-change scenarios in diverse decision and planning processes. This trend is strongest for planning and decisions concerned with climate-change impacts and adaptation. The trend reflects advances in scientific understanding of climate change, gradual maturation of models and analytic tools, and increasing recognition by decision-makers of the potential importance of climate change. Given the high general concern about climate change and the advance of background scientific knowledge, we expect this trend to continue, and to broaden to other types of climate-related decisions.

Scenarios of global emissions and resultant changes in atmospheric trace-gas concentrations and climate are a core requirement shared by many diverse climate-related decisions. Although climate-change decision-makers and their particular information needs are highly diverse, many will need scenarios of global emissions and resultant climate change, and many more will need information that depends on these. Consistent scenarios of global emissions and climate change, provided centrally at the national or international level, can serve these diverse needs – if they are presented with enough transparency and documentation of their underlying reasoning and assumptions.

Beyond global emissions and resultant climate change, decision-makers' needs from scenarios are highly diverse. Different climate-change decision-makers will have highly variable needs from scenarios, in the factors and variables included, the time and spatial scale at which they are provided, and the nature of uncertainties represented. The means for meeting these additional needs will likely be diverse, too. Some will call for separate, specialized scenario production capabilities. A major distinction in scenario-related needs can be drawn between impacts and adaptation managers, mitigation policy-makers, and energy resource and technology managers.

Impacts and adaptation managers need scenarios that project impacts relevant to their specific responsibilities, and the major determinants of vulnerability and adaptive capacity. Impacts and adaptation managers include both national officials and others responsible for more specific domains of impact. These decision-makers need climate-change scenarios, driven by specified global emissions scenarios, to provide information about potential climate-related stresses on their areas of responsibility. In addition, they need other environmental and socio-economic information specific to their areas of responsibility, at appropriate spatial and temporal scales. Meeting these needs will require both easy access to centrally produced climate scenario information with associated tools and support, and development of decentralized capabilities for developing and applying additional scenario-related information. Many of these specific information needs are likely to be similar in character for many particular locations and types of impact.

Meeting information needs for impacts and adaptation requires a cross-scale organizational structure. These decisions' combination of centralized and decentralized information needs suggests the need for a linked network of institutions at national and sub-national levels to develop scenarios. Such a structure would combine central provision of globally consistent emissions and climate change scenarios; decentralized elaboration of these scenarios with additional variables required for regional impacts and adaptation analyses; and provision of tools and resources to support development and use of scenarios.

Scenarios for impact and adaptation managers should be based on emissions assumptions that include a likely range of mitigation interventions, now and in the future. The emissions assumptions underlying scenarios for impacts managers should be based on the likely range of future global emissions trajectories, including explicit assumptions about what degrees of further mitigation effort are likely over time. This will typically imply a narrower range of emission futures than is considered in scenarios to support mitigation decisions.

Mitigation policy-makers need scenarios that project alternative emissions trends in their own jurisdiction and others, and the major factors that will influence mitigation opportunities, constraints, and costs. Mitigation policy-makers are usually officials who make national policy and participate in international negotiations, but this group also includes sub-national officials when they share mitigation responsibilities or undertake mitigation initiatives. Serious pursuit of greenhouse-gas mitigation will require major policy innovations that carry significant risks of many kinds, including the effectiveness and cost of the policies but also their effects on government budgets, competitiveness of particular industries, opportunities for national technological capabilities, etc. Decision-makers considering such policies will need scenarios of global and national emissions trends, resultant climate change, and aggregate impacts. In addition, they will need to consider many factors specific to their jurisdictions – e.g., national policies, institutions, economic structure, technological capabilities, and the detailed structure of national emissions – and information about the policy environment for their choices, including alternative scenarios of other nations' mitigation strategies, international mitigation decisions, and implementation and compliance.

Scenarios for mitigation decisions should include a wide range of baseline emissions assumptions and should not prejudge the likely level of mitigation effort. Scenarios used to inform mitigation decisions should consider the full range of potential mitigation choices on the agenda, defined relative to baseline assumptions that, as much as possible, reflect only efforts already enacted or committed, including a range of reasonable assumptions about implementation and compliance. This assumption typically implies a wider range of emissions futures than is considered in scenarios to support impacts and adaptation decisions.

Mitigation decision-makers can use target-driven scenarios for backcasting. Mitigation decision-making may also benefit from scenarios that impose explicit future environmental targets such as limits on emissions or atmospheric concentrations, together with assumptions about policy and implementation elsewhere, and reason backwards to explore alternative paths to, and implications and requirements of attaining that goal, including feasibility, costs, and tradeoffs. These must be defined in ways relevant to the level of decision-making being informed, i.e., alternative national targets to inform national policy-making, in the broader context of alternative global baselines or global targets.

Informing mitigation decisions requires capacity for scenario development at the national level. While core scenarios of global emissions and climate-change can provide some of the required input into mitigation decisions, these decisions require additional information that must be provided at the national or sub-national level where the decisions are being considered, generated in consultation with relevant decision-makers or their surrogates.

Energy resource and technology managers need scenarios that represent the political and economic environment for energy investments, including mitigation policies. Energy resource and technology managers concerned with private responses to mitigation policy primarily need scenarios that represent alternative policy regimes. Emissions and climate change underlie these as influences on policy decisions, but do not capture the most important uncertainties for these decision-makers. While many actors may wish to generate these scenarios privately to keep their assumptions and analyses confidential, there may also be value in multi-party collaborative scenario-building exercises in which today's policy-makers and corporate planners jointly examine what range of policy, economic, and energy regimes is plausible or likely over the 30- to 50-year time horizons relevant for investment and technology-development decisions.

Scenarios must be periodically revised and updated. For all types of decisions and decision-makers, developing scenarios, applying them to inform decisions, and refining scenario methods, are iterative processes. Limitations to present scenarios or methods do not in general justify delaying consideration of such decisions, any more than scientific uncertainties do. Still, scenarios must be periodically updated, based on new knowledge, experience, and priorities, as well as further developments in scenario-related methods. Such updates are needed much more frequently than the decision time horizons.

5.2. Use of Scenarios in Climate-Change Assessments

Large-scale, official assessments are the major use for scenarios at present and are likely to remain an important use. Large-scale, official assessments represent the most prominent demand for climate-related scenarios at present, and are likely to remain major users, particularly for coordinated scenarios of global emissions and resultant climate-change.

Within assessments, scenarios mainly serve to support further analysis, modeling, and assessment. When scenarios are used in assessments, certain users are clearly identified: e.g., climate modelers are major users of emissions scenarios, while impacts assessors and modelers are major users of climate-change scenarios. These users have specific scenario needs, and close consultation is possible between scenario producers and users to meet these needs. Substantial progress has been made in providing useful scenarios for these groups, at both the national and international level. These efforts should be continued and expanded.

The presentation of scenarios in assessments leads to many additional, unforeseen uses. Scenarios presented in large-scale assessments gain prominent dissemination that results in their being put to many uses their developers did not foresee. Scenarios should pursue clarity of documentation and transparency about underlying reasoning and assumptions, to improve the ease of use and reduce the risk of misunderstanding in such derivative uses, although they cannot anticipate all information needs of an open-ended set of diverse potential uses.

In assessments, scenarios can strongly in fluence issue framing. Also because of their prominent dissemination, scenarios presented in major assessments can exercise substantial influence over the framing of policy discussions or provide simple, widely used metrics of the seriousness of the issue. They may consequently exercise broad influence over many decisions that depend upon such aggregate perceptions of seriousness. The prospect of such influence further heightens the responsibility for transparency in production of scenarios.

Scenarios contain unavoidable elements of judgment in both their production and use. Although they draw on relevant data, knowledge, and analysis, scenarios inevitably contain elements of judgment. In addition to putting serious responsibilities onto scenario developers, this implies that there is no authoritative way to resolve arguments over whether a scenario is plausible or not. When a wide enough range of potential futures is considered, some scenarios are likely to draw criticism, in part motivated by opposition to their foreseeable implications for action. Any scenario can be attacked as unreasonable, speculative or unlikely, and close enough scrutiny of any scenario can usually reveal inconsistencies, but these do not provide sufficient basis for excluding a scenario from consideration. Indeed, scenarios designed to represent extreme events, or to lie near one end of a distribution of potential outcomes, should by definition appear unlikely. The most productive response to such criticisms lies in transparency about the process, reasoning, and assumptions used to produce scenarios. Such transparency can shift arguments to underlying uncertainties, and help limit biases in the production of scenarios.

5.3. Characteristics of "Core" Emissions and Climate Scenarios

Centrally provided scenarios of emissions and climate change can help inform mitigation and adaptation decisions at national and sub-national scale, but these will usually require additional information as well. Central scenarios can provide information about trends in world emissions, underlying socio-economic conditions at the scale of major world regions, and the large-scale pattern of global policy response. They can also provide access to climate-model scenario output, plus tools, data, and support for producing finer-scale scenario information needed for particular impact and adaptation applications. Mitigation and adaptation decisions and associated assessments at national or smaller spatial scale will need more detailed and finer-scale climate and socio-economic information than can be provided by centralized scenarios, so these must be extended and/or modified by national and sub-national scenario processes.

Scenarios of emissions and resultant climate change should be global in scope and century-scale in time horizon. Core emissions and climate-change scenarios should specify major climate-relevant emissions and other perturbations, globally and for major world regions. They should extend over a time horizon of at least 100 years (including some that extend 200-3 00 years to support assessments of sea level rise), with interim results at roughly decadal resolution.

Emissions scenarios of several distinct logical types will be needed to serve diverse purposes. These will include some combination of alternative baselines, alternative levels of incremental stringency of mitigation effort, and specified future targets to support backcasting and feasibility analysis.

For some uses, emissions scenarios should be coupled to explicit scenarios of alternative socio-economic futures. For these scenarios, the range of potential socio-economic and policy futures considered should be wider than has been considered to date, including scenarios of policy failure and conflict, and a wide range of stringency and timing of mitigation effort. For example, what if development stagnates in major world regions? What if world emissions grow sharply for several decades with little control effort, followed by a subsequent shift to stringent mitigation efforts? What if part of the world makes a lot of effort and part makes very little? Considering such varied future histories is crucial for considering long-term risks and opportunities from major mitigation choices.

Scenarios should reflect various explicit degrees of coordination, depending on their intended uses. Some uses will require groups of simple coordinated scenarios to provide standardized inputs for downstream modeling and analysis – e.g., standard emissions scenarios as inputs to climate models and standard climate scenarios as inputs to impact assessments – for exploring present uncertainties and tracking developments of knowledge over time. Other scenarios should be based on multiple models using common input assumptions. Non-standardized scenarios produced at the initiative of researchers and modelers should also be produced, which explore alternative assumptions or meet specific user needs, provided these meet basic standards of quality control, transparency, and documentation.

Some scenarios should seek to link qualitative and quantitative elements. Some scenarios of socio-economic conditions, whether produced to support global emissions scenarios or impacts assessments, should include both qualitative and quantitative elements and sustained analytic efforts to link the two. Qualitative or narrative scenario elements can provide a vehicle to explore major historical uncertainties with large implications for global emissions, climate change, and vulnerability to climate impacts; provide a coherent rationale and logical structure to connect assumed trajectories for multiple variables, including both quantitative and qualitative ones; and provide guidance to other analysts or users who may wish to extend the scenarios by elaborating additional detail. Achieving these benefits will require more sustained effort to integrate model-based projections of quantitative variables with qualitative and narrative scenario elements, to iterate between these, and to critically examine each element in light of the other, than has been made thus far. These efforts should seek to connect alternative qualitative and narrative scenarios not just to alternative parameter values in quantitative models, but also to alternative forms of causal relations and model structures. Generating multiple alternative model quantifications based on the same narrative and associated causal logic is one promising route to enriching understanding of uncertainties in key quantitative variables such as future economic output and emissions.

5.4. Scenario Process: Developer-User Interactions

There is value in collaboration between scenario developers and users, particularly at the beginning and ending stages of a scenario exercise. The appropriate degree and means of this collaboration vary substantially among scenario exercises. User engagement is most important in the initial scoping and design of a scenario exercise, and in the evaluation and application of the scenarios generated. The value of user engagement in details of scenario development, quantification, elaboration, and checking, depends on the specific case.

The ease of achieving such collaboration and its value are likely to be greater when scenario users are clearly identified, few in number, and similar in their interests and perspectives. When potential scenario users are identified, relatively few, and relatively homogenous, close and intensive collaboration between users and developers is likely to be most productive. When potential users are numerous and diverse, intensive engagement may be infeasible and more structured processes for consultation, representation, and information exchange are needed. While progress has been made in new methods to allow larger numbers to participate in scenario exercises, further development of such methods is needed.

5.5. Communication of Scenarios

Effective communication of scenarios is essential, in forms useful to audiences of diverse interests and technical skills. Scenarios must be communicated effectively to their potential users, including both technical and non-technical audiences. In addition to the contents or outputs of scenarios, communication should include associated documentation, tools, and support for their use. Various methods should be used to promote broad dissemination of scenario information; for instance, presentations, reports, websites, and centralized data distribution centers. To facilitate user understanding of results, various methods should be used to communicate numerical and technical information, including multiple tabular, summary, and graphical formats, ideally with user-interactive capabilities.

Transparency of underlying reasoning and assumptions is crucial. Scenario communication should include transparent disclosure of underlying assumptions, models, and reasoning used to produce the scenarios, to support the credibility of scenarios, to alert potential users to conditions under which they might wish to use or modify them, and to inform criticism and improvement of scenarios. This should include explicit identification of the major uncertainties represented in each scenario and the sources of underlying information, whether drawn from the scientific literature, formal expert-elicitation exercises, or informal judgments of the scenario team. It is possible in virtually all cases to formulate simple, accessible, honest descriptions of why a scenario was undertaken, why it was necessary, what was done, how and why, and why it merits respect as a reasonable judgment.

5.6. Consistency and Integration in Scenarios

Any scenario should be internally consistent in its assumptions and reasoning, to the extent this can be established given present knowledge. Carefully pursuing consistency within individual scenarios can be an intensive and time-consuming process, but is crucial to avoid problems that can discredit a scenario exercise.

In scenario exercises that use multiple models to explore potential uncertainties in future conditions, consistency between models should be pursued primarily through coordination of inputs, not outputs. Use of multiple models in parallel to produce alternative descriptions of future conditions can improve understanding of uncertainties, if models are run under consistent assumptions about exogenous inputs. Forcing models to generate consistent trajectories for endogenous outputs poses several risks, including suppressing variation from alternative causal structures that could provide valuable insights into uncertainties, and encouraging over-confidence from spurious precision. For quantities that are exogenous in some models and endogenous in others, the appropriate treatment varies case by case, but it is not generally desirable to force multiple models to convergent values of such variables without more detailed examination of the underlying uncertainties.

Imposing consistent outputs in multi-model exercises can be useful, however, when these outputs represent common goals for policy evaluation. For example, consistent constraints on some environmentally relevant target such as emissions, atmospheric concentrations, or radiative forcings, can be used to examine inter-model uncertainties in the technological, economic, and resource conditions associated with meeting the specified targets.

Transparency in reporting scenario and model differences as well as underlying assumptions and reasoning can help mitigate the effects of inconsistencies among scenarios. Ideally, multiple scenarios in an exercise should differ only on those elements intentionally chosen to distinguish them, and be consistent in all other factors. However, this is not always possible, particularly when scenarios are produced using different models. Pursuing maximal transparency about the models, assumptions, and reasoning underlying each scenario perhaps by issuing detailed diagnostic reports that include explicit discussion of points of weakness, uncertainty, and disagreements, and the means used to resolve them – can mitigate any resultant confusion.

5.7. Treatment of Uncertainty in Scenarios

Some scenario exercises should include more explicit characterization of likelihood judgments than has been practiced so far. The advantages of being more explicit about the probability judgments that underlie scenario exercises are likely to outweigh the disadvantages. Such specification should be pursued further than has been done in major global-change scenario exercises to date, although not necessarily in all scenario exercises. The means available to express these judgments are of widely varying specificity, ranging from agreed terminology[202] to explicitly quantified probability distributions. All such judgments should include explicit acknowledgement of their inevitably subjective elements and appropriate caveats to help users avoid mistaking them as objectively true.

Explicit probability judgments are easiest to produce and least controversial in scenarios generated using quantitative models of climate change or specific impact domains. Scenarios generated using such models can be conditioned on specific assumed values for socio-economic inputs such as emissions, and can represent explicitly and quantitatively the effects of specified variation in initial conditions or unknown parameter values. These devices are also available, although in less widespread use, in economic models used to project emissions. These devices aid in constructing distributions of key quantitative characteristics, such as measures of global or regional climate, or of prominent quantitative impact measures, such as changes in river flows or sea level, although they neither capture all relevant uncertainty nor avoid the inevitably subjective nature of such probability judgments. Explicit probability judgments are more difficult and controversial when they involve socio-economic factors for which quantitative models are not available, and that do not depend in well understood ways on identified quantitative parameters. Such factors include major technological innovations, large-scale changes in attitudes or norms, or policy response.

Attempting to include explicit probability judgments is likely to be most useful and successful when key variables are few, quantitative outcomes are needed, and potential users are numerous and diverse. The case for assigning explicit likelihood judgments is strongest when scenarios' most salient components are quantitative projections of a few key variables, such as emissions or average temperature change over the globe or some region, because the technical barriers to assigning probabilities are least severe in this case. The case is strongest when a primary purpose of the scenario exercise is to provide inputs to other quantitative assessment activities, or to inform decisions that primarily depend on one or a few key quantitative variables, because such uses are most likely to require probability judgments. The case is strongest when the set of potential scenario users and uses is large and heterogeneous, because this situation provides the least opportunity for informal communication of implicit judgments of likelihood or priority through intense, sustained collaboration between scenario developers and users.

Attempting to include explicit probability judgments is likely to be least useful and successful when scenarios specify multiple characteristics, including prominent narrative or qualitative components; when the purpose of a scenario exercise is sensitivity analysis or heuristic exploration; and when potential users are few, similar, and known. When scenarios are primarily construed as rich, qualitative narratives that present major alternative historical and socio-economic trajectories, the technical obstacles to explicit probability assignment are greatest and the likely confidence in scenario developers' subjective probability judgments lowest. When the main purpose of a scenario exercise is to stimulate critical or creative thought, to probe the limits of a model or decision strategy through sensitivity analysis, or to explore ways of meeting a specified target, explicit probability assignment provides little or no benefit. When users are few, similar, and specifically identified, they can be intensively involved in scenario production, allowing effective informal communication of likelihood judgments without stating them explicitly. Under these conditions, scenario exercises can also be structured to engage users in the potentially instructive activity of assigning and discussing their own probability judgments, rather than putting that responsibility exclusively on the researchers or analysts developing scenarios.

Centrally provided scenarios of global emissions and climate change should attempt to include explicit probability judgments. Because of the large, diverse set of users for these scenarios, explicit likelihood judgments should be provided for ranges of key quantitative outputs, including global emissions and global-average temperature change. Scenarios should typically include several paths that span a wide range of judged uncertainty, e.g., 95 percent to 99 percent. The associated probability judgments may include some that are unconditional and some that are conditioned on specific assumed prior conditions. Such estimates should be provided by multiple groups using diverse methods. As for all such probability judgments, their unavoidably subjective nature and the specific assumptions on which they are conditioned should be stated explicitly and prominently.

Providing explicit probability judgments allows scenario users to choose whether to use them or not. Some users may choose to use these explicitly in their subsequent analysis or decision support, while others may use them only to help decide which scenarios to use, and still others may disregard them entirely. Users may select a different group of scenarios or a different subset of the uncertainty range for various reasons, including differences in risk aversion, differences in the scope of their decision authority, or differences in their assumptions about decisions by other actors (present or future). Presenting complete descriptions of scenarios together with underlying assumptions and reasoning, including probability judgments, preserves all these options for users.

Scenario exercises should give more attention to extreme cases. Some uses of scenarios require consideration of low-probability, high-consequence extreme cases, such as loss of a major continental ice sheet or major changes in meridional ocean circulation. Consequently, such scenarios should be included in large, general-purpose scenario exercises producing emissions or climate-change scenarios, together with more likely middle-case scenarios. When extreme scenarios are included in an exercise, it is especially critical to be explicit and transparent about the reasoning and assumptions underlying each scenario, and scenario developers' judgments of relative likelihoods.

In addition to enhancing the utility of scenario outputs, probabilistic methods can contribute throughout the scenario development process. Developing scenarios requires making many judgments about unknown characteristics and developing many arguments and pathways to link these. Scenarios based on quantitative models typically require specifying many exogenous inputs and parameters. Even narrative scenarios require specifying values of multiple characteristics, both qualitative and quantitative. Explicit discussion of uncertainties and associated probabilities can help structure and facilitate many aspects of the scenario development process, including deciding appropriate ranges of variables to consider, defining boundaries of what outcomes are considered plausible, elaborating associated causal mechanisms and linkages, discussing and integrating knowledge and judgments from multiple participants, and clarifying disagreements. Explicit conversation about probabilities can support insights throughout these processes, in addition to supporting communication of scenario judgments to users.

5.8. Expanding and Sustaining Capacity for Production and Use of Scenarios[203]

Present scenario capacity is inadequate. Although scenario-based reasoning is required for many aspects of global change assessment and decision support, the present capacity to produce, disseminate, apply, evaluate, and adapt scenarios is inadequate. There has not been enough continuity to enable effective learning, because scenarios are typically produced *de novo* for each major application. There has not been enough transparency about methods, reasoning, and assumptions. Constructing integrated scenarios and exploring alternative methods has been difficult, in part because scenario exercises have tended to be dominated by use of quantitative models, separated along disciplinary lines. Inadequate resources have been devoted to methods development, for scenarios and related decision-support tools Finally, there has been no systematic evaluation and critique of scenarios or their application.

To Help Fulfill these Presently Unmet Needs, CCSP Should Establish a Program to :

- ***Commission scenarios for use in assessments and decision-support activities.*** This task includes facilitating agreement among relevant producers and users on standard scenarios in cases where multiple assessment activities require standard scenarios for comparability, and convening and supporting a diverse collection of more extensive and detailed scenario-related exercises, by multiple groups using a wide range of models and approaches.
- ***Disseminate scenarios with associated documentation, tools, and guidance materials.*** This task includes multiple forms of support and program-building for diverse groups seeking to apply, modify, and extend existing scenario-based information at various regional and sectoral scales, through providing data, models, tools, resources, and associated documentation and technical support, in multiple forms and through multiple media.
- ***Commission various groups to evaluate scenarios and their applications and to develop improved methods.*** This task includes defining and promulgating standards

for quality control – which, given the need for diverse approaches, would principally concern matters of process such as transparency, documentation, evaluation, and dissemination of results and supporting information. The task also includes broad efforts to develop scenario-related skills, tools, and methods, e.g., by providing resources for methods development and evaluation projects; conducting and establishing procedures for evaluation of particular activities and programs; and convening workshops, conferences, etc., to evaluate progress overall, or in particular methodological areas.

- ***Archive results and documentation related to all these tasks, to provide historical perspective and institutional memory for future scenario-related activities.*** This task includes preserving for retrospective scrutiny a wide range of materials: not just the methods, contents, and results of scenario exercises, but also the progressive evaluations of particular activities and approaches, and of the entire program. In its ongoing development and evaluation of methods, the program should not draw rigid boundaries between scenario development and application and other related methods of assessment and decision support for long-term global change issues.

Several conditions in the design and management of this new program would be required to ensure its effectiveness.

- ***The program should build and maintain strong connections with outside relevant expertise, and analytic and modeling capability.*** While the program must develop enough internal expertise to be a full participant in debates over scenarios and assessment methods, it cannot and should not attempt to impose a unilateral vision of preferred scenarios, methods, or approaches. Rather, it must build and maintain close collegial connections with outside networks of researchers and analysts in multiple fields of expertise, including emissions modelers, climate scientists and modelers, impacts researchers, and resource managers. These relationships would be facilitated by establishing governing mechanisms, such as a senior advisory board, drawn from the broad communities of researchers, modelers, and analysts who are developing and using scenarios and related methods. Although established as a US national program, it must also support, collaborate, and coordinate with parallel activities in other nations and internationally, and with relevant sub-national activities.
- ***The program should integrate and balance goals and criteria related to scientific and technical quality, and those related to utility and relevance to users.*** This balance is needed for the program to support promising but speculative activities, to encourage creativity and diversity of approach, to avoid being captured by any particular discipline or modeling approach, and to be willing to make and explain judgments about quality and promise that reflect both scientific and practical considerations. To achieve this, the program needs broad discretion over the type of projects supported, including sponsoring fairly sharply targeted activities, supporting speculative activities, and investing to develop and assess capabilities that do not yet exist.
- ***The program should be insulated from political control.*** For the scenarios and analyses based on them to be perceived as credible by their diverse users, the

program needs enough insulation from political control, at both the national or international level, to prevent scenarios from becoming proxies for conflict over near-term policies, and to allow exploration of the implications of alternative futures that represent plausible risks but that some political actors would find objectionable.

- ***The program should strive for maximum transparency in its own activities, in addition to demanding it from activities it supports***. The program should strive for maximal transparency regarding inputs, models, assumptions, and reasoning employed in developing scenarios, as well as any significant disagreements that arose and how they were resolved and any remaining weaknesses recognized by the developers. The broader and more diverse the collection of intended uses and users, the more crucial is transparency of the scenario-production process – because different users may require scenarios produced using different underlying assumptions, and they must be able to track the underlying logic to exercise this choice. This would enhance credibility in the scenario-development process. While calls for such transparency are widely made, experience suggests it is difficult to achieve, particularly for such matters as disagreements or recognized weaknesses that may risk professional embarrassment. Still, achieving more transparency and more widely informed debate on such matters is essential for advancing scenario methods.
- ***The program will require the authorityand resources necessary to artic-ulate and promulgate standards fortransparency, consistency (e.g., ofunits and formats), and quality control.***This task involves facilitatingdiscussions among the communityand formulating persuasive guide-lines and supporting arguments. Italso requires use of incentives suchas seals of approval, access to participation in particular processes, access to particular disseminationoutlets, and access to resources. Aweak "clearinghouse" that solicits,supports, or publicizes scenarios butcannot exercise quality control,propose and stimulate new directions, or convene critical reviews, ofthe whole enterprise and of particular exercises, is not an adequatemodel for the program.
- ***The program will require an adequatesustained resources.***The programmust build and maintain a sophisticated analytic capability, and develop skills and institutional memory regarding prior experiences,successes, and failures. This requirement precludes the programbeing a series of ad hocone-time activities or a part-time, unfunded burden imposed on people and organizations with other fulltime responsibilities.

ACRONYMS

A1, A2, B1, B2, A1FI

names of specific emissions scenarios published in the IPCC Special Report on Emissions Scenarios

CCSP Climate Change Science Program
CCTP Climate Change Technology Program

CFCs	chlorofluorocarbons
DDC	Data Distribution Center (IPCC)
DOD	Department of Defense (U.S.)
DWR	Department of Water Resources (California)
EMF	Energy Modeling Forum
ENSO	El Ni o/Southern Oscillation
EPA	Environmental Protection Agency
GBN	Global Business Network
GCM	general circulation model
GDP	gross domestic product
GFDL	Geophysical Fluid Dynamics Laboratory, US National Oceanographic and Atmos pheric Administration (NOAA), and climate model produced by this laboratory
GtC	gigatonnes (billion metric tons) of carbon
HadCM2	UK Hadley Centre climate model, Version 2
IIASA	International Institute for Applied Systems Analysis, Laxenburg, Austria
IPCC	Inter-governmental Panel on Climate Change
IS92	series of emissions scenarios produced by the IPCC in 1992, including specific scenarios named IS92a through IS92f
LRTAP	Convention on Long-Range Transboundary Air Pollution
MEA	Millennium Ecosystem Assessment
MEC	Metropolitan East Coast assessment of US National Assessment
MER	market exchange rates
NAPAP	National Acid Precipitation Assessment Program
NAST	National Assessment Synthesis Team
NRC	National Research Council (U.S.)
NYCDEP	New York City Department of Environmental Protection
NYCHP	New York Climate and Health Project
OECD	Organization for Economic Co-operation and Development
ONA	Office of Net Assessments (Office of US Secretary of Defense)
OTA	Office of Technology Assessment, United States Congress
PDO	Pacific Decadal Oscillation
PPP	purchasing-power parity
SRES	Special Report on Emissions Scenarios (IPCC)
TGICA	Task Group on Data and Scenario Support for Impact and Climate Analysis (IPCC)
UKCIP	United Kingdom Climate Impts Program
VEMAP	Vegetation Ecosystem Mapping and Analysis Project
VOC	olatile organic compounds The U.S. Climate Change Science Program Acronyms

REFERENCES

Alcamo, J., Bouwman, A., Edmonds, J., Gruebler, A., Morita, T. & Sugandhy, A. (1995). An Evaluation of the IPCC IS92 Emissions Scenarios." In: *Climate Change 1994, Radiative Forcing of Climate Change and an Evaluation of the IPCC IS92 Emission Scenarios* [J. T., Houghton, L. G., Meira Filho, J., Bruce, H., Lee, B. A., Callander, E., Haites, N. Harris, & K. Maskell, (eds.)], Cambridge University Press, Cambridge, UK.

Allen, M., Raper, S., et al., (2001). Climate Change-Uncertainty in the IPCC's Third Assessment Report. *Science*, *293*, 5529, 430.

Alley, R. B., Sowers, T., Mayewski, P. A., Stuiver, M., Taylor, K. C. & Clark, P. U. (1997). Holocene climate instability: a prominent, widespread event 8,200 years ago. *Geology*, *26*, 6.

Alley, R. B. (2004). Abrupt climate change. *Scientific American*, November, 62-69.

Arnell, N. W., Livermore, M. J. L., Kovats, S., Levy, P. E., Nicholls, R., Parry, M. L. & Gaffin, S. R. (2004). Climate and socio-economic scenarios for global-scale climate change impact assessments: characterizing the SRES storylines. *Global Environmental Change*, *14*, 3-20.

Barras, J., Beville, S., Britsch, D., Hartley, S., Hawes, S., Johnston, J., Kemp, P., Kinler, Q., Martucci, A., Porthouse, J., Reed, D., Roy, K., Sapkota, S. & Suhayda, J. (2003). *Historical and Projected Coastal Louisiana Land Changes: 1978-2050*. Open File Report 03-334, US Geological Survey, Washington, DC, USA.

Barney, G. O. (1981). *Global 2000 Report to the President*. Council on Environmental Quality, Washington, DC, USA.

Barrow, E., Gachon, P. & Caya, D. (2004). Creating climate change scenarios for Canada. In: *Climate Change and Variability in Canada: Past, Present and Future*. Environment Canada, Ottawa.

Berkhout, F. & Hertin, J. (2000). Socio-economic futures for climate impact assessment. *Global Environmental Change*, *10*, 165-168.

Berkhout, F. & Hertin, J. (2001). Foresight Futures 2001: Revised Scenario and User Guidance. *Final Report of the Review and Revise Environmental Futures study, SPRU*, Brighton, UK.

Berkhout, F., Hertin, J., Lorenzoni, I., Jordan, A., Turner, K., O'Riordan, T., Cobb, D., Ledoux, L., Tinch, R., Palutikof, J., Hulme, M. & Skea, J. (2001). Presentation of the UKCIP socioeconomic scenarios for climate impact assessment. In: UKCIP, *Socio-Economic Scenarios for Climate Impact Assessment: A Guide to Their Use in the UK*, www.ukcip.org.uk/ resources/publications/pub_dets.asp?ID=34.

Berkhout, F., Hertin, J. & Jordan, A. (2002). Socio-economic futures in climate impact assessment: using scenarios as learning machines. *Global Environmental Change*, *12*, 83-95.

Boer, G. J., Flato, G. M., Reader, M. C. & Ramsden, D. (1999a). A transient climate change simulation with historical and projected greenhouse gas and aerosol forcing: experimental design and comparison with the instrumental record for the 20th century. *Climate Dynamics*, *16*, 405-426.

Boer, G. J., Flato, G. M. & Ramsden, D. (1999b). A transient climate change simulation with historical and projected greenhouse gas and aerosol forcing: projected climate for the 21st century. *Climate Dynamics*, *16*, 427-450.

Bracken, P. (1977). Unintended consequences of strategic gaming. *Simulation and Games*, *8*, 283-318 (3 September).

Bracken, P. (1990). Gaming in hierarchical defense organizations. In: *Avoiding the Brink: Theory and Practice in Crisis Management* [A. C., Goldberg, D. Van Opstal, & J. H. Barkley, (eds.)]. Brassey's, London.

Bretherton, F. P., Bryan, K. & Woods, J. D., et al. (1990). Time-dependent greenhouse-gas-induced climate change. In: *Climate Change: The IPCC Scientific Assessment* [J. T., Houghton, G. J. Jenkins, & J. J. Ephraums, (eds.)]. Cambridge University Press, Cambridge, UK.

Brewer, G. D. (1986). Methods for synthesis: policy exercises. In: *Sustainable Development of the Biosphere* [W. C. Clark, & R. E. Munn, (eds.)]. Cambridge University Press, Cambridge, UK.

Brewer, G. D. (1990). Discovery is not prediction. In: Avoiding the Brink: *Theory and Practice in Crisis Management* [A. C., Goldberg, D. Van Opstal, & J. H. Barkley, (eds.)]. Brassey's, London.

Brewer, G. D. & Shubik, M. (1979). The War Game: a Critique of Military Problem-Solving. Harvard University Press, Cambridge, MA.

Burkett, V. R., Wilcox, D. A., Stottlemeyer, R., Barrow, W., Fagre, D., Baron, J., Price, J., Nielsen, J., Allen, C. D., Peterson, D. L., Ruggerone, G. & Doyle, T. (2005). Nonlinear dynamics in ecosystem response to climate change: case studies and policy implications. *Ecological Complexity*, *2*, 357-394.

California DWR, (Dept of Water Resources). (2005). *California Water Plan Update 2005*. Bulletin 160-05. Available at http://www.waterplan.water.

Castles, I. & Henderson, D. (2003a). The IPCC emissions scenarios: an economic-statistical critique. *Energy and Environment*, *14(2-3)*, 159-185.

Castles, I. & Henderson, D. (2003b). Economics, emissions scenarios, and the work of the IPCC. *Energy and Environment*, *14(4)*, 415-435.

CCSP. (2007). *Scenarios of Greenhouse Gas Emissions and Atmospheric Concentrations*. A report by the U.S. Climate Change Science Program and the subcommittee on Global Change Science Research [L., Clarke, J., Edmonds, J., Jacoby, H., Pitcher, J. Reilly, & R. Richels, (Eds.)]. *Department of Energy, Office of Biological & Environmental Research*, Washington, DC, USA, *102*.

Chess, C. & Purcell, K. (1999). Public P and the environment: do we know what works? *Environmental Science & Technology*, *33*, 2685-2691.

Cowling, E. B. (1992). The performance and legacy of NAPAP. *Ecological Applications*, *2(2)*, 111-116.

Cook, C. W. (1939). *Scenery of Florida Interpreted by a Geologist*. The State Geological Survey, Tallahassee, FL, USA.

Curry, R. & Mauritzen, C. (2005). Dilution of the northern North Atlantic Ocean in recent decades. *Science*, *308*, 5729, 1772-1774.

Cubasch, U., Meehl, G. A., Boer, G. J., Stouffer, R. J., Dix, M., Noda, A., Senior, C. A., Raper, S. C. B. & Yap, K. S. (2001). Projections of future climate change. In: *Climate*

Change 2001: The Scientific Basis [J. T., Houghton, G. J. Jenkins, & J. J. Ephraums, (eds.)]. Cambridge University Press, Cambridge, UK.

Davis, G. (2003). Meeting future energy needs: choices and possibilities. *The Bridge*, *33(2)*, 16-21.

Chesnaye, F. de la & Weyant, J. (eds.), (2006). Multi-greenhouse gas mitigation and climate policy. *The Energy Journal*, special issue.

DeWeerd, H. (1967). *Political Military Scenarios*, P-3535. Rand Corporation, Santa Monica, CA.

DeWeerd, H. (1975). A contextual approach to scenario construction. *Simulation and Games*, *5*, 403-414.

Dickson, B., Yashayaev, I., Meincke, J., Turrell, B., Dye, S. & Hoffort. J. (2002). Rapid freshening of the deep North Atlantic Ocean over the past four decades. *Nature*, *416*, 832.

Doyle, T. W., Day, R. H. & Biagas, J. M. (2003). Predicting coastal retreat in the Florida Big Bend Region of the Gulf Coast under climate change induced sea-level rise. In: *Integrated Assessment of the Climate Change Impacts on the Gulf Coast Region - Foundation Document* [A. H., Ning, R. E., Turner, T. Doyle, & K. Abdollahi, (eds.)]. LSU Press, Baton Rouge, USA.

Duncan, N. & Wack, P. (1994). Scenarios designed to improve decision making. *Planning Review*, *22(4)*, 18-25.

The Economist, (2003a). Hot potato. *The Intergovernmental Panel on Climate Change had better check its calculations*. February 13.

The Economist, (2003a). Hot potato revisited: *a lack-of-progress report onthe Intergovernmental Panel on Climate Change*. November 6.

EMF (Energy Modeling Forum), (2006). EMF 22: *Greenhouse Gas Stabilization. Prospectus. www.stanford.edu/group/EMF/projects/* EMF22/emf22prospectus.pdf.

Envision Sustainability Tools, (1999). Lower Fraser Basin QUEST Model Structure, Ver 1.0. Vancouver BC. www.sdri.ubc.ca/downloads/ Models.pdf

Fairbanks, R. G. (1989). A 17,000-year glacioeustatic sea level record: influence of glacial melting rates on the Younger Dryas and deep ocean circulation. *Nature*, *342*, 7, 637-642.

Gagosian, R. B. (2003). Abrupt climate change: *should we be worried? Presented to World Economic Forum, Davos*, Switzerland, January 27.

GBN (Global Business Network), (2004a). *Executive Summary, An Abrupt Climate Change Scenario and Its Implications for US National Security*, Oct. 2003. www.e ms.org/climate

GBN (Global Business Network), (2004b). Abrupt Climate Change. Press Release, February. www.gbn.com/ArticleDisplayServlet.srv?aid=26231.

Giorgi, F., Hewitson, B., Christensen, J., Hulme, M., von Storch, H., Whetton, P., Jones, R., Mearns, L. & Fu, C. (2001). Regional climate information: evaluation and projections. In: *Climate Change, 2001: The Scientific Basis* [J. T., Houghton, G. J. Jenkins, & J. J. Ephraums, (eds.)]. Cambridge University Press, Cambridge, UK.

Godet, M. & Roubelat. F. (1996). Creating the future: the use and misuse of scenarios. *Long Range Planning*, *29*, 2, 164-171.

Gosseink, J. G. (1984). *The Ecology of Delta Marshes of Coastal Louisiana: a Community Profile*. FWS/OBS-84/09, U.S. Fish and Wildlife Service, Washington, DC, USA.

Gough, C., Castells, N. & Funtowicz, S. (1998). Integrated assessment: an emerging methodology for complex issues. *Environmental Modeling and Assessment*, *3*, 19-29.

Greenberger, M., Brewer, G. D., Hogan, W. W. & Russell. M. (1983). *Caught Unawares: The Energy Decade in Retrospect*. Ballinger, Cambridge, MA, USA.

Gregory, R. & McDaniels, T. (2005). Improving environmental decision processes. Appendix B in: NRC 2005, *Decision Making for the Environment*. National Academy Press, Washington DC, USA.

Grübler, A. & Nakicenovic, N. (2001). Identifying dangers in an uncertain climate. *Nature*, *412*, 5-15.

Grübler, A., Naki enovi, N., Alcamo, J., Davis, G., Fenhann, J., Hare, B., Mori, S., Pepper, B., Pitcher, H., Riahi, K., Rogner, H., La Rovere, E. L., Sankovski, A., Schlesinger, M., Shukla, R. P., Swart, R., Victor, N. & Jung, T. Y. (2004). Emission scenarios: a final response. *Energy & Environment*, *15*, 1, 11-24.

Hansen, B., Turrell, W. & Østerhus, S. (2001). Decreasing overflow from the Nordic Seas into the Atlantic Ocean through the Faroe Bank Channel since 1950. *Nature*, *411*, 927-930.

Hausrath, A. H. (1971). *Venture Simulation in War, Business, and Politics*. McGraw-Hill, New York, USA.

Herrick, C. (2002). Atmospheric Science and the Constitution of Public Policy: The Case of the National Acid Precipitation Assessment Program (NAPAP). *Case study prepared for the American Meteorological Society*, 2002 Summer Policy Colloquium. http://www.ametsoc.org/atmospolicy/PolicyCaseStudies.html.

C. S. Holling, (ed.), (1978), *Adaptive Environmental Assessment and Management*. Wiley, New York, USA.

Holman, I. P., Loveland, P. J., Nicholls, R. J., Shackley, S., Berry, P. M., Rounsevell, M. D. A., Audsley, E., Harrison, P. A. & Wood, R. (2002). *REGIS - Regional Climate Change Impact Response Studies in East Anglia and North West England*. UK Climate Impacts Programme, UK.

Holtsmark, B. J. & Alfsen, K. H. (2005). PPP Correction of the IPCC emissions scenarios: does it matter? *Climatic Change*, *68*, 1, 11-19.

Huss, W. R. (1988). A move toward scenario analysis. *International Journal of Forecasting*, *4*, 377-388.

IPCC (Intergovernmental Panel on Climate Change). (1990). *Climate Change: The IPCC Scientific Assessment* [J. T., Houghton, G. J. Jenkins, & J. J. Ephraums, (eds.)]. Cambridge University Press, Cambridge, UK.

IPCC (Intergovernmental Panel on Climate Change). (1996a). *Climate Change (1995), The Science of Climate Change* [J. T., Houghton, L. G., Meira Filho, B. A., Calander, N., Harris, A. Kattenberg, & K. Maskell, (eds.)]. Cambridge University Press, Cambridge, UK.

IPCC (Intergovernmental Panel on Climate Change). (1996b). *Climate Change (1995), Economic and Social Dimensions of Climate Change* [J. P., Bruce, H. Lee, & E. F. Haites, (eds.)]. Cambridge University Press, Cambridge, UK.

IPCC (Intergovernmental Panel on Climate Change). (2001a). *Climate Change 2001, The Scientific Basis* [J. T., Houghton, Y., Ding, D. J., Griggs, M., Noguer, P. J. van der Linden, & D. Xiaosu, Cambridge University Press, Cambridge, UK.

IPCC (Intergovernmental Panel on Climate Change). (2001b). *Climate Change 2001: Impacts, Adaptations, and Vulnerability* [J. J., McCarthy, O. F., Canziani, N. A., Leary, D. J. Dokken, & K. S. White, Cambridge University Press, Cambridge, UK.

IPCC (Intergovernmental Panel on Climate Change). (2001c). *Climate Change 2001: Mitigation*, [B., Metz, O., Davidson, R. Swart, & J. Pan, (eds.)]. Cambridge University Press, Cambridge, UK.

IPCC-TGCIA (Intergovernmental Panel on Climate Change-Task Group on Scenarios for Climate Impact Assessment). (1999). *Guidelines on the Use of Scenario Data for Climate Impact and Adaptation Assessment*, Version 1, [T. R., Carter, M. Hulme, & M. Lal, (eds.)].

Johns, T. C., Carnell, R. & Crossly, J., et al. (1997). The second Hadley Centre Coupled Ocean-Atmosphere GCM: model description, spinup, and validation. *Climate Dynamics 13*, 225-237.

Jones, W. M. (1985). *On Free-Form Gaming*. Rand Note N-2322-RC, *Rand Corporation, Santa Monica*, CA, USA.

Kahn, H. & Wiener, A. J. (1967). *The Year 2000: a Framework for Speculation on the Next Thirty-Three Years*. Macmillan, New York.

Kalkstein, L. S. & Greene, J. S. (1997). An evaluation of climate/mortality relationships in large U.S. cities and the possible impacts of a climate change. *Environmental Health Perspectives*, *105*, 84-93.

Keith, D. W. (2000). Geoengineering the climate: history and prospect. *Annual Review of Energy and Environment*, *25*, 245-284.

Keith, D. W., Ha-Duong, M. & Stolaroff, J. K. (2006). Climate strategy with CO_2 capture from the air. *Climatic Change*, *74*, 1-3.

Kemp-Benedict, E., Heaps, C. & Raskin, P. (2002). Global Scenario Group Futures: Technical Notes. *Stockholm Environment Institute PoleStar Series Report* no. 9.

Kinney, P., et al., (2005). *Assessing Potential Public Health and Air Quality Impacts of Changing Climate and Land Use in Metropolitan New York*. Columbia Earth Institute, New York, USA.

Kinney, P. L., Rosenthal, J. E., Rosenzweig, C., Hogrefe, C., Solecki, W., Knowlton, K., Small, C., Lynn, B., Civerolo, K., Ku, J. Y., Goldberg, R. & Oliveri, C. (2006). Assessing potential public health impacts of changing climate and land uses: the New York Climate & Health Project. In: *Climate Change and Variability: Impacts and Responses* [M., Ruth, K., Donaghy, & P. Kirshen, (eds.)]. New Horizons in Regional Science, Edward Elgar, Cheltenham, UK.

Kittel, T. G. F., Rosenbloom, N. A., Painter, T. H., Schimel, D. S. & VEMAP Participants, (1995). the VEMAP Integrated Database for Modeling United States Ecosystem/ Vegetation Sensitivity to Climate Change. *Journal of Biogeography*, *22*, 4-5, 857-862.

Leggett, J., Pepper, W., Swart, R. J., Edmonds, J. A., Meira Filho, L. G., Mintzer, I., Wang, M. X. & Watson, J. (1992). Emissions scenarios for the IPCC, an update: assumptions, methodology, and results. In: *Climate Change, 1992: The Supplementary Report to the IPCC Scientific Assessment* [J. T., Houghton, B. A. Callander, & S. K. Varney, (eds.)]. Cambridge University Press, Cambridge, UK.

Lempert, R. J., Groves, D. G., Popper, S. W. & Bankes, S. C. (2006). A general, analytic method for generating robust strategies and narrative scenarios. *Management Science*, *52*, *4*, 5 14-528.

Levine, R. A. (1964a). Crisis games for adults. In: Crisis Games 27 Years Later: Plus C'est Déjà vu. P-7719, *RAND Corporation*, Santa Monica, CA, USA.

Levine, R. A. (1964b). Crisis games: a rejoinder to Tom Schelling and to some extent to Bill Jones. *Crisis Games 27 Years Later: Plus C'est Déjà vu.* P-7719, *RAND Corporation*, Santa Monica, CA, USA.

Levy, M. A. (1995). International co-operation to combat acid rain. In: *Green Globe Yearbook of International Co-operation on Environment and Development, 1995* [H., Bergesen, G. Parmann, & Ø. Thommessen, (eds.)]. Oxford University Press, Oxford, UK.

Lofgren, B. M., Quinn, F. H., Clites, A. H., Assel, R. A. & Eberhardt, A. J. (2000). Water resources. In: *Preparing for a Changing Climate: Great Lakes Overview*.

Lomborg, B. (2001). *The Skeptical Environmentalist.* Cambridge University Press, Cambridge, UK.

London Observer, (2004). Now the Pentagon tells Bush: climate change will destroy us. February 22:1. observer.guardian. 153513,00.html,

Lorenzoni, I., Jordan, A., Hulme, M., Turner, R. K. & O'Riordan, T. (2000). A co-evolutionary approach to climate change impact assessment: part i. integrating socio-economic and climate change scenarios. *Global Environmental Change, 10*, 1, 57-68.

MacCracken, M. C., Barron, E. J., Easterling, D. R., Felzer, B. S. & Karl, T. R. (2003). Climate change scenarios for the U.S. National Assessment. *Bulletin of the American Meteorological Society, 84*, 12, 1711- 1723.

MacCracken, M. C., Barron, E., Easterling, D., Felzer, B. & Karl, T. (2001). Scenarios for climate variability and change. See NAST 2001.

McLean, R. F., Tsyban, A., Burkett, V., Codignotto, J., Forbes, D., Ittekkot, V., Mimura, N. & Beamish, R. J. (2001). Coastal zones and Marine ecosystems. In: *Climate Change: Impacts, Adaptation, and Vulnerability* [J. J., McCarthy, O. F., Canziani, N. A., Leary, D. J. Dokken, & K. S. White, (eds.)]. Cambridge University Press, Cambridge, UK.

Manabe, S. & Stouffer, R. J. (1979). A CO_2-climate sensitivity study with a mathematical model of the global climate. *Nature, 282*, 491-493.

Manabe, S., Stouffer, R. J., Spelman, M. J. & Bryan, K. (1991). Transient responses of a coupled ocean-atmosphere model to gradual changes of atmospheric CO_2. part i: annual mean response. *Journal of Climate, 4, 8*, 785-818.

Manabe, S. & Wetherald, R. T. (1967). Thermal equilibrium of the atmosphere with a given distribution of relative humidity. *Journal of the Atmospheric Sciences, 24, 3*, 241-259.

Manne, A., Richels, R. & Edmonds, J. A. (2005). Market exchange rates or purchasing power parity: does the choice make any difference to the climate debate? *Climatic Change, 71*, 1-2, 1-8.

MEA (Millennium Ecosystem Assessment). (2005). *Ecosystems and Human Well-Being: Synthesis Report*. Island Press, Washington, DC, USA.

MEA (Millennium Ecosystem Assessment).(2006). *Ecosystems and Human Well-Being: Scenarios*. Island Press, Washington, DC, USA.

Meadows, D. H., Meadows, D. L., Randers, J., Behrens III, W. W. (1972). *Limits to Growth: A Report to the Club of Rome*. Universe Books, New York, USA.

Mearns, L. O., Rosenzweig, C. & Goldberg, R. (1992). Effect of changes in interannual climatic variability on CERES-wheat yields: sensitivity and 2 x CO_2 general circulation model studies. *Agricultural and Forest Meteorology, 62*, 159-189.

Mearns, L. O., Rosenzweig, C. & Goldberg, R. (1996). The effect of changes in daily and interannual climatic variability on CERESwheat: a sensitivity study. *Climatic Change*, *32*, 257-292.

Mearns, L. O., Hulme, M., Carter, T. R., Leemans, R., Lal, M. & Whet- ton, P. (2001). Climate scenario development. In: *Climate Change 2001: The Scientific Basis* [J. T., Houghton, Y., Ding, D. J., Griggs, M., Noguer, P. J., van der Linden, X., Dai, K. Maskell, & C. A. Johnson, (eds.)]. Cambridge University Press, Cambridge, UK.

Michaels, P. J. (2003a). Science or political science: an assessment of the US National Assessment of the potential consequences of climate variability and change. In: *Politicizing Science: the Alchemy of Policymaking* [M. Gough, (ed.)]. Publication 517, Hoover Institution Press, Stanford, CA, USA.

Michaels, P. (ed.), (2003b). Bad math. *World Climate Report*, *13*, 10.

Miller, S. S. (1990). NAPAP: a unique experience. *Environmental Science and Technology*, *24*, 12, 1781-1782.

Mitchell, J. F. B., Manabe, S., Meleshko, V. & Tokioka, V. (1990). Equilibrium climate change and its implications for the future. In: *Climate Change: The IPCC Scientific Assessment* [J. T., Houghton, G. J. Jenkins, & J. J. Ephraums, (eds.)]. Cambridge University Press, Cambridge, UK.

Mitchell, R. B., Clark, W. C., Cash, D. W. & Dickson, N. M. (eds.), (2006). *Global Environmental Assessments: Information and Influence*. MIT Press, Cambridge, MA, USA.

Morita, T. & Lee, H. C. (1998). IPCC SRES Database. *Database prepared for IPCC Special Report on Emissions Scenarios*. http:wwwcger.nies.go.jp/cger-e/db/ipcc.html.

Morita, T., Robinson, J., Adegbulugbe, A., Alcamo, J., Herbert, D., La Rovere, E. L., Nakicenovic, N., Pitcher, H., Raskin, P., Riahi, K., Sankovski, A., Sokolov, V., de Vries, B. & Dadi, Z. (2001). Greenhouse gas emissions: mitigation scenarios and implications. In: *Climate Change, 2001: Mitigation* [B., Metz, O., Davidson, R. Swart, & J. Pan, (eds.)]. Cambridge University Press, Cambridge, UK.

Morgan, M. G. & Keith, D. W. (1995). Subjective judgments by climate experts. *Environmental Science & Technology*, *29*, 468-476.

Morgan, M. G., Kandlikar, M., Risbey, J. & Dowlatabadi, H. (1998). Why conventional tools of policy analysis are often inadequate for problems of global change. *Climatic Change*, *41*, 271-281.

Morgan, M. G., Cantor, R., Clark, W. C., Fisher, A., Jacoby, H. D., Janetos, A. C., Kinzig, A. P., Melillo, J., Street, R. B. & Wilbanks, T. J. (2005). Learning from the U.S. National Assessment of Climate Change. *Environmental Science & Technology*, *39*, 9023-9032.

Morton, R. A., Buster, N. A. & Krohn, M. D. (2002). Subsurface controls on historical subsidence rates and associated wetland loss in southcentral Louisiana. *Transactions - Gulf Coast Association of Geological Societies*, *52*, 767-778.

Moss, R. H. & Schneider, S. H. (2000). Uncertainties in the IPCC TAR: recommendations to lead authors for more consistent assessment and reporting. In: Cross Cutting Issues Guidance Papers, *Intergovernmental Panel on Climate Change*, Geneva, Switzerland.

Mote, P. W. & Parson, E. A., et al., (2003), Climate and the water, forests, and salmon of the Pacific Northwest. *Climatic Change*, *61*, 1-2, 45-8 8.

Naki enovi, N. & Swart, R. (eds.). (2000). Special Report on Emissions Scenarios. Cambridge University Press, Cambridge, UK. http://www.grida.no/climate

Naki enovi, N., Grübler, A., Gaffin, S., Jung, T. T., Kram, T., Morita, T., Pitcher, H., Riahi, K., Schlesinger, M., Shukla, P. R., van Vuuren, D., Davis, G., Michaelis, L., Swart, R. & Victor, N. (2003). IPCC SRES revisited: a response. *Energy & Environment*, *14*, 2-3, 187-2 14.

NAPAP (National Acid Precipitation Assessment Program. (1982). *Annual Report: National Acid Precipitation Assessment Program*. National Acid Precipitation Assessment Program, Washington, DC, USA.

Nash, J. F. (1950). Equilibrium points in n-person games. *Proceedings of the National Academy of Sciences*, *36(1)*, 48-49.

NAST (National Assessment Synthesis Team). (2000). *Climate Change Impacts on the United States: The Potential Consequences of Climate Variability and Change, Overview*. Cambridge University Press, Cambridge, UK.

NAST (National Assessment Synthesis Team). (2001). *Climate Change Impacts in the United States: Potential Consequences of Climate Change and Variability and Change*. Cambridge University Press, Cambridge, UK.

NRC (US National Research Council). (1996). *Understanding Risk: Informing Decisions in a Democratic Society*. [P. C. Stern, & H. V. Fineberg, (eds.)]. Committee on Risk Characterization. National Academy Press, Washington, DC, USA.

NRC (US National Research Council). (1999). *Our Common Journey: A Transition toward Sustainability*. Board on Sustainable Development. National Academy Press, Washington, DC, USA.

NRC (US National Research Council). (2002). *Abrupt Climate Change: Inevitable Surprises*. Committee on Abrupt Climate Change. National Academy Press, Washington, DC, USA.

NRC (US National Research Council). (2005). *Decision Making for the Environment: Social and Behavioral Science Research Priorities*. Committee on Human Dimensions of Global Change. National Academy Press, Washington, DC, USA.

Parikh, J. K. (1992). IPCC strategies unfair to the South. *Nature*, *360*, 507-508.

Parikh, J. K. (1998). The emperor's new clothes: long-range energy-use scenarios by IIASA-WEC and IPCC. *Energy*, *23(1)*, 69-70.

Parson, E. A. (1996). What can you learn from a game? In: *Wise Choices: Games, Decisions, Negotiations* [R., Zeckhauser, R. L. Keeney, & J. K. Sebenius, (eds.)]. Harvard Business School Press, Boston, USA.

Parson, E. A. (1997). Informing global environmental policy-making: a plea for new methods of assessment and synthesis. *Environmental Modeling and Assessment*, *2(4)*, 267-279.

Parson, E. A. (2003). *Protecting the Ozone Layer: Science and Strategy*. Oxford University Press, New York, USA.

Parson, E. A. (2006). Reflections on air capture: the political economy of active intervention in the global environment. *Climatic Change*, *74*, 1-3.

Parson, E. A. & Fisher-Vanden, K. (1997). Integrated assessment models of global climate change. *Annual Review of Energy and the Environment*, *22*, 589-628.

Parson, E. A. & Morgan, M. G., with Janetos, A., Joyce, L., Miller, B., Richels, R. & Wilbanks, T. (2001). Socioeconomic context for climate impact assessment. In: *Climate Change Impacts on the United States*, NAST. Cambridge University Press, Cambridge, UK.

Parson, E. A. & Corell, R. W., et al., (2003). Understanding climate impacts, vulnerabilities, and adaptation in the United States: building a capacity for assessment. *Climatic Change, 57(1)*, 9-42.

Payne, J. T., Wood, A. W., Hamlet, A. F., Palmer, R. N. & Lettenmaier, D. P. (2004). Mitigating the effects of climate change on the water resources of the Columbia River Basin. *Climatic Change, 62(1-3)*, 233-256.

Perhac, R. M. (1991). Usable science: lessons from acid rain legislation, NAPAP. *Power Engineering, 95*, 10, 26-29.

Pitcher, H. M. (2005). Downscaling: something for nothing? Presentation to workshop on global-change scenarios, S*nowmass Colorado*, July 26.

Pittock, A. B., Jones, R. N. & Mitchell, C. F. (2001). Probabilities will help us plan for climate change. *Nature, 413*, 249.

Providence Journal, (2004). Pentagon report plans for climate catastrophe. March 3, A1.

Raisanen, J. & Palmer, T. (2001). A probability and decision-model analysis of a multi-model ensemble of climate change simulations. *Journal of Climate, 14*, 3212-3226.

Raskin, P., Banuri, T., Gallopin, G., Gutman, P., Hammone, A., Kates, R. & Swart, R. (2002). *Great Transition: the Promise and Lure of the Times Ahead*. Global Scenario Group, Stockholm Environment Institute, Boston.

Reilly, J., Stone, P. H., et al., (2001). Climate change – uncertainty and climate change assessments. *Science, 293*, 430.

Renn, O., Webler, T. & Wiedemann, P. (1995). *Fairness and Compensation in Citizen Participation*. Kluwer, Dordrecht, The Netherlands.

Roberts, L. (1991). Learning from an acid rain program. *Science, 251*, 1302-1305.

Robinson, J. B. (1982). Energy backcasting: a proposed method of policy analysis. *Energy Policy, 10(4)*, 33 7-45.

Robinson, J. B. (2003). Future subjunctive: backcasting as social learning. *Futures, 35*, 839-856.

Rosenberg, N., Crosson, P., Frederick, K., Easterling, W., McKenney, M., Bowes, M., Sedjo, R., Darmstadler, J., Katz, L. & Lemon, K. (1993). The MINK methodology: background and baseline. *Climatic Change, 24*, 7-22.

Rosenzweig, C. & Solecki, W. D. (eds.). (2001). *Climate change and a global city*: the potential consequences of climate variability and change-metro east coast. Columbia Earth Institute, New York, USA.

Rosenzweig, C. & Major, D. C. (2006). *Climate Impact Assessment of Environmental Infrastructure Systems: Phase I Final Report: Scoping for Phase II*. Center for Climate Systems Research, Columbia University, New York, USA.

Rothman, D., Robinson, J. & Biggs, D. (2003). Signs of life: linking indicators and models in the context of QUEST. In: *Implementing Sustainable Development- Integrated Assessment and Participatory Decision-Making Processes* [H. Abaza, & A. Baranzini, (eds.)]. Edward Elgar, Cheltenham, UK.

Rubin, E. S. (1991). Benefit-cost implications of acid rain controls: an evaluation of the NAPAP integrated assessment. *Journal of the Air and Waste Management Association, 41(7)*, 914-921.

Ruosteenoja, K., Carter, T. R., Jylha, K. & Tuomenvirta, H. (2003). *Future Climate in World Regions: An Intercomparison of Model-Based Projections for the New IPCC Emissions Scenarios*. Finnish Environment Institute, Helsinki.

Russell, M. (1992). Lessons from NAPAP. *Ecological Applications, 2(2)*, 107-110.

San Francisco Chronicle. (2004). Pentagon-sponsored climate report sparks hullabaloo in Europe. February *25*, A5.

Schelling, T. C. (1964). An uninhibited pitch for crisis games. In: *Crisis Games 27 Years Later: Plus C'est Deja* Vul. P-7719, RAND Corporation, Santa Monica, CA, USA.

Schelling, T. C. (1994). Nuclear history program oral history transcript. NHP Berlin Crisis Oral History Project, Interview with Thomas Schelling. University of Maryland Center for International Security Studies, *Nuclear History Program*, College Park, MD, USA.

Schelling, T. C. (1983). Climate change: implications for welfare and policy. In: *Changing Climate*, Report of the Carbon Dioxide Assessment Committee, *US National Research Council*. National Academy Press, Washington DC, USA.

Schneider, S. H. (2001). What is "dangerous" climate change? *Nature*, *411*, 17-19.

Schneider, S. H. (2002). Can we estimate the likelihood of climatic changes at 2100? *Climatic Change*, *52(4)*, 441-451.

Schoemaker, P. J. H. (1995). Scenario planning: a tool for strategic thinking. *Sloan Management Review*, Winter.

Schultz, R. L. & Sullivan, E. M. (1972). Developments in simulation in social and administrative science. In: *Simulation in the Social and Administrative Sciences* [H., Guetzkow, P. Kotler, & R. L. Schultz, (eds.)]. Prentice-Hall, Englewood Cliffs, NJ, USA.

Schwartz, P. (1991). *The Art of the Long View: Planning for the Future in an Uncertain World*. Currency Doubleday, New York, USA.

Semenov, M. A. & Porter, J. R. (1995). Climatic variability and the modeling of crop yields. *Agricultural and Forest Meteorology*, *73*, 265-283.

Shell International. (2001). *Energy Needs, Choices, and Possibilities: Scenarios to 2050*. Global Business Environment.

Shell International. (2003). *Scenarios: an Explorer's Guide. Global Business Environment. www-static.shell.com/static/royal-en/downloads/* scenarios_explorersguide.pdf.

Shubik, M. (1975). *The Uses and Methods of Gaming*. Elsevier, New York, USA.

Shinkle, K. D. & Dokka, R. K. (2004). *Rates of Vertical Displacement at Benchmarks in the Lower Mississippi Valley and the Northern Gulf Coast*. NOAA Technical Report NOS/NGS 50, US Department of Commerce, National Oceanic and Atmospheric Administration, Washington, DC, USA.

Smil, V. (2005). *Energy at the Cross Roads*. MIT Press, Cambridge, MA, USA.

Southeast Regional Assessment Team, (2002). *Preparing for a Changing Climate: Potential Consequences of Climate Variability and Change, Southeast*. Publication 8-40002, Global Hydrology and Climate Center, University of Alabama in Huntsville, USA.

Stipp, D. (2004). The Pentagon's weather nightmare: the climate could change radically, and fast. *Fortune*, February 9, *100*.

Stockholm Environment Institute. (1999). *POLESTAR: System Manual for Version 2000*. POLESTAR Series Report No. 2., Stockholm Environment Institute, Sweden.

Stouffer, R. J., Manabe, S. & Bryan, K. (1989). Interhemispheric asymmetry in climate response to a gradual increase of CO_2. *Nature*, *342*, 660-662.

Svedin, U. & Aniansson, B. (eds.), (1987). *Surprising Futures. Notes from an International Workshop on Long-term World Development*, Friibergh Manor, Sweden, January 1986. Report 87, 1, Swedish Council for Planning and Coordination of Research, Stockholm.

Swart, R. J., Raskin, P. & Ribonson, J. (2004). The problem of the future: sustainability science and scenario analysis. *Global Environmental Change*, *14*, 137-146.

Tebaldi, C., Mearns, L. O., Smith, R. L. & Nychka, D. (2004). Regional probabilities of precipitation change: a Bayesian analysis of multi-model simulations. *Geophysical Research Letters*, *31*, L24213.

Tebaldi, C., Smith, R. L., Nychka, D. & Mearns, L. O. (2005). Quantifying uncertainty in projections of regional climate change: a Bayesian approach to the analysis of multimodel ensembles. *Journal of Climate*, *18(10)*, 1524-1540.

Terleckyj, N. E. (1999a). *Analytic Documentation of Three Alternate Socioeconomic Projections*, *199 7-2050*. NPA Data Services, Washington, DC, USA.

Terleckyj, N. E. (1999b). *Development of Three Alternate National Projection Scenarios, 1997-2050*. NPA Data Services, Washington, DC, USA.

Toth, F. & Wilbanks, T. W. (2004). *Considering the Technical and Socioeconomic Assumptions Embedded in the SRES Scenario Families*. Guidance Paper, Fourth Assessment Report, IPCC Working Group 2. September 2004.

Tversky, A. & Kahneman, D. (1974). Judgment under uncertainty: heuristics and biases. *Science*, *185*, 1124-1131.

UKCIP (UK Climate Impacts Programme). (1998). *Climate Change Scenarios for the United Kingdom*. UKCIP Technical Report No. 1, Oxford, UK. http://www.ukcip .org.uk/resources pub_ dets.asp?ID=1 1.

UKCIP (UK Climate Impacts Programme). (2000). *Climate Change: Assessing the Impacts, Identifying the Responses* [M., McKenzie Hedger, M., Gawith, I., Brown, R. Connell, & T. E. Downing, (eds.)]. UKCIP Short Report, Oxford, UK. http://www.ukcip.org.uk/re-sources

UKCIP (UK Climate Impacts Programme). (2001). *Socio-Economic Scenarios for Climate Change Impact Assessment: A Guide to Their Use in the UKCIP*. UKCIP, Oxford, UK. ukcip.org.uk/resources/

UKCIP (UK Climate Impacts Programme). (2005). *Measuring Progress: Preparing for Climate Change through the UK Climate Impacts Programme* [West, C. and M. Gawith (eds.)]. UKCIP, Oxford, UK. http://www.ukcip.org.uk/resources

UK House of Lords. (2005). *The Economics of Climate Change*. Select Committee on Economic Affairs. 6 July. The Stationery Office Limited, London, UK.

UK Office of Science and Technology. (2002). *Foresight Futures 2020: Revised Scenarios and Guidance*. Department of Trade and Industry, London, UK.

US Army Corps of Engineers. (2004). *Louisiana Coastal Area (LCA), Louisiana, Ecosystem Restoration Study*. New Orleans District, Louisiana, USA.

US Climate Change Science Program and Subcommittee on Global Change Research. (2003). *Strategic Plan for the U.S. Climate Change Science Program*. Washington, DC, USA.

US EPA (US Environmental Protection Agency). (1989). *The Potential Effects of Global Climate Change on the United States* [J. Smith, & D. Tirpak, (eds.)]. EPA-230-05-89-050, Washington, DC, USA.

US OTA (US Congress, Office of Technology Assessment). (1993). *Preparing for an Uncertain Climate* (2 vols.). OTA-O-567 and -568, US Government Printing Office, Washington, DC, USA.

Van der Heijden, K. (1996). *Scenarios: The Art of Strategic Conversation*. John Wiley and Sons, Chichester, UK.

van Notten, P. W. F., Rotmans, J., van Asselt, M. B. A. & Rothman, D. S. (2003). An updated scenario typology. *Futures*, *35*, 423-443.

van't Klooster, S. A. & van Asselt, M. B. A. (2006). Practicing the scenario-axes technique. *Futures*, *38*, 15-30.

Vaughan, D. G. & Spouge, J. R. (2002). Risk estimation of collapse of the West Antarctic Ice Sheet. *Climatic Change*, *52*, 65-91.

VEMAP Members. (1995). Vegetation/Ecosystem Modeling and Analysis Project (VEMAP): comparing biogeography and biogeochemistry models in a continental-scale study of terrestrial ecosystem responses to climate change and CO_2 doubling. *Global Biogeochemical Cycles*, *9(4)*, 407-437.

von Neumann, J. & Morgenstern, O. (1944). *Theory of Games and Eco-nomic Behavior*. Princeton University Press, Princeton, NJ, USA.

Wack, P. (1985a). Scenarios: uncharted waters ahead. *Harvard Business Review*, *63(5)*, 73-89.

Wack, P. (1985b). Scenarios: shooting the rapids. *Harvard Business Review*, *63(6)*, 139-150.

Wallsten, T. S. & Whitfield, R. G. (1986). *Assessing the Risks to Young Children of Three Effects Associated with Elevated Blood Lead Levels*. ANL/AA-32, Argonne National Laboratory, Argonne, IL, USA.

Washington, W. M. & Meehl, G. A. (1989). Climate sensitivity due to increased CO_2: experiments with a coupled atmosphere-ocean general circulation model. *Climate Dynamics*, *4(1)*, 1-38.

WBCSD (World Business Council on Sustainable Development). (1997). *Exploring Sustainable Development: WBCSD Global Scenarios, 2000-2050. www.wbcsd.org/DocRoot/FFiAJwjBGGNjlawOAipD/exploringscenarios.pdf.*

Webster, M. D. (2003). Communicating climate change uncertainty topolicy-makers and the public. *Climatic Change*, *61*, 1-2, 1-8.

WEC/IIASA (World Energy Council, International Institute for Applied Systems Analysis). (1995). *Global Energy Perspectives to 2050 and Beyond*. World Energy Council, London, UK.

Weyant, J., Davidson, O., Dowlatabadi, H., Edmonds, J. A., Grubb, M., Parson, E. A., Richels, R., Rotmans, J., Shukla, P. R., Tol, R. S. J., Cline, W. & Fankhauser, S. (1996). Integrated assessment of climate change: an overview and comparison of approaches and results. In: *Climate Change 1995, Economic and Social Dimensions of Climate Change*, [J. P., Bruce, H. Lee, & E. F. Haites, (eds)]. Cambridge University Press, Cambridge, UK.

[1] Weyant, J. P. & Hill, J. N. (1999). Introduction and overview, the costs of the Kyoto Protocol: a multi-model evaluation. *The Energy Journal*, Special Issue, vii-xliv.

[2] J. P. Weyant, (ed.). (2004). EMF 19: alternative technology strategies for climate change policy. *Energy Economics*, *26*, 4.

[3] Wigley, T. M. L., Richels, R. & Edmonds, J. A. (1996). Economic and environmental choices in the stabilization of atmospheric CO_2 concentrations. *Nature*, *379*, 240-243.

[4] Wilby, R. L. & Wigley, T. M. L. (1997). Downscaling general circulation model output: a review of methods and limitations. *Progress in Physical Geography*, *21*, 530-548.

[5] Williams, K. L., Ewel, K. C., Stumpf, R. P., Putz, F. E. & Workman, T. W. (1999). Sea-level rise and coastal forest retreat on the west coast of Florida. *Ecology*, *80(6)*, 2045-2063.

End Notes

[2] Morgan et al. 1998.
[3] Lempert et al. 2006.
[4] See, e.g., Lomborg 2001; Michaels 2003b; Castlesand Henderson 2003a, b; UK House of Lords 2005.
[5] IPCC 2001b:149.
[6] Shell International 2003.
[7] Nakicenovic and Swart 2000:62.
[8] IPCC 2001a:741.
[9] van der Heijden 1996:5.
[10] Berkhout et al. 2001:i.
[11] MEA 2006:xvii.
[12] Brewer and Shubik 1979.
[13] Von Neumann and Morgenstern 1944, Nash 1950.
[14] Levine 1964a,b; Schelling 1964; DeWeerd 1967,1975; Brewer and Shubik 1979.
[15] For relevant history, see Hausrath 1971, Shubik 1975,Greenberger et al. 1983, Huss 1988, Schoemaker 1995,Schultz and Sullivan 1972, Schwartz 1991, Shell International 2003.
[16] Parson 2003:9; Mitchell et al. 2006.
[17] NRC 1996; Hausrath 1971; Brewer 1986; Shubik1975; Svedin and Aniansson 1987; Schultz and Sullivan 1972; Jones 1985; Parson 1996, 1997.
[18] Crisis-response exercises are often an exception, presenting one scenario at a time showing a novel challenge, to which participants must respond and whichis implicitly contrasted to the status quo.
[19] See, e.g., Duncan and Wack 1994, Godet and Roubelat 1996, van Notten et al. 2003.
[20] Several such errors are collected and discussed inBracken 1977, 1990; and Brewer 1990.
[21] Brewer 1990.
[22] Shell International 2001, 2003. For example, in a1960s crisis exercise on a Soviet invasion of Iran, oneparticipant realized the local supply of jet fuel available to support a rapid US response was ten timeslarger than had been thought, because kerosene – anacceptable substitute – was used for domestic cookingand heating (Schelling 1994).
[23] This dimension is presented by Van Notten et al. 2003as “exploration” versus “decision support.”
[24] See, for example, the simple scenario exercise inNRC (1999:161-176) that posited specific targets to reduce world hunger and greenhouse-gas emissions byyear 2050, or the scenarios of the Global ScenarioGroup, which included some defined by specifiedtrends and others back-cast from normatively specifiedtargets for 2050 (Kemp-Benedict et al. 2002, Raskin etal. 2002).
[25] Robinson 1982, 2003.
[26] This approach does not preclude such misuse: if agoal is strongly desired, scenarios are at risk of conscious or unconscious bias to make it look easy. Japanese war-games of the Battle of Midway provide strikingexamples (Bracken 1977).
[27] E.g., note the predominance of scenarios on the “exploration,” rather than the “decision support” side inthe survey of Van Notten et al. 2003.
[28] Alternative interpretations of this matrix structure arediscussed in van’t Klooster and van Asselt 2006.
[29] Van der Heijden 1996; Wack 1985a,b; Schwartz1991; Shell International 2003.
[30] Wack 1985a:74.
[31] See, e.g., Meadows et al. 1972, Barney 1981; summary of early ozone assessments in Parson 2003; andsummary history of scenarios in global-change applications in Swart et al. 2004. What was the earliest scenario work in global change depends, of course, on howthe boundaries of global change are defined. Kahn andWiener 1967 might be considered an early example.
[32] While this categorization has frequently been criticized for neglecting actions with overlapping effectsand the third category of direct interventions in the climate system (Schelling 1983, Keith 2000, Keith et al.2006, Parson 2006), it remains a useful approximationfor most currently proposed responses.
[33] Weyant et al. 1996, Parson and Fisher-Vanden 1997.
[34] e.g., Manabe and Wetherald 1967, Manabe andStouffer 1979.
[35] Washington and Meehl 1989, Manabe et al. 1991,IPCC 1996a.
[36] Cubash et al. 2001.

[37] de la Chesnaye and Weyant 2006, EMF 2006, CCSP2007.
[38] Wigley et al. 1997.
[39] See, e.g., Weyant and Hill 1999; Weyant 2004; de laChesnaye and Weyant 2006; EMF 2006.
[40] CCSP 2007.
[41] IPCC – TGCIA 1999, Barrow et al. 2004.
[42] Mearns et al. 2001.
[43] e.g., Mearns et al. 1992, 1996; Semenov and Porter 1995.
[44] E.g., Kalkstein and Greene 1997.
[45] E.g., Rosenberg et al. 1993.
[46] Giorgi et al. 2001.
[47] Wilby and Wigley 1997.
[48] Mearns et al. 2001, Giorgi et al. 2001.
[49] IPCC 2001a.
[50] Burkett et al. 2005.
[51] Vaughan and Spouge 2002.
[52] MEA 2005.
[53] Parson et al. 2001, 2003; Arnell et al. 2004.
[54] Berkhout et al. 2001, citing UNEP 1994 guidelines.
[55] Toth and Wilbanks 2004, Pitcher 2005.
[56] Berkhout et al. 2002.
[57] Berkhout et al. 2001, Parson et al. 2001
[58] The scenarios were mentioned in a 1-page Appendixto the Working Group 1 report. The one non-equilibrium run available was a preliminary transient runusing 1 percent annual CO_2 concentration increase.See Mitchell et al. 1990, Bretherton et al. 1990, IPCC1990.
[59] Leggett et al. 1992.
[60] The 1 percent scenario was similar to IS92a, but gavetotal radiative forcing about 20 percent greater by 2100.Washington and Meehl 1989; Stouffer et al. 1989;Bretherton et al. 1990:180-182.
[61] Nakicenovic and Swart 2000.
[62] Alcamo et al. 1995.
[63] Parikh 1992, 1998.
[64] Nakicenovic and Swart 2000: 324, Appendix I (termsof reference).
[65] Morita and Lee 1998.
[66] See, e.g., WEC/IIASA 1995, WBCSD 1997.
[67] Minutes, Lead Authors Meeting, Paris, April 13-15,1997.
[68] Arnell et al. 2004; Minutes, Lead Authors Meeting,Paris, April 13-15, 1997.
[69] Minutes, informal modelers meeting, Berkeley, Feb7-8.
[70] Draft minutes, informal modelers meeting, Berkeley,Feb 7-8:4.
[71] Minutes of the Laxenburg meeting, July 2-3, 1998,reporting results of June 29-July 1 IPCC ScopingMeeting, Bonn.
[72] In July 1998, members decided that A1F or A2 couldbe the requested high-emissions scenario (with emissions of ~ 30 GtC in 2100), B2 or A1B a central case(~15 GtC in 2100, with two different SO2 profiles),and B1 or an A1 variant called A1R a stabilization case(at about 550 ppm) (Laxenburg report, July 2-3,1998:1).
[73] Because markers were produced by different modelswith different time steps, the interim years to be harmonized differed for each scenario.
[74] For example, discussions in Beijing reconfirmed thatallowed deviation from markers at 4-region level wouldbe 5 percent for GDP and 10 percent for final energy,but substantial discrepancies in base-year energy couldnot be harmonized due to time constraints (SRES modelers meeting report, Beijing, October 6-7, 1998:2).
[75] SRES modelers meeting report, Beijing, October 6-7, 1998:4. At this meeting, B1 was also proposed forremoval, but was retained based on a decision that noneof the many policy interventions it presumed was anexplicit greenhouse-gas limitation, so it was consistentwith the terms of reference.
[76] A1FI was the gas-intensive scenario, A1G, with revisions to methane emissions and additional non-CO_2 gases added from the A1 run of the MESSAGE model.
[77] Morita et al. 2001.
[78] E.g., Minutes of London meeting, March 1999.
[79] Draft minutes of the Washington, DC, meeting, April29-30, 1998:6.
[80] Schneider 2001, 2002; Pittock et al. 2001; Allen etal. 2001; Reilly et al. 2001.
[81] Grubler and Nakicenovic 2001.
[82] Castles and Henderson 2003a, 2003b; The Economist2003a,b; Michaels 2003b.
[83] Nakicenovic et al. 2003, McKibben et al. 2004,Holtsmark and Alfsen 2005, Manne et al. 2005, Grbleret al. 2004.
[84] Minutes of the Beijing meeting, October 6-7,1998:10.
[85] Draft minutes of the Bilthoven meeting, September17-19, 1997:7-8; draft minutes of the Berkeley meeting, February 7-8, 1997:6; draft minutes of the Washington, DC, meeting, April 29-30, 1998.

[86] While names proposed for the "1" storylines suggestsubstantial common understanding (A1 was called"High Growth," "Productivity," and "Golden Economic Age," B1 was "Green" and "Sustainable development"), names proposed for the "2" scenarios,particularly B2, do not (A2 was called "Regional Consolidation," Divided World," and "Clash of Civilizations"; B2, "Regional Stewardship," "Small IsBeautiful," "Dynamics as Usual," "Gradually Better,"and "Muddling Through") (draft minutes of theBilthoven meeting, September 17-19:7-8; draft minutes of the Berkeley meeting, February 7-8, 1997;UKCIP 1998 report summarizing SRES progress;Pitcher 1998 presentation slides.

[87] Minutes of the Bilthoven meeting, September 17-19,1997:11.

[88] Alcamo et al. 1995.

[89] Draft minutes of the Bilthoven meeting, September17-19, 1997:5.

[90] See, e.g., discussion with Mike Hulme on behalf ofTGICA, draft minutes of the Washington, DC, meeting, April 29-30, 1998:9.

[91] January 1998, meeting with Richard Moss, WG2Technical Support Unit, described in draft meetings ofthe Berkeley meeting, February 7-8, 1997.

[92] Pitcher 2005.

[93] There had been two previous assessments of US climate impacts. The US EPA (1989) did a preliminaryassessment for five representative US regions and fivesectors (agriculture, forests, water resources, health,and coasts), while the US OTA (1993) examined impacts for six sectors – coasts, water, agriculture, wetlands, protected areas, and forests.

[94] NAST 2000, 2001.

[95] NAST 2001:25. It is arguable whether or not the inverse approach involves scenarios by the definition wehave adopted here, because it does not stipulate specified future climate conditions, but attempts to identifythem from presumed thresholds or breakpoints. However, we are following the usage of the NAST reportsin calling these approaches three types of scenarios.

[96] NAST 2001:31-32; MacCracken et al. 2001; Mac-Cracken et al. 2003:1714.

[97] Johns et al. 1997; Boer et al. 1999a, 1999b; Mac-Cracken et al. 2003.

[98] Cubasch et al. 2001, Table 9.1:538-540; and Table9A.1:577.

[99] Ensembles of climate-model runs are repeated simulations with small variations in initial conditions whichimprove the characterization of climate variability. TheCanadian group had completed only one ensemble runat this time. The Hadley Center had completed three,but the National Assessment was only able to use one.

[100] The IS92a scenario is described in section 3.1. Therewere small differences among climate-modelinggroups in the way they converted emissions trajectories into atmospheric concentrations and radiative forcings, making the actual scenarios driving each modelrun very close, but not quite identical.

[101] See www.usgcrp.gov/usgcrp/nacc/background/sce-narios/emissions.html for further detail on emissionsscenarios used in the National Assessment.

[102] VEMAP members 1995, Kittel et al. 1995.

[103] NAST 2001:36, Table 2.

[104] Cubasch et al. 2001, Figure 9.5a:541. While theCanadian model lies at the high end, it is not an outlier.The GFDL model (which was more responsive thanthe Canadian model, with a climate sensitivity of 4.2°C) projected higher global warming than the Canadianmodel in this scenario for the first few decades of thecentury, but only had results through 2060 in time forthe TAR.

[105] The seven models for which these results were available clustered at the top and the bottom. Three of them– the Canadian, GFDL, and Hadley 3 models – lay veryclose together at the high end, the Canadian the highest by a fraction of a degree; three others lay close together at the low end, Hadley 2 the highest of them bysomewhat less than a degree. A seventh model,ECHAM4, tracked the high group through 2050, thelast year for which its results were available. Sincethese comparisons usually reflect only one ensemblerun of each model, small differences between runs mayreflect consistent intermodel differences, or noise reflected in a single ensemble run. See NAST 2001:547,Figure 7.

[106] NAST 2001:545, Figure 8.

[107] NAST 2001:552, Figure 16 and 18.

[108] Lofgren et al. 2000; NAST 2001:175.

[109] NAST 2001:256.

[110] E.g., Mote et al. 2003, Southeast Regional Assessment Team 2002.

[111] Rosenberg et al. 1993.

[112] For an example of such efforts, see the AIACC (Assessments of Impacts and Adaptations to ClimateChange) project, information at http://www.aiaccproject.org.

[113] Parson et al. 2001:102-103.

[114] Terleckyj 1999a,b.

[115] The high-growth scenario was roughly comparablewith A1, medium with B1, and low with A2 and B2.

[116] Rosenzweig and Solecki 2001.

[117] Morgan et al. 2005.

[118] Lorenzoni et al. 2000, Berkhout and Hertin 2000,Parson et al. 2003.
[119] Congressional Record, June 16, 2001, Statements ofSenators Hagel (page S5292) and Craig (page S5294).
[120] MIT Integrated Assessment project, comments onNational Assessment, Aug 11, 2000:15.
[121] Michaels 2003a:171-192.
[122] Disparities between the two models' projections werethe basis of an unsuccessful lawsuit brought against theAssessment under the Federal Data Quality Act (SeeCompetitive Enterprise Institute, "Complaint for Declarative Relief," http://www.cei.org/pdf/3595.pdf, atparagraph 24.)
[123] UKCIP 1998:13-15.
[124] UKCIP 2000.
[125] UKCIP 1998:iv.
[126] UKCIP 2001.
[127] UKCIP, 2001.
[128] Berkhout et al. 2001.
[129] Berkhout and Hertin 2001.
[130] UKCIP 2005.
[131] Holman et al. 2002.
[132] Regional (National) Enterprise was taken as UKCIPHigh (IPCC A2); Global Markets as UKCIP Medium-High (A1B); Regional (Local) Stewardship UKCIPMedium-Low (B2); and Global Sustainability UKCIPLow (B1).
[133] UK Office of Science and Technology 2002.
[134] West and Gawith 2005.
[135] E.g., the Convention on Biological Diversity, theConvention to Combat Desertification, the Conventionon Migratory Species, and the Ramsar Convention onWetlands.
[136] MEA 2006:xii.
[137] MEA 2006:153 (Table 6.1) and 304 (Table 9.2).
[138] MEA 2006:152.
[139] MEA 2006, Figure 5.2.
[140] MEA 2006, Ch 5.5.1
[141] This scenario was originally named "Fortress World"(report of first meeting of MEA global modelinggroup, Jan 7, 2003). The later name reflected participants' judgments that in such a decentralized worldpreoccupied with security concerns, maintaining globalorder would require democratic nations to be militarilystrong – i.e., it is a world of "realist" international affairs (MEA 2006:133)
[142] MEA 2006:131.
[143] MEA 2006:182.
[144] MEA 2006:304, Table 9.2.
[145] MEA 2006, Table S3.
[146] MEA 2005, Table S2.
[147] MEA 2006:155.
[148] MEA 2006, Table S3.
[149] MEA 2006:127.
[150] Report of the First Meeting of the MEA Global Modeling Group, 7 Jan 2003, at www.usf.unikassel.de/ma-gmgroup/dl/first_report.doc; Second Report of theMEA Global Modeling Group, 7 March 2003, available at www.usf.unikassel.de/ma-gmgroup/dl/sanjose_report.doc.
[151] This concern is particularly present regarding implications of the assumption that ecosystem managementis either proactive or reactive (See, e.g., MEA 2006, Ch8.4.2.1 and Ch 9).
[152] NAST 2001.
[153] Shell International 2001, Davis 2003.
[154] Morita et al. 2001, CCSP 2007.
[155] Ged Davis, personal communication. (posted expertreview comments).
[156] Rosenzweig and Solecki 2001; Kinney et al. 2005,2006; Rosenzweig and Major 2006.
[157] Gosselink 1984, Williams et al. 1999, Burkett et al.2005, Morton et al. 2002.
[158] Shinkle and Dokka 2004, Barras et al. 2003.
[159] US Army Corps of Engineers 2004.
[160] See, e.g., http://www.clear.lsu.edu/clear/web-content/index.html.
[161] Presentation by Randy Hanchey, Louisiana Department of Natural Resources, to Governor's AdvisoryCommission on Coastal Protection, Restoration andConservation, Baton Rouge, LA, June 22, 2006.
[162] Cook 1939, Doyle et al. 2003.
[163] California Water Plan home page, http://www.water-plan.water.ca.gov.
[164] California DWR 2005:4-32.
[165] Smil 2005, Greenberger et al. 1983.
[166] This example is drawn from Parson 2003.
[167] Mote et al. 2003, Payne et al. 2004.
[168] www.nwcouncil.org/energy/powerplan/plan.

[169] Chess and Purcell 1999; Gregory and McDaniels2005; Holling 1978; NRC 1996, 2005; Renn et al.1995.
[170] McLean et al. 2001.
[171] CCSP 2003:112.
[172] See, e.g., Envision Sustainability Tools 1999, Rothman et al. 2003, Stockholm Environment Institute1999.
[173] NAPAP 1982, Herrick 2002.
[174] Herrick 2002.
[175] Roberts 1991, Cowling 1992, Russell 1992, Miller1990, Perhac 1991, Rubin 1991.
[176] Gough et al. 1998.
[177] Levy 1995.
[178] See, e.g., Svedin and Aniansson 1987.
[179] Information on the TGICA is at ipcc-wg1.ucar.edu/wg1/wg1_tgica.html. The DDC is jointlyoperated by the UK Climatic Research Unit and theDeutsches Klimarechenzentrum, with several mirrorsites around the world. Data are provided via the webor CD-ROM. All data distributed are in the public domain.
[180] Ruosteenoja et al. 2003.
[181] Raisanen and Palmer 2001; Tebaldi et al. 2004, 2005.
[182] Note that this is not the case if the purpose of scenarios is to explore the implications of specified limitson future emissions. If an emission constraint is assumed to be imposed by policy, then different modelscan be used to explore the implications of that constraint for costs, technologies, and other impacts. Inthis case, caution is needed in deciding what othermodel variables, if any, should be constrained.
[183] Weyant and Hill 1999.
[184] When a scenario exercise uses just one scenario, thisusually presents some specific threat or challengeposed to existing procedures or decision-makers. Inthese cases, uncertainty is still represented by differences among scenarios, but the single scenario is implicitly contrasted to the status quo.
[185] E.g., Wack (1985a:74) states that such a scenario isjust "quantification of a clearly recognized uncertainty."
[186] Morgan and Keith 1995.
[187] Tversky and Kahnemann 1974, Wallsten and Whitfield 1986.
[188] Initially A2 and B2 were most widely used. More recent work has used A2 and B1, sometimes with A1B.
[189] NRC 2002.
[190] Shell International 2003.
[191] Webster 2003.
[192] NRC 2002.
[193] See, e.g., Dickson et al. 2002, Hansen et al. 2001,Gagosian 2003, Curry and Mauritzen 2005, Fairbanks1989.
[194] Global Business Network 2004.
[195] Alley et al. 1997.
[196] Note: these regional projections are 5-10 times fasterthan the IPCC's projected global 21st-century warming.
[197] GBN 2004a:7, 2004b; Schwartz interview.
[198] Stipp 2004 (released, January 26, 2004).
[199] London Observer2004.
[200] San Francisco Chronicle2004, Providence Journal2004.
[201] E.g., Alley 2004 cites it as a useful worst-case assessment.
[202] See, e.g., the consistent uncertainty language proposed for IPCC reports by Moss and Schneider 2000.
[203] Recommendations made in this chapter regarding programmatic and organizational changes, and the adequacy of current budgets, reflect the judgment of thereport's authors and the CPDAC and are not necessarily the views of the U.S. Government.

INDEX

C

D

E

F

G

H

I

N

O

P

Q

R

S

T

U

V

W

Y